Mathematical Modelling and Numerical Analysis in Electrical Engineering

Editors

Udochukwu B. Akuru
Ogbonnaya I. Okoro
Yacine Amara

Basel • Beijing • Wuhan • Barcelona • Belgrade • Novi Sad • Cluj • Manchester

Editors

Udochukwu B. Akuru
Department of
Electrical Engineering
Tshwane University
of Technology
Pretoria
South Africa

Ogbonnaya I. Okoro
Department of Electrical and
Electronic Engineering
Michael Okpara University
of Agriculture
Umudike
Nigeria

Yacine Amara
GREAH
Université Le
Havre Normandie
Le Havre
France

Editorial Office
MDPI AG
Grosspeteranlage 5
4052 Basel, Switzerland

This is a reprint of articles from the Special Issue published online in the open access journal *Mathematics* (ISSN 2227-7390) (available at: https://www.mdpi.com/si/mathematics/CO34J81013).

For citation purposes, cite each article independently as indicated on the article page online and as indicated below:

Lastname, A.A.; Lastname, B.B. Article Title. *Journal Name* **Year**, *Volume Number*, Page Range.

ISBN 978-3-7258-1773-3 (Hbk)
ISBN 978-3-7258-1774-0 (PDF)
doi.org/10.3390/books978-3-7258-1774-0

© 2024 by the authors. Articles in this book are Open Access and distributed under the Creative Commons Attribution (CC BY) license. The book as a whole is distributed by MDPI under the terms and conditions of the Creative Commons Attribution-NonCommercial-NoDerivs (CC BY-NC-ND) license.

Contents

About the Editors vii

Dazhi Wang, Wenhui Li, Jiaxing Wang, Keling Song, Yongliang Ni and Yanming Li
A Parameterized Modeling Method for Magnetic Circuits of Adjustable Permanent Magnet Couplers
Reprinted from: *Mathematics* 2023, 11, 4793, https://doi.org/10.3390/math11234793 1

Chiweta E. Abunike, Udochukwu B. Akuru, Ogbonnaya I. Okoro and Chukwuemeka C. Awah
Sizing, Modeling, and Performance Comparison of Squirrel-Cage Induction and Wound-Field Flux Switching Motors
Reprinted from: *Mathematics* 2023, 11, 3596, https://doi.org/10.3390/math11163596 19

Kyung-Hoon Jang, Sang-Won Seo and Dong-Jin Kim
A Study on Electric Potential and Electric Field Distribution for Optimal Design of Lightning Rod Using Finite Element Method
Reprinted from: *Mathematics* 2023, 11, 1668, https://doi.org/10.3390/math11071668 43

Aleksey Paramonov, Safarbek Oshurbekov, Vadim Kazakbaev, Vladimir Prakht and Vladimir Dmitrievskii
Investigation of the Effect of the Voltage Drop and Cable Length on the Success of Starting the Line-Start Permanent Magnet Motor in the Drive of a Centrifugal Pump Unit
Reprinted from: *Mathematics* 2023, 11, 646, https://doi.org/10.3390/math11030646 54

Vladimir Dmitrievskii, Vladimir Prakht and Vadim Kazakbaev
Design Optimization of a Synchronous Homopolar Motor with Ferrite Magnets for Subway Train
Reprinted from: *Mathematics* 2023, 11, 589, https://doi.org/10.3390/math11030589 72

Nikita V. Martyushev, Boris V. Malozyomov, Svetlana N. Sorokova, Egor A. Efremenkov and Mengxu Qi
Mathematical Modeling of the State of the Battery of Cargo Electric Vehicles
Reprinted from: *Mathematics* 2023, 11, 536, https://doi.org/10.3390/math11030536 89

Abdolhamid Mazloumi, Alireza Poolad, Mohammad Sadegh Mokhtari, Morteza Babaee Altman, Almoataz Y. Abdelaziz and Mahmoud Elsisi
Optimal Sizing of a Photovoltaic Pumping System Integrated with Water Storage Tank Considering Cost/Reliability Assessment Using Enhanced Artificial Rabbits Optimization: A Case Study
Reprinted from: *Mathematics* 2023, 11, 463, https://doi.org/10.3390/math11020463 108

Aleksey Paramonov, Safarbek Oshurbekov, Vadim Kazakbaev, Vladimir Prakht and Vladimir Dmitrievskii
Study of the Effect of Throttling on the Success of Starting a Line-Start Permanent Magnet Motor Driving a Centrifugal Fan
Reprinted from: *Mathematics* 2022, 10, 4324, https://doi.org/10.3390/math10224324 129

Katleho Moloi and Innocent Davidson
High Impedance Fault Detection Protection Scheme for Power Distribution Systems
Reprinted from: *Mathematics* 2022, 10, 4298, https://doi.org/10.3390/math10224298 144

Antony Plait and Frédéric Dubas
A 2D Multi-Layer Model to Study the External Magnetic Field Generated by a Polymer Exchange Membrane Fuel Cell
Reprinted from: *Mathematics* **2022**, *10*, 3883, https://doi.org/10.3390/math10203883 . . . . **163**

Mahmoud Hamouda, Fahad Al-Amyal, Ismoil Odinaev, Mohamed N. Ibrahim and László Számel
A Novel Universal Torque Control of Switched Reluctance Motors for Electric Vehicles
Reprinted from: *Mathematics* **2022**, *10*, 3833, https://doi.org/10.3390/math10203833 . . . . **178**

About the Editors

Udochukwu B. Akuru

Udochukwu B. Akuru received his B.Eng. and M.Eng. degrees from the University of Nigeria, Nsukka, in August 2008 and June 2013, respectively, and his Ph.D. degree in electrical engineering from Stellenbosch University, South Africa, in December 2017. He held a Postdoctoral Research Fellowship at Stellenbosch University between 2018 and 2019, and worked as a lecturer at the University of Nigeria, Nsukka, from 2011 to 2019. He is currently a Senior Lecturer in the Department of Electrical Engineering, Tshwane University of Technology, South Africa. His core research interest is the design and analysis of electrical machines for electrical power and renewable energy technologies. He has published extensively and is a reviewer for several journals and conferences. Dr. Akuru is a Registered Engineer of the Council for the Regulation of Engineering in Nigeria (COREN); a Senior Member of IEEE and SAIEE; a South African National Research Foundation (NRF)-Rated Researcher; and a volunteer for various societies and committees such as IEEE IAS EMC and IEEE IES EMTC. He has been the recipient of several grants and awards, including the prestigious Jorma Luomi Student Forum Award at the 2018 International Conference on Electrical Machines (ICEM) and the 2020 TWAS-DFG Research Fellowship. Dr. Akuru is a Guest Editor of the Special Issue "Mathematical Modelling and Numerical Analysis in Electrical Engineering" in the journal *Mathematics* published by MDPI, as well as the current Chairperson of the IEEE South Africa Section.

Ogbonnaya I. Okoro

Ogbonnaya I. Okoro received his B.Eng and M.Eng. degrees in Electrical Engineering from the University of Nigeria, Nsukka in 1991 and 1997, respectively. He holds a Ph.D. in Electrical Machines from the University of Kassel, Germany, under the DAAD scholarship program. He is a professor of electrical power and machines, a registered Electrical Engineer (COREN) and a Senior Member of the IEEE. He is former Dean of the College of Engineering and Engineering Technology and former Head of Department of Electrical/Electronic Engineering, Michael Okpara University of Agriculture, Umudike, Nigeria. He is the current Dean of the Postgraduate School at Michael Okpara University of Agriculture, Umudike. Prof. Okoro has published extensively in reputable international journals. His research interests include the dynamic simulation and control of induction machines, as well as the thermal and dynamic analysis of AC machines. He is the author of two Textbooks published by JUTA (South Africa): *Concise Higher Electrical Engineering* and *The Essential Matlab/Simulink for Engineers and Scientists*. Prof Okoro is a Guest Editor of the Special Issue "Mathematical Modelling and Numerical Analysis in Electrical Engineering" in the journal *Mathematics*, published by MDPI, as well as an Editorial Board Member of the Nigerian Journal of Technology, published by the Faculty of Engineering, University of Nigeria, Nsukka, Nigeria.

Yacine Amara

Yacine Amara graduated as an electrical engineer from the "Ecole Nationale Polytechnique" of Algiers (Algeria) in 1997. In 1998, he began his doctoral thesis work on hybrid excited synchronous machines at the SATIE laboratory. After obtaining his Ph.D. on Applied Physics from the "Université de Paris-Sud" in December 2001, he joined the university as a temporary lecturer from January to August 2002. From October 2002 to March 2003, he held a post-doctoral position at SATIE, working on the Power Optimized Aircraft European Project. He joined the University of Sheffield

(UK) (Electrical Machines and Drives (EMD) group) as a research associate in March 2003. Here, he worked on the Free Piston Energy Converter European Project. In September 2004, he joined the "Université de Technologie de Belfort-Montbéliard" (France) as a lecturer in the Department of Electrical Engineering, and conducted research at the L2ES laboratory. In February 2008, he joined the "Université du Havre", where he taught electrical engineering om the Department of Science and Technology and performed research at the GREAH laboratory. He became Professor of electrical engineering in September 2017. His research is mainly focused on electrical machines and the modelling, analysis, and design of actuators. He has published many papers on the design, analysis, and control of radial field hybrid excited synchronous machines.

Article

A Parameterized Modeling Method for Magnetic Circuits of Adjustable Permanent Magnet Couplers

Dazhi Wang [1], Wenhui Li [1,*], Jiaxing Wang [1], Keling Song [2], Yongliang Ni [2] and Yanming Li [2]

1. College of Information Science and Engineering, Northeastern University, Shenyang 110819, China
2. China Northern Vehicle Research Institute, Beijing 100072, China
* Correspondence: wenhuilio@stumail.neu.edu.cn; Tel.: +86-133-2406-1559

Abstract: The contactless transmission between the conductor rotor and the permanent magnet (PM) rotor of an adjustable permanent magnet coupler (APMC) provides the device with significant tolerance for alignment errors, making the performance estimation complicated and inaccurate. The first proposal of an edge coefficient in this paper helps to describe the edge effect with better accuracy. Accurate equivalent magnetic circuit (EMC) models of the APMC are established for each region. Models of magnetic flux, magnetic resistance, and eddy current density are established by defining the equivalent dimensional parameters of the eddy current circuit. Furthermore, the concept of magnetic inductance is proposed for the first time, parameterizing eddy currents that are difficult to describe with physical models and achieving the modeling of the dynamic eddy current circuit. The magnetic resistance is subdivided into two parts corresponding to the output and slip according to the power relationship. Furthermore, eddy current loss and dynamic torque models are further derived. The method proposed in this paper enables the APMC to be modeled and calculated in a completely new way. The correctness and accuracy of the model have been fully demonstrated using finite element simulation and an experimental prototype. In addition, the limitations of the proposed method and the reasons are fully discussed and investigated.

Keywords: adjustable permanent magnet coupler; equivalent magnetic circuit; electromagnetic field analytical modeling; parametric expression of eddy current circuit; performance estimation; finite element analysis; prototype verification

MSC: 00A06

1. Introduction

1.1. Types of Adjustable Permanent Magnet Couplers

APMCs have played an irreplaceable role in reducing the vibration and friction losses in various fields of transmission systems due to their advantages of no mechanical connection, low maintenance costs, and strong environmental adaptability. However, the hysteresis in air gap adjustment makes it difficult to accurately estimate and fully utilize their performance, and disadvantages such as the losses caused by eddy current heating have become the main factors hindering the development of APMCs toward higher power levels [1]. Therefore, it is necessary to conduct satisfactory performance estimations to improve their overall behavior.

APMCs can be roughly divided into axial and radial types, as shown in Figure 1, according to the excitation directions of the PMs [2]. From the perspective of practical applications, an axial configuration is selected in this paper due to the smaller axial space required between the motor and the load and higher axial alignment tolerance, and the difficulty of aligning the PM rotor and conductor can be reduced. Axially magnetized PMs are arranged alternately on the back-iron and connected to the load. Each pair of magnetic poles produces the same electromagnetic characteristics and mechanical performance. The primary side is connected to the prime mover using a copper back-iron structure with high conductivity and permeability [3]. It is evident that this connection mode enables absolute mechanical isolation between the prime mover and the load. Once the load is exceeded during operation, the active and passive shafts will automatically decouple, which can be restored after relieving the load, without any damage to the device. Therefore, based on its self-protection mechanism, the APMC is particularly suitable for transmission linkages in high-pressure systems, absolutely sealed systems (e.g., hazardous, corrosive, high cleanliness, etc.), and high-vacuum systems. Electromagnetic torque is generated by the interaction between the reaction field of the induced current and the permanent magnetic field, which are excited by the copper and PMs, respectively [4].

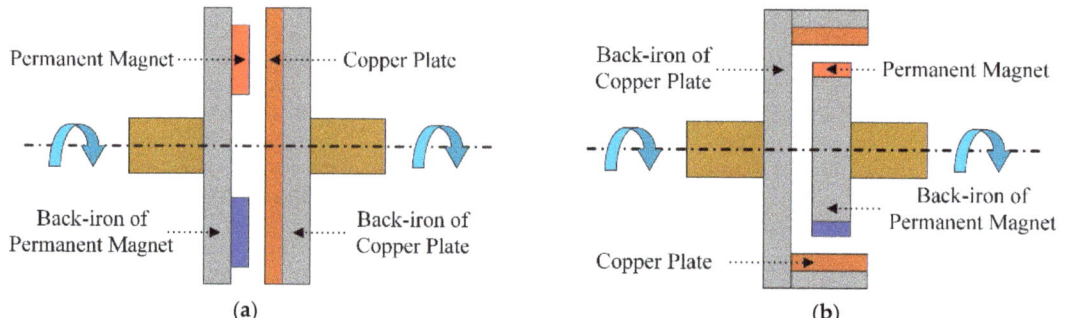

Figure 1. Two types of APMC: (**a**) excitation in axial direction; (**b**) excitation in radial direction.

1.2. Methodologies in Adjustable Permanent Magnet Coupler Modeling

APMCs can be studied numerically or analytically. The numerical methods discretize the continuous model and mainly include the finite element method (FEM), finite difference method, and boundary element method. Currently, the FEM is widely used in the research on APMCs [5,6]. It can be found that the FEM can simulate the actual working conditions and thus obtain extremely accurate analysis results. However, this method requires precise meshing, involves massive computational efforts, and imposes harsh requirements in terms of the computer hardware. It is not suitable for the initial design processes, which require frequent parameter modifications. Instead, it is mostly used as a verification tool [7–10].

Compared to numerical methods, the analytical methods are more flexible and convenient to modify parameters and involve smaller computational efforts, which are more suitable for the initial design of the device. The analytical methods commonly applied for the performance calculations of APMCs include the layer model method and the EMC method [11]. The former divides the research object into different regions based on the consistency of the material properties and establishes corresponding governing equations. These governing equations typically consist of Poisson's equation, Laplace's equation, and diffusion equations [12]. These partial differential equations are solved using the separation of variables method, and the unknown constants in the general solution are determined according to the boundary conditions. The layer model method has been a focus of research for many scholars both domestically and internationally, playing an important role in the analytical modeling and analysis of magnetic fields, eddy current fields, and torque performance of APMC [13–15]. However, the two-dimensional (2D) model often requires three-dimensional (3D) correction. Although the accuracy of the 3D layer model method is significantly improved, the parameters of the model are complicated, which increases the difficulty of the solution process [16–18].

In the past decade, EMCs have been widely used in the analytical modeling of APMCs. This method regards PMs as the magnetic sources, calculates the magnetic resistance in each magnetic circuit branch, establishes the EMC of the whole device, and calculates the magnetic flux in a static state, just like solving the circuit current [19]. Through this operation, abstract and complex magnetic field calculation problems are transformed into concrete circuit analysis problems. Currently, EMCs have been applied in the analysis and design of electromagnetic devices such as transformers, switched reluctance motors, induction motors, and wire-wound synchronous motors [20–22]. Compared to the layer model method, this method has clear physical meanings and simple calculations, making it more suitable for the magnetic circuit design of APMCs. Nevertheless, EMCs can only provide results under static conditions (i.e., the conductor rotor is assumed to be purely resistive). Therefore, the inductive properties of the conductor rotor may not be considered [23].

1.3. Summary

The main objective of this paper is to explore an EMC model that parameterizes the eddy current circuit. Different from the existing research methods, the resistance and inductance characteristics of the conductor, as well as the eddy currents and their magnetic flux, are effectively parameterized in the EMC model. A new EMC model is established to flexibly consider the magnetic resistance of the PM–back-iron at different positions and its influence on magnetic flux. The magnetic resistance of the adjacent PMs, PM–back-iron, and air gap–copper are also considered, while the magnetic flux of each branching magnetic circuit is calculated. The comprehensive modeling of these magnetic resistances and magnetic fluxes significantly improves the computational accuracy of the model. Additionally, the concept of the edge coefficient is proposed to reduce the edge effects near the inner and outer radii of the sector-shaped PMs. Furthermore, the concept of magnetic inductance is proposed for the first time, making eddy currents that are difficult to describe using physical models parameterized and modeling the dynamic eddy current circuit. The magnetic resistance is subdivided into two parts corresponding to the output and slip according to the power relationship, and the eddy current loss and the dynamic torque models are further derived. The method utilizes the electromagnetic coupling regularities followed by the device to achieve the parameter solutions, representing a bold attempt and innovation in this field. The strong comparative results involving analysis, finite element analysis, and experiments have demonstrated the feasibility and superiority of the proposed method.

2. Theories and Methods

2.1. Assumptions and Definition

The specifications of the APMC studied in this paper are summarized in Table 1. It is noteworthy that the sector-shaped PMs configured in the APMC have no centrifugal force and they are mechanically protected by the frame.

Table 1. The specifications of APMC.

Parameter	Value	Unit
PM remanence B_{PM}	1.25	T
PM inner radius r_{PM1}	105	mm
PM outer radius r_{PM2}	135	mm
Copper inner radius r_{cop1}	90	mm
Copper outer radius r_{cop2}	150	mm
Copper conductivity σ_{cop}	58	MS
Number of pole pairs p	9	/
Back-iron thickness of copper h_{copb}	20	mm
Back-iron thickness of PMs h_{PMb}	20	mm
Copper thickness h_{cop}	10	mm
PM thickness h_{PM}	30	mm

The goal of this paper is to minimize the reliance on natural material properties to reduce the impact of model assumptions on the performance estimation and improve the accuracy of analytical modeling. The essence of modeling the dynamic process of the

APMC is to solve the eddy current problem in moving conductors. To mathematically describe the strong coupling process between the dynamic eddy current field, induced magnetic field, and rotating magnetic field, the following appropriate assumptions should be made:

1. The relative velocity between the conductor rotor and the PM rotor is v_{re};
2. Each layer is homogeneous with uniform material properties, and the air regions outside the two back-iron layers are not considered;
3. The thickness of the rotor back-iron is sufficient to prevent magnetic saturation;
4. The eddy current effect of the PMs is not considered.

2.2. Geometries and Topologies

The geometrical parameters and a cross-section schematic of the APMC extending along the circumference are shown in Figure 2. The structural parameters h, r, and n in Figure 2 denote thickness, radius, and rotational speed, respectively. The subscripts *cop*, *PM*, *air*, *copb*, *PMb*, *cop1*, *cop2*, *PM1*, and *PM2* represent the copper plate, PM, air gap, copper plate back-iron, PM back-iron, copper plate inner side, copper plate outer side, PM inner side, and PM outer side, respectively. Moreover, α_{PM} is the angle of the PMs in the circumferential direction, N and S are the north and south poles of the PMs, and θ_{PM} is the angle of the 1/2 pole pitch.

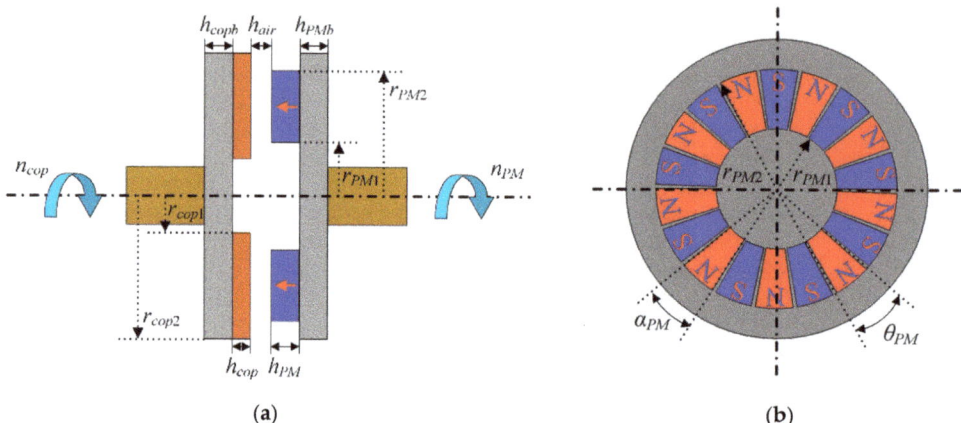

Figure 2. Geometrical parameters and cross-section schematic of APMC: (**a**) axial topology of APMC; (**b**) circumferential distribution of PMs.

2.3. Programming of Magnetic Flux Paths

The 2D finite element qualitative simulation results for the APMC are presented in Figure 3 to intuitively visualize the magnetic flux distribution and rationally plan the magnetic circuit. As shown in Figure 3, the APMC exhibits a typical axisymmetric structure, so only the magnetic flux paths under the pair of poles need to be planned: the main magnetic circuit (solid green line) passes through the copper rotor back-iron, copper rotor, air gap, PM rotor, and PM back-iron, forming a closed loop. The magnetic flux paths consisting of the adjacent PMs, PM, and its back-iron and the copper rotor and its back-iron are all leakage magnetic flux paths (dashed blue line). In particular, the induced eddy current within the copper rotor is indicated in Figure 3 with red dots and cross symbols. The magnetic flux paths generated by the eddy current effect are represented by the long dashed red line, which passes through the conductor region and forms a closed magnetic circuit within the air gap and back-iron. The magnitude of the eddy current reaction varies with slip. In the case of high slip, the proportion of magnetic flux generated by the eddy currents is significant to the total leakage magnetic flux [4,24].

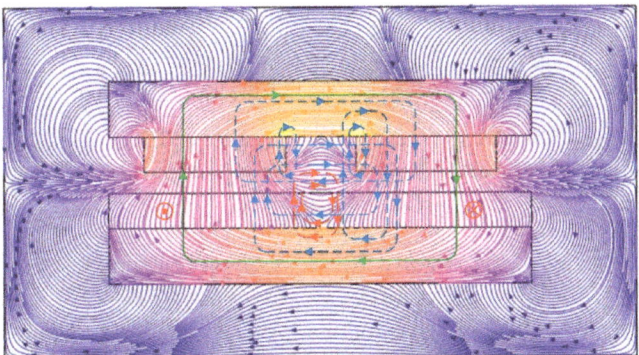

Figure 3. Planning of magnetic flux paths.

The sector-shaped PMs can not only provide a larger magnetic flux cross-sectional area S_{PM} but also ensure that the pole-arc coefficients at different radii are the same, thus eliminating numerous approximations and greatly improving the accuracy of the model. They can also be equivalent to other shapes of PMs with the same area and thickness, such as circular or rectangular PMs [2].

$$\alpha_{PM} = \frac{\pi \alpha}{p} = \alpha \theta_{PM} \tag{1}$$

$$S_{PM} = \frac{\alpha_{PM}}{2}(r_{PM2}^2 - r_{PM1}^2) \tag{2}$$

2.4. Equivalent Magnetic Circuit Method

The EMC can consider material properties such as iron saturation, remanence, and coercivity, as well as the complex geometry of the device. However, this method can only be derived under a steady state. Under dynamic conditions, the varying eddy current inside the conductor rotor excites alternating magnetic fields, causing inductive properties to be manifested, rather than purely resistive ones. Thus, a method based on sinusoidal steady-state circuits is no longer applicable.

2.4.1. Novel Equivalent Magnetic Circuit Model

Based on the symmetrical structure of the APMC and the magnetic circuit analysis, an initial EMC model for a half-pole pair shown in Figure 4 is established, where the red dashed line represents the branch magnetic circuit of the eddy current effect. The expressions for PM magnetic flux Φ_{PM} and magnetic resistance R_{PM} can be derived. The eddy current circuit contains two parts of magnetic resistance corresponding to the output power (defined as R_{out}) and slip power (defined as R_{slip}), as well as the magnetic inductance M_L corresponding to the changing magnetic field.

$$\Phi_{PM} = B_{PM} S_{PM} \tag{3}$$

$$R_{PM} = \frac{h_{PM}}{\mu_0 \mu_r S_{PM}} \tag{4}$$

where B_{PM}, μ_0, and μ_r are the residual magnetic flux density, vacuum permeability, and relative permeability of the PMs, respectively.

In Ref. [24], the magnetic resistance is neglected by assuming an infinite magnetic permeability for the PM back-iron. Although this assumption simplifies the analysis of the moving eddy current problem, it overlooks the influence of the back-iron magnetic flux on the device performance, thereby reducing the accuracy of the model [25].

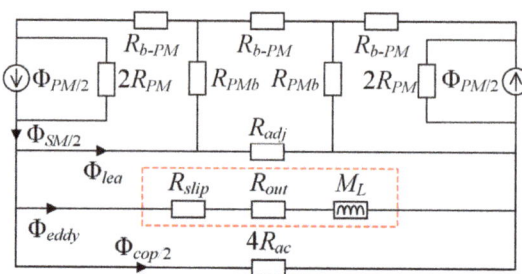

Figure 4. Initial EMC model.

The arrangement characteristics of axially magnetized PMs are fully considered in this paper, and the back-iron magnetic resistance corresponding to the PMs and the air gap are set, respectively. Based on the power relationships of the APMC in the eddy current path, the magnetic resistance is divided into two parts, corresponding to the output power and slip power [26]. Furthermore, by reasonably transforming the EMC shown in Figure 4, a more analytically favorable Figure 5 is obtained. In Figure 5, the red dashed box still represents the eddy current effect.

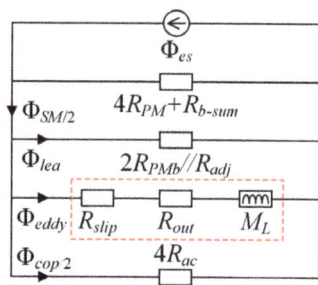

Figure 5. Simplified EMC model.

The EMC model established at the average radius treats the axial magnetic flux of APMC as a linear actuator, neglecting the curvature effect and the edge effect [23]. The curvature effect primarily manifests in the leakage flux between the adjacent PMs. In previous studies, researchers attempted to calculate the magnetic flux at each radius using modulation functions and then superimpose it in the radial direction. However, there are two drawbacks to employing modulation functions, which can be referred to in Ref. [27]. To avoid repetition, the calculation details will not be reiterated here.

The edge effect will appear near the inner and outer radii of the sector-shaped PMs, resulting in a reduction in the no-load magnetic flux, which is manifested as a saddle-shape defect in the magnetic density waveform at the inner and outer radii [27]. In this paper, the concept of an edge coefficient β is proposed to reduce the influence of the edge effect on the model accuracy, and a strict definition of β is required to avoid distorted magnetic flux paths caused by an excessive air gap length in the non-magnetic regions, which leads to prolonged magnetic circuits and a weakened effective magnetic flux.

$$\beta = 1 + \frac{(r_{cop2} - r_{cop1}) - (r_{PM2} - r_{PM1})}{4} \quad (5)$$

$$R_{ac} = \frac{h_{ac}}{\mu_0 \beta^2 S_{PM}} \quad (6)$$

where R_{ac} and h_{ac} are the magnetic resistance and thickness of the air gap–copper plate, respectively.

Based on the 2D finite element qualitative simulation results of the APMC as shown in Figure 3, the leakage magnetic flux paths and the integral paths are investigated. Figure 6 depicts the leakage magnetic flux between the adjacent PM surfaces and the leakage magnetic flux between the PM surface and its back-iron, respectively. It can be observed that the integral path of the leakage magnetic flux between the adjacent PM surfaces includes a straight segment and two arc segments within the air gap. The integral path of each leakage magnetic flux between the PM surface and its back-iron consists of a straight line segment and an arc segment in the air gap [28]. The magnetic resistance R_{adj} between the adjacent PMs and the magnetic resistance R_{PMb} between the PM surface and its back-iron are given in (7) and (8), where r_{av} is the average radius of the PMs.

$$\frac{1}{R_{adj}} = \frac{\mu_0(r_{PM2} - r_{PM1})}{\pi} \ln\left(\frac{\theta_{PM} + \alpha_{PM}}{\theta_{PM} - \alpha_{PM}}\right) \tag{7}$$

$$\frac{1}{R_{PMb}} = \frac{2\mu_0(r_{PM2} - r_{PM1})}{\pi} \ln\left(1 + \frac{\pi r_{av}\alpha_{PM}}{2(\pi h_{air} + h_{PM})}\right) \tag{8}$$

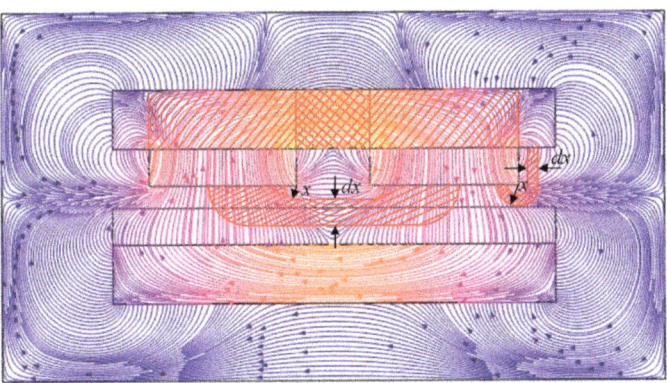

Figure 6. Magnetic flux paths of leakage flux.

Due to the tight connection between the PMs and the back-iron, the magnetic flux density distribution of the latter is significantly uneven. To effectively distinguish the magnetic flux in different areas, the magnetic resistance of the back-iron is equated to a series connection of three parts, as shown in Figure 6. Consequently, the total magnetic resistance and source flux of the parallel magnetic circuit in Figure 6 can be obtained smoothly.

$$R_{b-sum} = R_{b-air} + 2R_{b-PM} \tag{9}$$

$$R_{b-air} = \frac{r_{av}(\theta_{PM} - \alpha_{PM})}{\mu_0 \mu_{b-air} h_{PMb}(r_{PM2} - r_{PM1})} \tag{10}$$

$$R_{b-PM} = \frac{\alpha_{PM}\theta_{PM}r_{av}}{\mu_0 \mu_{b-PM}(\alpha_{PM}\theta_{PM}r_{av}/2 + h_{PMb})(r_{PM2} - r_{PM1})} \tag{11}$$

$$R_{par} = \frac{2R_{adj}R_{PMb}(4R_{PM} + R_{b-sum})}{2R_{adj}R_{PMb} + (4R_{PM} + R_{b-sum})(2R_{PMb} + R_{adj})} \tag{12}$$

$$\Phi_{es} = \frac{2\Phi_{PM}R_{ac}}{4R_{PM} + R_{b-sum}} \tag{13}$$

In (9)~(13), the magnetic resistances $R_{b\text{-}sum}$, $R_{b\text{-}air}$, $R_{b\text{-}PM}$, and R_{par} associated with the PMs correspond to the total back-iron, air gap–back-iron, PM–back-iron, and parallel magnetic resistance, respectively. Φ_{es} is the source magnetic flux of the PMs.

2.4.2. Eddy Current Circuit Modeling

According to Ampere's law, the magnetic flux Φ_{cop}, magnetic flux density B_{cop}, and average eddy current density J_{cop} within the copper can be expressed as follows [29].

$$\Phi_{cop} = \frac{2R_{par}C_{av}S_{PM}\Phi_{es}}{C_{av}S_{PM}(R_{par}+4R_{ac}) + \mu_0\mu_r S_{av}R_{par}\sigma_{cop}h_{cop}\alpha_{PM}r_{av}} \tag{14}$$

In Equation (14), σ_{cop}, s, and n_{cop} are the conductivity, slip, and rotational speed of the copper rotor, respectively. C_{av} and S_{av} correspond to the perimeter and area at the average radius of the sector-shaped PM.

$$B_{cop} = \frac{2\Phi_{cop}}{S_{PM}} \tag{15}$$

$$J_{cop} = \frac{2\sigma_{cop}v_{re}\Phi_{cop}}{S_{PM}} \tag{16}$$

$$S_{av} = \frac{\alpha_{PM}r_{av}^2}{2} \tag{17}$$

$$C_{av} = \pi(r_{PM2} + r_{PM1}) \tag{18}$$

$$v_{re} = \frac{C_{av}sn_{cop}}{60} \tag{19}$$

According to the previous analysis, it is known that there is a correspondence between the eddy current circuit and the PMs. Using the concept of the edge coefficient proposed in this paper to define the equivalent size of the eddy current circuit (equivalent area S_{eddy} in Equation (20) and equivalent length L_{eddy} in Equation (21) of the eddy current path), the eddy current, which is difficult to describe using a physical model, are expressed parametrically, so that the eddy current circuit can be modeled and calculated in a completely new way.

$$S_{eddy} = \beta^2 S_{PM} \tag{20}$$

$$L_{eddy} = \beta^2 L_{PM} \tag{21}$$

$$I_{eddy} = \frac{J_{cop}h_{cop}\alpha_{PM}r_{av}}{2} \tag{22}$$

$$\Phi_{eddy} = \frac{\mu_0\mu_r S_{eddy} I_{eddy}}{L_{eddy}} \tag{23}$$

The 3D finite element simulation results of the magnetic flux and corresponding eddy current density on the surface of the copper rotor at a slip speed of 100 rpm are shown in Figure 7.

It can be seen from the results that the eddy currents distributed along the radial direction in the copper plate excite the magnetic field and form a closed loop with the PMs in the main magnetic circuit to generate the output torque. The magnetic flux varies periodically with the arrangement of PMs, and the number of eddy current circuits is also the same as the number of PMs. The eddy currents distributed along the circumference, although they excite the magnetic field, do not contribute to the output torque but generate heat, corresponding to the eddy current losses.

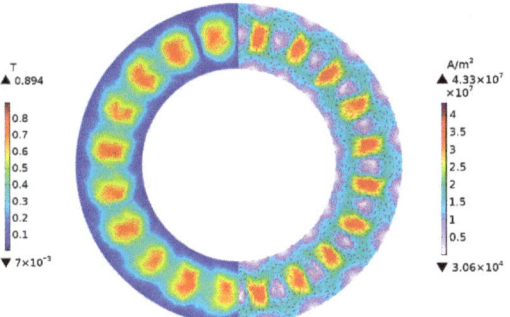

Figure 7. FEM results of magnetic flux and eddy current on the surface of copper rotor.

The relative velocity v_{re} can be expressed more intuitively as Equation (24). By applying Equations (16) and (20)–(24), the relationship between the magnetic fluxes Φ_{eddy} and Φ_{cop} can be obtained in Equation (25). Obviously, the magnetic flux Φ_{eddy} under steady-state conditions is also a function of the rotational speed difference (n_{cop}-n_{PM}), which is of great significance for solving the dynamic output torque.

$$v_{re} = \frac{30(n_{cop} - n_{PM})}{\pi r_{av}} \quad (24)$$

$$\Phi_{eddy} = \frac{30\mu_0\mu_r h_{cop}\alpha_{PM}\sigma_{cop}(n_{cop} - n_{PM})}{\pi L_{PM}}\Phi_{cop} \quad (25)$$

The magnetic resistance R_{slip} is established using the eddy current I_{eddy} and the magnetic flux Φ_{eddy} of the eddy current circuit based on Kirchhoff's law, and the magnetic resistance R_{out} is further obtained according to the power relationship.

$$R_{slip} = \frac{s(2\Phi_{cop}R_{ac} - I_{eddy})}{\Phi_{eddy}} \quad (26)$$

$$R_{out} = \frac{1-s}{s}R_{slip} \quad (27)$$

In fact, the existing research results are unable to describe the eddy currents accurately, which is a core factor limiting the modeling accuracy of the APMC. This paper attempts to model the eddy current circuit and parameterize the magnetic flux paths from the perspective of an EMC. According to Faraday's law of electromagnetic induction, eddy currents are excited on the conductor when they cut through the rotating magnetic field of the PM rotor. The changing magnetic field will generate an induced electromotive force to resist the variation in the closed loop, causing the inductive characteristics of the conductor. The inductive characteristics are an inherent property of the closed circuit and should not be ignored.

This paper combines both field and circuit theories to develop a quantity that can describe the inductive characteristics, called the magnetic inductance M_L. Equation (25) reflects the relationship between Φ_{eddy} and Φ_{cop} under steady-state conditions, while they will be redistributed into Φ_{eddy}' and Φ_{cop}' due to the presence of M_L under dynamic conditions. Ohm's law of magnetic circuits is introduced into Figure 5 and Equation (28) is given. The differential $d\Phi_{eddy}'/dt$ in Equation (28) can be converted into the differential of the rotational speed difference over time according to Equation (25). Moreover, this conversion can enable a measurable dynamic torque to be obtained.

$$4R_{ac}\frac{\Phi_{cop}'}{2} = (R_{slip} + R_{out})\Phi_{eddy}' + M_L\frac{d\Phi_{eddy}'}{dt} \quad (28)$$

The physical rules do not depend on the choice of measurement units, and the application of dimensional analysis allows for the exploration of this invariance, so as to deeply reflect the intrinsic relationships among physical quantities. Via dimensional analysis of Equation (28), the expression of M_L is reversely derived: M_L can be equivalent to the ratio of time t to inductance L.

Ref. [30] takes square inductors as an example to establish the closed form expression of inductance. The difference from other modeling methods is that the inductance calculation is based on the average segment interactions, rather than summing the individual segment interactions one by one. Hence, the total inductance consists of self-inductance and mutual inductance, including all negative and positive interactions between all segments. It should be noted that there are just the geometry parameters but no unphysical fitting factors in the model.

There is a lack of modeling for magnetic induction in existing EMCs. Therefore, the idea of modeling square inductors in Ref. [30] is extended to model the magnetic inductance of the APMC eddy current circuit shown in Figure 8a. The eddy current path is modeled as the n-turn polygonal symmetric spiral shape shown in Figure 8b. Considering the convenience of modeling, a single-turn eddy current is divided into eight segments at the average radius in Figure 8b. Of course, any position and any number of segments can be applied.

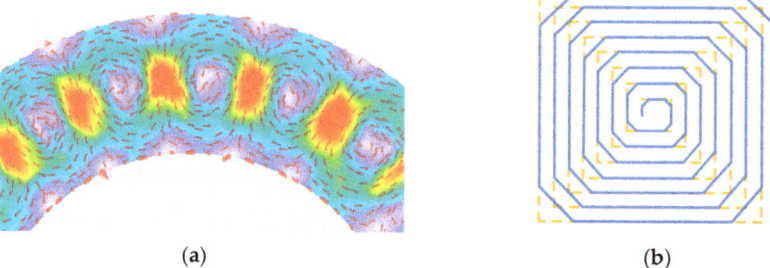

Figure 8. Eddy current path and octagonal spiral model of APMC: (**a**) eddy current path based on COMSOL; (**b**) octagonal spiral model.

At the average radius, the length of the octagonal spiral is C_{av}, and then the self-inductance of the straight segment shown in Equation (7) of Ref. [30] can be obtained. Equation (29) is defined as the total square inductor length in Equations (8) and (9) in Ref. [30], since 45°-inclined segments can be decomposed into their vertical and horizontal contributing components. Therefore, the total inductance L of an octagonal spiral with a length of C_{av} is Equation (30). The relationship between time and relative velocity shown in Equation (31) can be obtained according to Equation (28). Based on the above analysis, the expression of M_L shown in Equation (32) can be obtained from Equations (30) and (31).

$$l_{seg} = 4 \times \left(\frac{C_{av}}{8} + \frac{C_{av}}{8} \sin 45° + \frac{C_{av}}{8} \cos 45° \right) = 1.21 C_{av} \tag{29}$$

$$\begin{aligned} L &= \frac{\mu_0 C_{av}}{2\pi} \left(\ln \frac{C_{av}}{h_{cop} + r_{av}\alpha_{PM}/2} - 0.9 \right) - 0.47 \frac{\mu_0}{2\pi} l_{seg} \\ &= \frac{\mu_0 C_{av}}{2\pi} \left(\ln \frac{C_{av}}{h_{cop} + r_{av}\alpha_{PM}/2} - 0.5687 \right) \end{aligned} \tag{30}$$

$$t = \frac{r_{av} v_{re}}{n_{cop} - n_{PM}} = \frac{30}{\pi} \tag{31}$$

$$M_L = \frac{60}{\mu_0 C_{av} \left(\ln \frac{C_{av}}{h_{cop} + r_{av}\alpha_{PM}/2} - 0.5687 \right)} \tag{32}$$

The parallel magnetic resistance of the two branches in Figure 5 is defined as $R_{//}$ in Equation (33). The redistributed magnetic fluxes Φ_{eddy}' and Φ_{cop}' can be solved by constructing the magnetic circuit Equation (34) and associating it with Equation (28).

$$R_{//} = (4R_{PM} + R_{b-sum})//\left(2R_{PMb}//R_{adj}\right) \tag{33}$$

$$\left(\Phi_{es} - \Phi_{eddy}' - \Phi_{cop}'\right)R_{//} = 4R_{ac}\frac{\Phi_{cop}'}{2} \tag{34}$$

2.4.3. Calculation of Eddy Current Losses and Torque

It can be observed from Equation (28) that the magnetic inductance and magnetic resistance of a conductor represent two different energy exchange modes. The magnetic field generated by the induced current serves as the carrier of energy, making the magnetic inductance an energy storage element. The conductor plate, on the other hand, is a metal with a certain conductivity, resulting in heat generation and power losses, making the magnetic resistance an energy dissipation element. The eddy current loss P_{eddy} and output torque T can be further obtained by applying the above model, which does not ignore the contribution of the eddy currents in the back-iron. The detailed derivation process has been provided in Appendix A.

$$P_{eddy} = \frac{2p\sigma_{cop}v_{re}^2 h_{cop}\left(\Phi_{eddy}' + \Phi_{cop}'/2\right)^2}{S_{PM}} \tag{35}$$

$$T = \frac{2p\sigma_{cop}v_{re}r_{av}h_{cop}\left(\Phi_{eddy}' + \Phi_{cop}'/2\right)^2}{S_{PM}} \tag{36}$$

3. Verification and Discussion for Proposed Model

3.1. Description of 3D Finite Element Model

A 3D finite element model based on Table 1 is conducted to verify the accuracy of the EMC model, and the meshes are divided according to the materials of each part of the APMC and their influence on the magnetic field. Figure 9 shows the division of 969,944 elements using a fixed calculation method, which allows for the simulation of the magnetic field and output torque separately, collecting more actual results that are difficult to acquire experimentally.

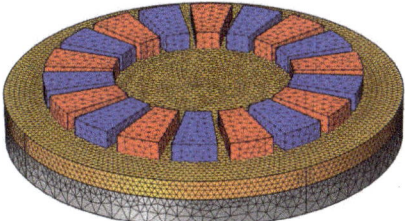

Figure 9. Mesh partition model based on FEM.

3.2. Parameter Sensitivity Analysis

In this section, the structural parameters of the APMC in Table 1 are used as benchmarks to investigate the effects of the conductor plate thickness, air gap thickness, and number of pole pairs on the device performance. The parameter sensitivity analysis on the one hand verifies the effectiveness and accuracy of the proposed method by establishing a 3D finite element model and an experimental prototype. On the other hand, it provides a research basis and reference data for optimizing the design.

3.2.1. Torque–Slip Speed Characteristics Related to Conductor Plate Thickness

The conductor plate serves as a carrier for eddy currents, which directly influence the eddy currents and torque. In practice, it is impossible to make the conductor plate infinitely large considering the material costs and device volume. Therefore, the thickness of the conductor plate becomes one of the important constraints in optimization tasks.

Figure 10 makes an arithmetic progression with 2 mm to 16 mm as the interval and 2 mm as the common difference to study the torque variation under different conductor plate thicknesses. In general, the torque increases significantly when increasing the slip speed and conductor plate thickness. At low slip, the effective magnetic field in the air gap mainly comes from the permanent magnetic field of the PMs, and the increase in the conductor plate thickness leads to an increase in the leakage magnetic flux, which reduces the effective magnetic field in the air gap. As the slip increases gradually, the induced magnetic field excited by eddy currents and the effective magnetic field in the air gap increase with the increase in the conductor plate thickness, manifested as the increase in torque. However, when the conductor plate thickness increases to a certain value, the growth rate of torque decreases at the same slip speed. Therefore, the relationship between the conductor plate thickness and the device volume cost should be considered comprehensively in the design and optimization of the APMC, so as to obtain the optimal matching scheme.

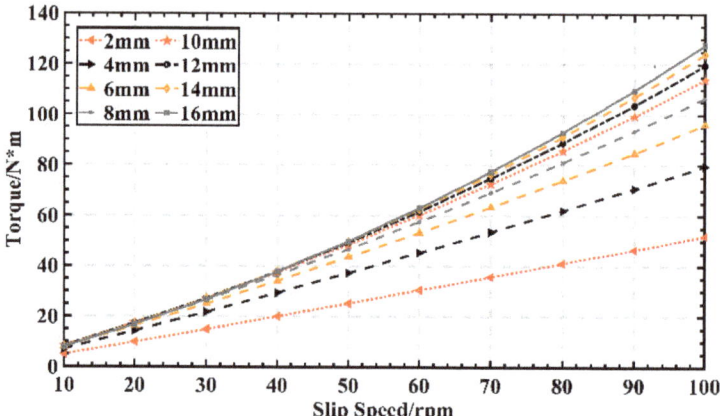

Figure 10. Torque–slip speed characteristics related to conductor plate thickness.

3.2.2. Torque–Slip Speed Characteristics Related to Air Gap Thickness

As mentioned above, the APMC realizes torque regulation by changing the thickness of the air gap between the conductor and PMs, which in turn regulates the output speed of the device, i.e., it is used as a governor. Figure 11 shows the analytical and finite element calculation results of the torque and errors at different slip speeds with air gap thicknesses of 1 mm, 3 mm, and 5 mm, respectively. It can be seen that the error between the analytical and finite element results is larger when the air gap thickness is 1 mm at low slip speeds. The reason is that when the air gap thickness and slip speed are both very small, the main magnetic circuit will experience relatively severe distortion, which makes the difference between the two calculation results larger. Therefore, in order to ensure a stable output performance in practical applications, the air gap thickness is generally controlled to be around 3 mm without being too small.

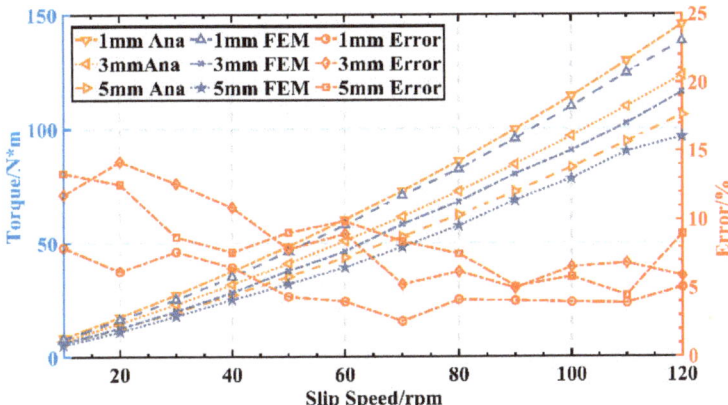

Figure 11. Torque–slip speed characteristics related to air gap thickness.

3.2.3. Eddy Current Loss–Slip Speed Characteristics Related to Air Gap Thickness

The results of the analytical and finite element calculations of the eddy current losses under different air gap thicknesses and the error between the two methods corresponding to Figure 11 are shown in Figure 12. The error is larger for an air gap thickness of 5 mm at low slip speed. The reason for this situation is that the eddy current mainly generates losses in the form of heat. When the air gap thickness is relatively large, the magnetic resistance of the air gap and the energy consumption increase, which ultimately leads to the deviation of the calculation results. Other than that, the errors of the analytical and finite element calculation results fluctuate within an acceptable small range, proving that the model proposed in this paper can calculate the eddy current losses accurately.

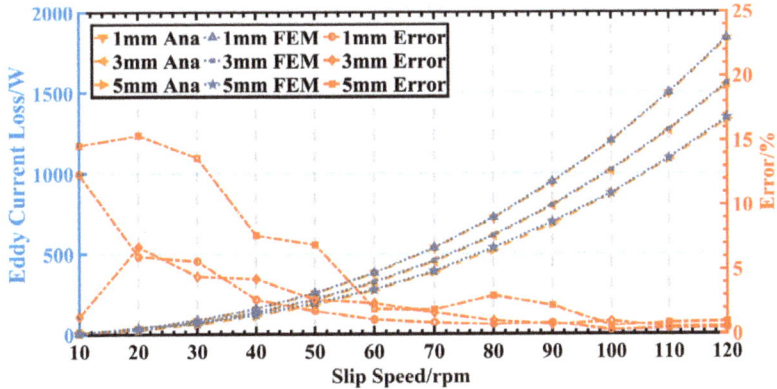

Figure 12. Eddy current loss–slip speed characteristics related to air gap thickness.

3.2.4. Torque–Slip Speed Characteristics Related to the Number of Pole Pairs

The establishment of the initial magnetic field and the generation of eddy currents in the APMC all originate from the PMs. The number of pole pairs of the PMs determines the distribution and periodicity of the magnetic field in the air gap and the eddy currents in the conductor plate, which plays a decisive role in the output performance: it is then one of the most important design parameters of the device. The output torque characteristic curves for the number of pole pairs vary from 6 to 11 and are given in Figure 13. It can be seen that the output torque increases with an increase in the number of pole pairs, but the growth rate decreases. In other words, there must be a critical pole pair value so that the torque no longer increases with the number of pole pairs. In addition, PMs cannot be used in large quantities due to their expensive prices in practical engineering applications.

Therefore, it is particularly important to balance the relationship between the number of PMs and output performance in the optimization of APMCs.

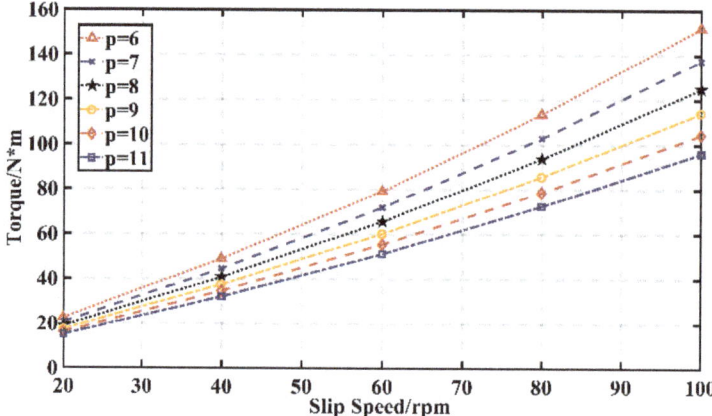

Figure 13. Torque–slip speed characteristics related to number of pole pairs.

3.2.5. Dynamic Torque Characteristics

The dynamic torque characteristics of the APMC at an air gap thickness of 2 mm and different slip speeds are shown in Figure 14. The startup process of the APMC is essentially the establishment and action process of the magnetic field. At the initial stage, there are no eddy currents inside the copper rotor. The increasing slip speed causes the eddy currents to gradually increase, and the output torque also increases accordingly. However, the torque will not reach its maximum immediately due to the magnetic resistance of the copper rotor. Ultimately, the output torque will achieve dynamic stability.

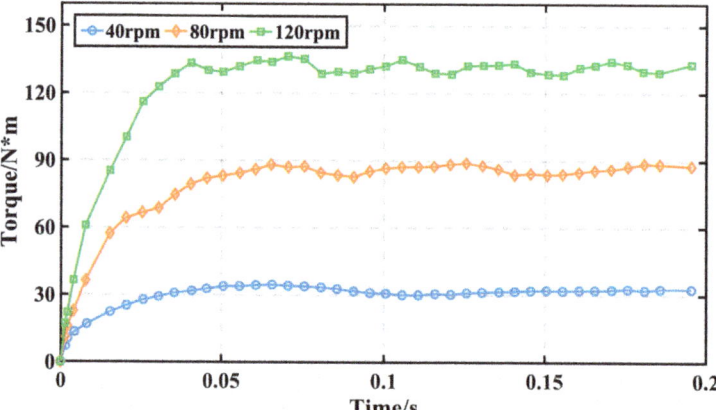

Figure 14. Dynamic torque characteristics.

3.3. Experimental Prototype Validation

The experimental prototype was manufactured based on the structural parameters shown in Table 1 to verify the accuracy of torque, and an experimental platform as shown in Figure 15 was built. The platform consists of a drag motor, an APMC prototype, and a generator. To measure the output torque of the drag motor and the rotational speed of the load, a torque–speed sensor is installed between the APMC and the load generator. The speed of the drag motor is measured using an encoder mounted on it. The measurement data collected by the torque–speed sensor and the encoder can be uploaded to a computer for storage.

Figure 15. The experimental platform of APMC.

The measured torque characteristic curves of the prototype at the air gap thicknesses of 2 mm and 4 mm and their error curves with analytical and finite element methods are given in Figure 16. According to the experimental results, it can be concluded that: (1) The increase in the air gap thickness significantly reduces the output torque at the same slip speed. (2) The error rate of the two methods is within 18%; although they are relatively accurate, there is still great room for improvement. (3) The 3D FEM is more consistent with the actual test results with an error rate of 5–8%. Therefore, in the absence of a measured prototype, the results of the 3D FEM can serve as a verification benchmark, which provides convenience for the optimization of the device. (4) The possible sources of the errors are mainly the influence of the conductor plate temperature rise on the characteristics of the APMC and the deviation generated by the sensor data.

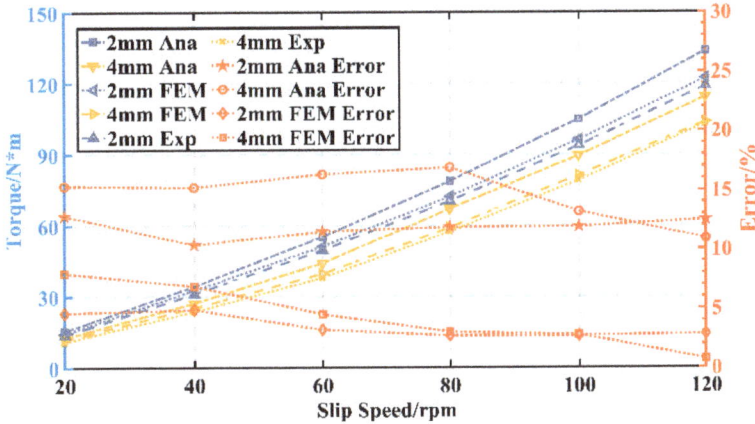

Figure 16. Torque-slip speed characteristics related to air gap thickness.

4. Conclusions and Future Highlights

A novel parameterized EMC modeling method for an APMC is proposed in this paper. According to the electromagnetic coupling characteristics of the APMC, the resistance and inductance characteristics of the conductor eddy current loop are introduced into the EMC model as a branch circuit. The analytical method based on the EMC makes the performance prediction for the APMC simple and practical. The concept of the edge coefficient effectively reduces the edge effect near the inner and outer radii of the sector-shaped PMs without 3D correction. The magnetic flux, magnetic resistance, and eddy current density of each branch magnetic circuit are established under steady-state conditions. The concept of magnetic inductance is first proposed to parameterize eddy currents that are difficult to accurately describe using physical models. On this basis, Ohm's law of a dynamic eddy current circuit is established to obtain the dynamic eddy current model. In addition, the magnetic resistance is subdivided into two parts corresponding to the output and slip according to the power relationship, from which the eddy current loss and dynamic torque models are further derived. The method proposed in this paper is verified by comparing the analytical prediction with the finite element calculation. It can achieve satisfactory results within a reasonable range of slip speed and structural parameters. The method proposed in this paper is suitable for different PM geometries, thus providing a theoretical basis for the design and optimization of the APMC. Compared with existing analytical methods, the EMC is easy to program and implement, and the calculation process is short, making it very suitable for the initial design stage of devices. The processing technology regarding the eddy current branch also provides a feasible method for effectively evaluating the EMC modeling, including the induced eddy current problem of the moving conductors.

The method proposed in this paper is a bold attempt and breakthrough of an EMC for an APMC. The parameter sensitivity analysis verifies the validity and accuracy of the method in terms of the conductor plate thickness, air gap thickness, and number of pole pairs. The 3D finite element simulation model and experimental prototype further demonstrate the reliability of the analytical method in modeling the APMC. However, concerns are expressed about the shortcomings of its generalization capability, which will be the motivation of and main work for further exploration in the future.

Author Contributions: Conceptualization, D.W. and W.L.; methodology, W.L.; software, K.S.; validation, Y.L.; formal analysis, J.W.; investigation, K.S.; resources, Y.N.; data curation, W.L.; writing—original draft preparation, W.L.; writing—review and editing, W.L.; visualization, J.W.; supervision, D.W.; project administration, W.L.; funding acquisition, D.W. All authors have read and agreed to the published version of the manuscript.

Funding: This research was funded by the National Natural Science Foundation of China under grant 52077027.

Data Availability Statement: Data are contained within the article.

Conflicts of Interest: The authors declare no conflict of interest.

Appendix A

The dynamic transition process is the intermediate process of the speed changing from one equilibrium state to another, so it can be regarded as the instantaneous steady-state value at each moment. Therefore, Φ_{eddy} and Φ_{cop} in Equation (25) can also be perceived as the steady-state value at each moment of the dynamic process. According to the above description, (A1) and (A2) can be obtained.

$$\Phi_{eddy}' = \frac{30\mu_0\mu_r h_{cop}\alpha_{PM}\sigma_{cop}(n_{cop} - n_{PM})}{\pi L_{PM}} \Phi_{cop}' \tag{A1}$$

$$\frac{d\Phi_{eddy}'}{dt} = \frac{30\mu_0\mu_r h_{cop}\alpha_{PM}\sigma_{cop}\Phi_{cop}'}{\pi L_{PM}} \frac{d(n_{cop} - n_{PM})}{dt} \tag{A2}$$

By substituting (A1) and (A2) into Equation (28), (A3) is derived.

$$2R_{ac}\Phi_{cop}' = \frac{30\mu_0\mu_r h_{cop}\alpha_{PM}\sigma_{cop}\Phi_{cop}'}{\pi L_{PM}} \left[(R_{slip} + R_{out})(n_{cop} - n_{PM}) + M_L \frac{d(n_{cop} - n_{PM})}{dt} \right] \tag{A3}$$

The above relationships are substituted into Equation (34) to realize the analytical expressions for eddy and cop. Then, (A4)–(A10) can be smoothly deduced.

$$2R_{ac}\Phi_{cop}' = \left[\Phi_{es} - \left(\Phi_{eddy}' + \Phi_{cop}'\right)\right] R_{//} \tag{A4}$$

$$\Phi_{eddy}' + \Phi_{cop}' = \frac{30\mu_0\mu_r h_{cop}\alpha_{PM}\sigma_{cop}(n_{cop} - n_{PM}) + \pi L_{PM}}{\pi L_{PM}} \Phi_{cop}' \tag{A5}$$

$$\Phi_{es} - \left(\Phi_{eddy}' + \Phi_{cop}'\right) = \frac{30\mu_0\mu_r h_{cop}\alpha_{PM}\sigma_{cop}\Phi_{cop}'}{\pi L_{PM} R_{//}} \left[\left(R_{slip} + R_{out}\right)\left(n_{cop} - n_{PM}\right) + M_L \frac{d(n_{cop} - n_{PM})}{dt}\right] \quad (A6)$$

$$\Phi_{es} = \frac{\Phi_{cop}'}{\pi L_{PM} R_{//}} \left\{ \left[\left(n_{cop} - n_{PM}\right) + \frac{\left(R_{slip} + R_{out}\right)\left(n_{cop} - n_{PM}\right) + M_L \frac{d(n_{cop} - n_{PM})}{dt}}{R_{//}}\right] \begin{array}{c} 30\mu_0\mu_r h_{cop}\alpha_{PM}\sigma_{cop} \times \\ \\ \end{array} + \pi L_{PM} \right\} \quad (A7)$$

$$L_1 = \frac{1}{\pi L_{PM} R_{//}} \left\{ \left[\left(n_{cop} - n_{PM}\right) + \frac{\left(R_{slip} + R_{out}\right)\left(n_{cop} - n_{PM}\right) + M_L \frac{d(n_{cop} - n_{PM})}{dt}}{R_{//}}\right] \begin{array}{c} 30\mu_0\mu_r h_{cop}\alpha_{PM}\sigma_{cop} \times \\ \\ \end{array} + \pi L_{PM} \right\} \quad (A8)$$

$$\Phi_{cop}' = \frac{\Phi_{es}}{L_1} \quad (A9)$$

$$\Phi_{eddy}' = \frac{30\mu_0\mu_r h_{cop}\alpha_{PM}\sigma_{cop}(n_{cop} - n_{PM})}{\pi L_{PM}} \frac{\Phi_{es}}{L_1} \quad (A10)$$

At this point, Φ_{eddy}' and Φ_{cop}' have been expressed as intermediate variables for solving the torque. Substituting (A9) and (A10) into Equations (35) and (36), respectively, can further determine the eddy current loss and torque of the APMC. It can be found that the output torque is a quantity related to the change in speed over time, i.e., acceleration, which is in line with objective facts and laws.

References

1. Ye, L.; Li, D.; Ma, Y.; Jiao, B. Design and Performance of a Water-Cooled Permanent Magnet Retarder for Heavy Vehicles. *IEEE Trans. Energy Convers.* **2011**, *26*, 953–958. [CrossRef]
2. Wang, J.; Lin, H.; Fang, S.; Huang, Y. A General Analytical Model of Permanent Magnet Eddy Current Couplings. *IEEE Trans. Magn.* **2014**, *50*, 1–9. [CrossRef]
3. Atallah, K.; Wang, J. A Brushless Permanent Magnet Machine with Integrated Differential. *IEEE Trans. Magn.* **2011**, *47*, 4246–4249. [CrossRef]
4. Yang, X.; Liu, Y.; Wang, L. Nonlinear Modeling of Transmission Performance for Permanent Magnet Eddy Current Coupler. *Math. Probl. Eng.* **2019**, *2019*, 1–14. [CrossRef]
5. Shin, H.-J.; Choi, J.-Y.; Jang, S.-M.; Lim, K.-Y. Design and Analysis of Axial Permanent Magnet Couplings Based on 3D FEM. *IEEE Trans. Magn.* **2013**, *49*, 3985–3988. [CrossRef]
6. Potgieter, J.H.J.; Kamper, M.J. Optimum Design and Comparison of Slip Permanent-Magnet Couplings with Wind Energy as Case Study Application. *IEEE Trans. Ind. Appl.* **2014**, *50*, 3223–3234. [CrossRef]
7. Gay, S.E.; Ehsani, M. Parametric Analysis of Eddy-Current Brake Performance by 3-D Finite-Element Analysis. *IEEE Trans. Magn.* **2006**, *42*, 319–328. [CrossRef]
8. Canova, A.; Cavalli, F. Design Procedure for Hysteresis Couplers. *IEEE Trans. Magn.* **2008**, *44*, 2381–2395. [CrossRef]
9. Zhang, B.; Wan, Y.; Li, Y.; Feng, G. Optimized Design Research on Adjustable-Speed Permanent Magnet Coupling. In Proceedings of the 2013 IEEE International Conference on Industrial Technology (ICIT), Cape Town, South Africa, 25–28 February 2013; IEEE: New York, NY, USA, 2013; pp. 380–385.
10. Erasmus, A.S.; Kamper, M.J. Analysis for Design Optimisation of Double PM-Rotor Radial Flux Eddy Current Couplers. In Proceedings of the 2015 IEEE Energy Conversion Congress and Exposition (ECCE), Montreal, QC, Canada, 20–24 September 2015; IEEE: New York, NY, USA; pp. 6503–6510.
11. Canova, A.; Vusini, B. Design of Axial Eddy-Current Couplers. *IEEE Trans. Ind. Appl.* **2003**, *39*, 725–733. [CrossRef]
12. Canova, A.; Vusini, B. Analytical Modeling of Rotating Eddy-Current Couplers. *IEEE Trans. Magn.* **2005**, *41*, 24–35. [CrossRef]
13. Shin, H.-J.; Choi, J.-Y.; Cho, H.-W.; Jang, S.-M. Analytical Torque Calculations and Experimental Testing of Permanent Magnet Axial Eddy Current Brake. *IEEE Trans. Magn.* **2013**, *49*, 4152–4155. [CrossRef]
14. Park, M.-G.; Choi, J.-Y.; Shin, H.-J.; Jang, S.-M. Torque Analysis and Measurements of a Permanent Magnet Type Eddy Current Brake with a Halbach Magnet Array Based on Analytical Magnetic Field Calculations. *J. Appl. Phys.* **2014**, *115*, 17E707. [CrossRef]
15. Lubin, T.; Rezzoug, A. Steady-State and Transient Performance of Axial-Field Eddy-Current Coupling. *IEEE Trans. Ind. Electron.* **2015**, *62*, 2287–2296. [CrossRef]
16. Lubin, T.; Rezzoug, A. 3-D Analytical Model for Axial-Flux Eddy-Current Couplings and Brakes Under Steady-State Conditions. *IEEE Trans. Magn.* **2015**, *51*, 1–12. [CrossRef]

17. Lubin, T.; Rezzoug, A. Improved 3-D Analytical Model for Axial-Flux Eddy-Current Couplings with Curvature Effects. *IEEE Trans. Magn.* **2017**, *53*, 1–9. [CrossRef]
18. Jang, G.-H.; Koo, M.-M.; Kim, J.-M.; Choi, J.-Y. Torque Characteristic Analysis and Measurement of Axial Flux-Type Non-Contact Permanent Magnet Device with Halbach Array Based on 3D Analytical Method. *AIP Adv.* **2017**, *7*, 056647. [CrossRef]
19. Lubin, T.; Mezani, S.; Rezzoug, A. Two-Dimensional Analytical Calculation of Magnetic Field and Electromagnetic Torque for Surface-Inset Permanent-Magnet Motors. *IEEE Trans. Magn.* **2012**, *48*, 2080–2091. [CrossRef]
20. Sheikh-Ghalavand, B.; Vaez-Zadeh, S.; Hassanpour Isfahani, A. An Improved Magnetic Equivalent Circuit Model for Iron-Core Linear Permanent-Magnet Synchronous Motors. *IEEE Trans. Magn.* **2010**, *46*, 112–120. [CrossRef]
21. Qu, R.; Lipo, T.A. Analysis and Modeling of Air-Gap and Zigzag Leakage Fluxes in a Surface-Mounted Permanent-Magnet Machine. *IEEE Trans. Ind. Appl.* **2004**, *40*, 121–127. [CrossRef]
22. Momen, M.F.; Datta, S. Analysis of Flux Leakage in a Segmented Core Brushless Permanent Magnet Motor. *IEEE Trans. Energy Convers.* **2009**, *24*, 77–81. [CrossRef]
23. Aberoomand, V.; Mirsalim, M.; Fesharakifard, R. Design Optimization of Double-Sided Permanent-Magnet Axial Eddy-Current Couplers for Use in Dynamic Applications. *IEEE Trans. Energy Convers.* **2019**, *34*, 909–920. [CrossRef]
24. Wang, J.; Zhu, J. A Simple Method for Performance Prediction of Permanent Magnet Eddy Current Couplings Using a New Magnetic Equivalent Circuit Model. *IEEE Trans. Ind. Electron.* **2018**, *65*, 2487–2495. [CrossRef]
25. Lubin, T.; Mezani, S.; Rezzoug, A. Simple Analytical Expressions for the Force and Torque of Axial Magnetic Couplings. *IEEE Trans. Energy Convers.* **2012**, *27*, 536–546. [CrossRef]
26. You, S.-A.; Wang, S.-H.; Tsai, M.-C.; Mao, S.-H. Characteristic Analysis of Slotted-Type Axial-Flux Permanent Magnetic Couplers. In Proceedings of the 2012 15th International Conference on Electrical Machines and Systems, Sapporo, Japan, 21–24 October 2012; p. 5.
27. de la Barriere, O.; Hlioui, S.; Ben Ahmed, H.; Gabsi, M.; LoBue, M. 3-D Formal Resolution of Maxwell Equations for the Computation of the No-Load Flux in an Axial Flux Permanent-Magnet Synchronous Machine. *IEEE Trans. Magn.* **2012**, *48*, 128–136. [CrossRef]
28. Chen, J.T.; Zhu, Z.Q. Influence of the Rotor Pole Number on Optimal Parameters in Flux-Switching PM Brushless AC Machines by the Lumped-Parameter Magnetic Circuit Model. *IEEE Trans. Ind. Appl.* **2010**, *46*, 1381–1388. [CrossRef]
29. Yang, F.; Zhu, J.; Yang, C.; Ding, Y.; Hang, T. A Simple Method to Calculate the Torque of Magnet-Rotating-Type Axial Magnetic Coupler Using a New Magnetic Equivalent Circuit Model. *IEEE Trans. Magn.* **2022**, *58*, 1–12. [CrossRef]
30. Jenei, S.; Nauwelaers, B.; Decoutere, S.; Nacm, A. Closed form inductance calculation for integrated spiral inductor compact modeling. In Proceedings of the 2000 Topical Meeting on Silicon Monolithic Integrated Circuits in Rf Systems, Digest of Papers, Garmisch Partenki, Germany, 26–28 April 2000.

Disclaimer/Publisher's Note: The statements, opinions and data contained in all publications are solely those of the individual author(s) and contributor(s) and not of MDPI and/or the editor(s). MDPI and/or the editor(s) disclaim responsibility for any injury to people or property resulting from any ideas, methods, instructions or products referred to in the content.

Article

Sizing, Modeling, and Performance Comparison of Squirrel-Cage Induction and Wound-Field Flux Switching Motors

Chiweta E. Abunike [1,2], Udochukwu B. Akuru [3,4,*], Ogbonnaya I. Okoro [1] and Chukwuemeka C. Awah [1]

1. Department of Electrical and Electronic Engineering, Michael Okpara University of Agriculture, Umudike 440101, Abia State, Nigeria; abunike.emmanuel@mouau.edu.ng (C.E.A.); okoro.ogbonnaya@mouau.edu.ng (O.I.O.); ccawah@ieee.org (C.C.A.)
2. School of Engineering, University of Aberdeen, Aberdeen AB24 3UE, UK
3. Department of Electrical Engineering, Tshwane University of Technology, Pretoria 0183, South Africa
4. Department of Electrical Engineering, University of Nigeria, Nsukka 410001, Enugu State, Nigeria
* Correspondence: akuruub@tut.ac.za; Tel.: +27-01-382-5605

Citation: Abunike, C.E.; Akuru, U.B.; Okoro, O.I.; Awah, C.C. Sizing, Modeling, and Performance Comparison of Squirrel-Cage Induction and Wound-Field Flux Switching Motors. *Mathematics* 2023, 11, 3596. https://doi.org/10.3390/math11163596

Academic Editor: Jacques Lobry

Received: 20 July 2023
Revised: 16 August 2023
Accepted: 17 August 2023
Published: 19 August 2023

Copyright: © 2023 by the authors. Licensee MDPI, Basel, Switzerland. This article is an open access article distributed under the terms and conditions of the Creative Commons Attribution (CC BY) license (https://creativecommons.org/licenses/by/4.0/).

Abstract: In this study, the analytical design and electromagnetic performance comparison of a squirrel-cage induction motor (SCIM) and a wound-field flux switching motor (WFFSM) for high-speed brushless industrial motor drives is undertaken for the first time. The study uses analytical sizing techniques and finite element analysis (FEA) to model and predict the performance of both motors at a 7.5 kW output power. This study includes detailed equations and algorithms for sizing and modeling of both types of motors, as well as performance calculations that aid in motor selection, design optimization, and system integration. The main findings show that the SCIM has superior torque performance for starting and overload conditions, while the WFFSM offers advantages in power factor, efficiency over a wide operating range, and potential for higher peak power output. To this end, the WFFSM is capable of high-speed and high-efficiency operation while the SCIM is suitable for applications requiring variable speed operation. The validation study shows good agreement between analytical and FEA calculations for both motors. The results provide insights into the design and performance characteristics of both motors, enabling researchers to explore innovative approaches for improving their efficiency, reliability, and overall performance.

Keywords: analytical design; brushless; finite element analysis (FEA); modeling; sizing; squirrel-cage induction motor (SCIM); wound-field flux switching motor (WFFSM)

MSC: 78-10; 78M10; 97M50

1. Introduction

Electric motors play a critical role in global energy conservation efforts, accounting for about half of all electrical energy consumed globally [1]. In recent years, there has been increasing interest in high-speed motors with low or no permanent magnet (PM) materials due to issues such as high cost, uncontrollable flux, and limited operating temperature associated with PM machines. Some examples of non-PM motors are squirrel-cage induction motors (SCIMs) and switched reluctance motors (SRMs). SCIMs have been widely used historically due to their robustness, low cost, manufacturing simplicity, and easy adaptation to high-speed operation [2,3]. In contrast, a relatively new class of machines belonging to the flux modulation family has recently emerged, such as the wound-field flux switching motor (WFFSM), which is characterized by features such as high torque density, robust rotor structure, and high-speed operation [4].

SCIMs consist of short-circuit copper or aluminum rotor bars that fit into rotor slots, requiring less conductor material due to the absence of brushes, commutators, and sliprings compared with slip-ring induction motors (SRIMs). As a result, SCIMs have higher efficiency than SRIMs. SCIMs are commonly used in various industrial applications, such as centrifugal pumps, industrial drives (e.g., running conveyor belts), large blowers and fans, machine tools, lathes, and other turning equipment [1]. Despite the advantages offered by SCIMs, they suffer some drawbacks such as high starting current and low efficiency under partial loads [2,5]. Similarly, although WFFSMs offer some attractive features as mentioned earlier, they are nonetheless prone to low power factor due to high leakage flux and high torque ripple due to the double-salient structure [6,7].

Therefore, the aim of this study is to compare SCIMs and WFFSMs, which, as magnet-less and brushless motors, both offer low-cost and robust designs for high-speed industrial motor drives. While some studies have presented comparisons of WFFSMs to other machines, mainly in terms of wind generator and electric traction applications, there is little research on comparing SCIMs and WFFSMs to the family of three-phase low-power industrial motors. For example, WFFSMs have been compared to ferrite PM flux switching machines (PMFSMs) as wind generators [8] and switched reluctance machines (SRMs) as electric vehicle motors [9], with the ferrite PMFSM exhibiting higher torque density and efficiency and SRM displaying lower torque capability, saturation withstand capability and higher torque ripple, compared with WFFSM, respectively. On the other hand, SCIMs have been compared to other motors, such as line-start permanent magnet synchronous motors (LSPMSMs) [10], synchronous reluctance motors (SynRMs) [11], and SRMs [12], with the SCIM found to operate at a lower efficiency than all three motors.

To this end, the current study compares SCIMs and WFFSMs by using analytical sizing techniques to model, evaluate, and compare their electromagnetic performance based on the finite-element analysis (FEA) method. While both SCIMs and WFFSMs have distinct pros and cons, there has been limited research comparing their sizing, modeling, and performance specifically for high-speed industrial drive applications in the lower power range. SCIMs are commonly used in industry, but they face efficiency challenges at partial loads. WFFSMs show potential but need further evaluation across operating regimes. This study aims to provide a detailed comparison between SCIMs and WFFSMs based on analytical and FEA techniques. The results can guide design trade-offs and technology selection decisions for various industrial end-uses. This study considers the high-speed operating regimes of both machines at 7.5 kW output power, representing typical operating requirements. The analysis provides insights into the electromagnetic performance differences between the two machines. The rest of the paper is organized as follows: Section 2 presents the semi-analytical design and performance modeling techniques of the proposed SCIM and WFFSM. Section 3 reports the FEA no-load and on-load electromagnetic performance comparison, while Section 4 presents and discusses the electromagnetic performance comparison in terms of analytic and FEA calculations. The concluding remarks are provided in Section 5.

2. Sizing-Design Algorithms and Analytical Performance Calculations for SCIM and WF-FSM

SCIMs and WFFSMs represent very different motor technologies in terms of their operating principles. Their analytical designs are based on sizing equations that give the machine's outer diameter as a function of design specifications such as the leakage factor (k_e) and aspect ratio (k_L). Performance-related values, such as efficiency (η), power factor (cosϕ), maximum airgap flux density (B_{gmax}), and stator electrical loading (A_s), are then imposed based on the machine type.

2.1. Design Process of SCIM

The operating principle of SCIM is based on the interaction between the rotating mutual airgap magnetic field and the rotor currents. As a varying magnetic field is required to produce the rotor current, there is a torque-proportional speed difference, commonly known as slip, between the airgap field and the rotor. The design steps for SCIM are shown in Figure 1.

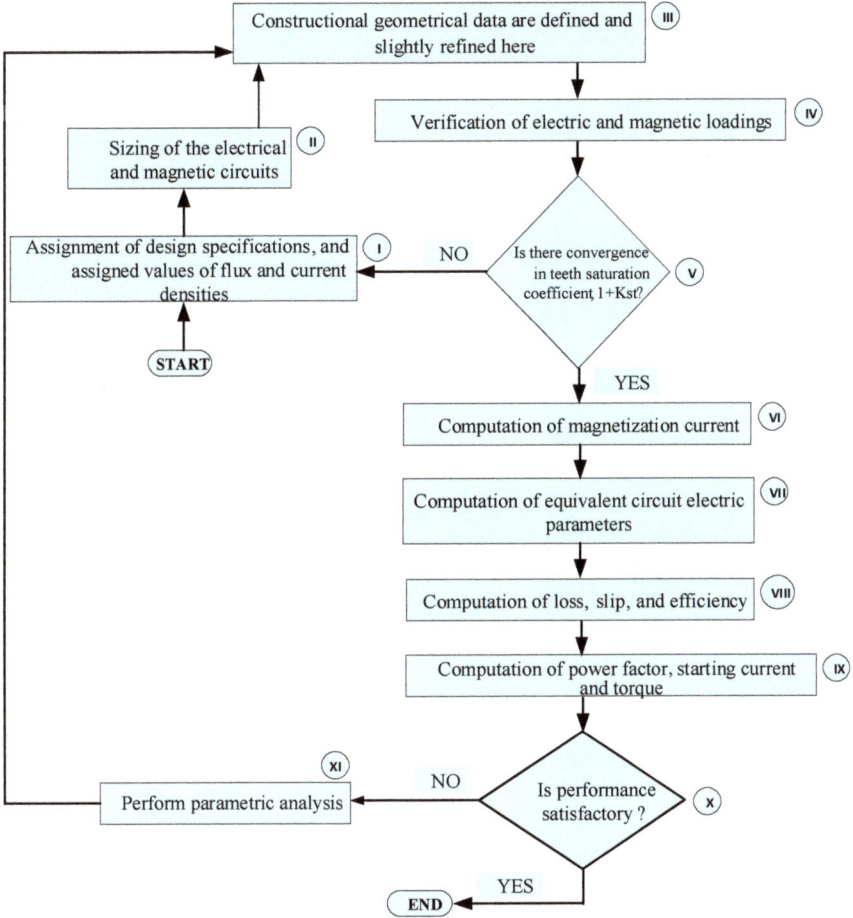

Figure 1. Flowchart for SCIM semi-analytical design process.

As shown in Figure 1, the design process for SCIMs begins with establishing the design specifications in I, while the assigned values of flux densities and current densities are calculated in II. The stator bore diameter (D_{is}), stack length, stator slots, stator outer diameter, and stator and rotor slot dimensions are then determined in III, with all dimensions adjusted to standardized values. Electric and magnetic loadings are verified in IV. In V, the design process loops back to I and the tooth flux density is adjusted until the magnetic saturation coefficient ($1 + K_{st}$) of the stator and rotor teeth are equal to prescribed values. Once this loop is completed, stages VI to IX are processed by computing the magnetization current in VI, equivalent circuit parameters in VII, losses, rated slip (S_n), and efficiency in VIII, and then determining the power factor, locked rotor current and torque, breakdown torque, and temperature rise in IX. In step X, all performance metrics are checked, initiating a potential process reset in I if any metrics are found unsatisfactory. During this reset, geometrical data are refined via parametric analysis. These new optimal values are then implemented, allowing the process to progress until reaching step X once again. The iterative cycle continues until the point when performance criteria are met satisfactorily, signifying the termination of the process.

2.1.1. Design Sizing for SCIM

The stator bore diameter of the SCIM is given as [13]:

$$D_{is(SCIM)} = \sqrt[3]{\frac{2p_1^2 S_{gap}}{\pi \lambda f C_0}}, \tag{1}$$

where p_1, S_{gap}, λ, f, and C_0 represent pole pairs, apparent airgap power, aspect ratio, supply frequency, and Esson's constant (which is 147×10^3 J/m^3), respectively.

The pole pitch, $\tau_{(SCIM)}$, stack length, $L_{st(SCIM)}$, and slot pitch, $\tau_{s(SCIM)}$, for dimensioning the SCIM are derived as follows:

$$\tau_{(SCIM)} = \frac{\pi D_{is(SCIM)}}{2p_1}, \quad L_{st(SCIM)} = \lambda \tau_{(SCIM)}, \quad \tau_{s(SCIM)} = \frac{\tau_{(SCIM)}}{3q}. \tag{2}$$

The aspect ratio, pole pitch, stator slots per pole, are respectively represented as λ, $\tau_{(SCIM)}$, and q, while the stator outer diameter, $D_{os(SCIM)}$, airgap length, and rotor outer diameter are approximated as follows:

$$D_{os(SCIM)} = \frac{D_{is(SCIM)}}{0.62}, \tag{3}$$

$$g_{(SCIM)} = \left(0.1 + 0.012 \times \sqrt[3]{P_n}\right), \tag{4}$$

$$D_{or(SCIM)} = D_{is(SCIM)} - g_{(SCIM)}. \tag{5}$$

The rated power is denoted as P_n. The determination of the number of turns per phase, W_1, depends on the airgap flux density, $B_{g(SCIM)}$, and is evaluated as follows:

$$W_1 = \frac{0.22 V_{SCIM}}{K_{w1} f \alpha_i \tau_{(SCIM)} L_{st(SCIM)} B_{g(SCIM)}}, \tag{6}$$

where V_{SCIM}, K_{w1}, and α_i, represent supply voltage, stator winding factor, and pole spanning coefficient, respectively. The number of conductors per slot, $n_{c(SCIM)}$, is given as:

$$n_{c(SCIM)} = \frac{a_1 W_1}{p_1 q}. \tag{7}$$

The number of current paths in parallel is denoted as a_1. It should be emphasized that an even number of slots is required in a double-layer winding since there are two separate coils per slot. The rated current, I_{in}, and wire gauge diameter, d_{co}, are defined as follows:

$$I_{in} = \frac{P_n}{\eta \cos(\phi) \sqrt{3} V_{SCIM}}, \tag{8}$$

$$d_{co} = \sqrt{\frac{2 I_{in}}{\pi a_p J_{(SCIM)}}}, \tag{9}$$

where a_p, $J_{(SCIM)}$, and η represent conductors in parallel, current density, and efficiency, respectively.

Assuming all the airgap flux passes through the stator teeth, the useful slot area, A_{su}, and stator tooth width, b_{ts}, are respectively given as:

$$A_{su} = \frac{\pi d_{co}^2 a_p n_{c(SCIM)}}{4 K_{fill}}, \tag{10}$$

$$b_{ts} = \frac{B_{g(SCIM)} \tau_{(SCIM)}}{0.96 B_{ts}}. \tag{11}$$

The fill factor and stator tooth flux density are represented by K_{fill} and B_{ts}, respectively. For stator slot sizing, the lower width, b_{s1}, higher width, b_{s2}, and slot height, h_s, are given as:

$$b_{s1} = \frac{\pi\left(D_{in(SCIM)} + 2h_{os} + 2h_w\right)}{N_s} - b_{ts}, \tag{12}$$

$$b_{s2} = \sqrt{4A_{su}\tan\frac{\pi}{N_s} + b_{s1}^2}, \tag{13}$$

$$h_{s2} = \frac{2A_{su}}{b_{s1} + b_{s2}}. \tag{14}$$

The lower slot height, wedge height, slot effective area, and number of stator slots are denoted as h_{os}, h_W, A_{su}, and N_s, respectively. If stator and rotor teeth produce the same effects, the teeth saturation factor $(1 + K_{st})$ is calculated as follows:

$$1 + K_{st} = \frac{1 + F_{mts} + F_{mtr}}{F_{mg}}, \tag{15}$$

where F_{mts}, F_{mtr}, and F_{mg} represent stator tooth, rotor tooth, and airgap MMFs, respectively. The end-ring current, I_{er}, and magnetization current, I_μ, are respectively calculated as:

$$I_{er} = \frac{I_b}{2\sin\left(\frac{\pi p_1}{N_r}\right)}, \tag{16}$$

$$I_\mu = \frac{\pi p_1\left(\frac{F}{2}\right)}{\sqrt[3]{2}W_1 K_{w1}}, \tag{17}$$

where I_b, N_r, and F are the rotor bar current, number of rotor slots, and the magnetization MMF, respectively.

For the rotor slot sizing, the rotor slot pitch, τ_r, and tooth width, b_{tr}, are respectively calculated as:

$$\tau_r = \frac{\pi\left(D_{in(SCIM)} - 2g_{(SCIM)}\right)}{N_r}, \tag{18}$$

$$b_{tr} = \frac{\tau_r B_{g(SCIM)}}{0.96 B_{tr}}. \tag{19}$$

The rotor tooth flux density is represented as B_{tr}. The rotor slot geometry is obtained by using the slot area, A_b, equation given as follows:

$$A_b = \frac{\pi}{8}\left(d_1^2 + d_2^2\right) + \frac{(d_1 + d_2)h_r}{2}. \tag{20}$$

The diameters d_1 and d_2 are obtained simultaneously from

$$d_1 = \frac{\pi(D_{re} - 2h_{or}) - N_r b_{tr}}{\pi + N_r} \text{ and } d_1 - d_2 = 2h_r \tan\frac{\pi}{N_r}. \tag{21}$$

The outer diameter of the rotor, rotor slot height, height of rotor back core, and lower height of the rotor slot are denoted as $D_{or(SCIM)}$, h_r, h_{cr}, and h_{or}, respectively. The maximum diameter of the shaft, D_{shaft}, is obtained as follows:

$$\left(D_{shaft}\right)_{max} \leq D_{is(SCIM)} - 2g_{(SCIM)} - 2\left(h_{or} + \frac{d_1 + d_2}{2} + h_r + h_{cr}\right). \tag{22}$$

The selection of the shaft diameter is determined by the rated torque, taking into account mechanical design considerations and previous knowledge. It is crucial to emphasize that these parameters will be recalculated if they do not meet the design objectives.

2.1.2. Performance Calculations for SCIM

The equivalent circuit and phasor diagrams of the SCIM, which are vital tools for understanding its performance behavior, are presented in Figure 2.

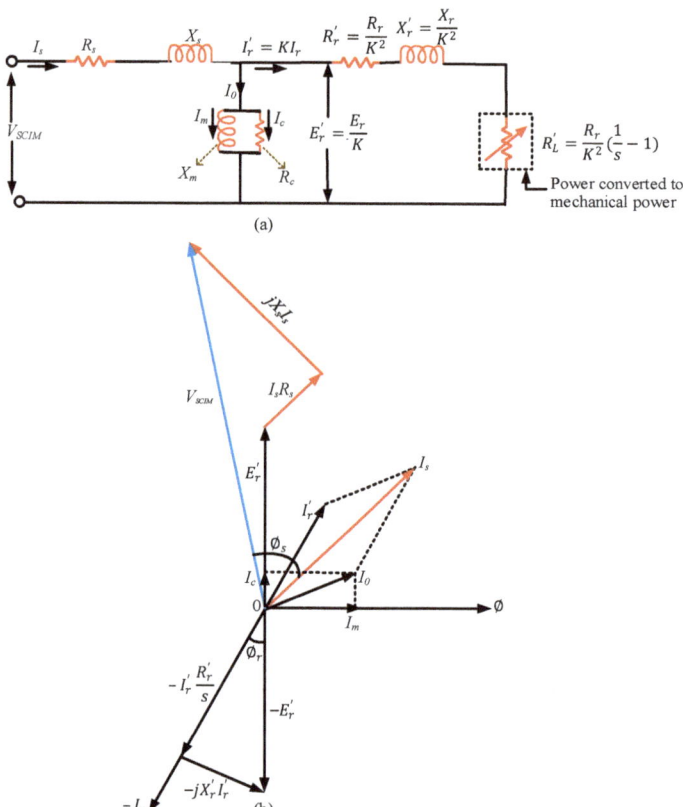

Figure 2. Model representation of SCIM: (**a**) per phase equivalent circuit and (**b**) phasor diagram.

The equivalent circuit in Figure 2a, simplifies the SCIM's electrical representation, enabling analysis of current flow, voltage drop, and power transfer. On the other hand, the phasor diagram of Figure 2b graphically depicts the motor's voltages and currents, facilitating visualization of phase relationships. In Figure 2, R_s, X_s, X_m, R_c, R'_r, R_r, X'_r, X_r, K, and R'_L represent stator resistance, stator inductive reactance, magnetizing reactance, core loss component, rotor resistance referred to stator, rotor resistance, rotor inductive reactance referred to stator, transformation ratio, and load resistance referred to the stator, respectively. These parameters aid in the calculation of performance characteristics such as power factor, rated slip, rated torque, efficiency, and magnetization behavior of the motor. The phase parameters of stator and rotor are respectively given as:

$$R_s = \frac{\rho_{co} L_c W_1}{A_{co} a_1}, \tag{23}$$

$$X_s = \frac{2\mu_0 \omega_1 L W_1^2 (\lambda_s + \lambda_{ds} + \lambda_{ec})}{p_1 q}, \tag{24}$$

$$R_r = \frac{4m(W_1 K_{w1})^2 R_{be}}{N_r}, \quad (25)$$

$$X_r = \frac{4m(W_1 K_{w1})^2 X_{be}}{N_r}, \quad (26)$$

while the magnetizing reactance is computed as:

$$X_m = \sqrt{\left(\frac{V_{SCIM}}{I_\mu}\right)^2 - R_s^2 - X_{sL}}. \quad (27)$$

The coil length, copper resistivity, permeability of free space, magnetic wire cross section, slot differential and end ring connection coefficients, equivalent rotor bar resistance and reactance, and phase rotor resistance are denoted as L_c, ρ_{co}, μ_o, A_{co}, λ_s, λ_{ds}, λ_{ec}, R_{be}, and X_{be}, respectively.

The rated slip s_n, starting current I_{str}, and starting torque T_{str} of the motor are respectively defined as:

$$s_n = \frac{P_{Al}}{P_n + P_{Al} + P_{mv} + P_{stray}}, \quad (28)$$

$$I_{str} = \frac{V_{SCIM} - E_r'}{jX_s} = \frac{V_{SCIM}}{\sqrt{R_s + R_r^{s=1\,2} + (X_s + X_r^{s=1})^2}}, \quad (29)$$

$$T_{str} = \frac{3p_1 R_r^{s=1} I_{str}^2}{\omega_n}, \quad (30)$$

where P_{Al}, P_{mv}, and P_{stray} represent rotor cage losses, mechanical/ventilation losses, and stray losses, respectively.

The power factor of the electrical motor is computed as:

$$PF = \frac{P_{in}}{3V_{rms} I_{rms}} = \cos(\theta_v - \theta_i), \quad (31)$$

where P_{in} is the input power of the motor, I_{rms} and V_{rms} are the root mean square (RMS) values of both the current (I_S) and voltage (V_{SCIM}), while θ_v and θ_i are the phase of voltage and current, respectively.

On the other hand, the rated shaft torque, T_n, and torque ripple, TR, are respectively computed as follows:

$$T_n = \frac{P_n}{\frac{2\pi f}{p_1}(1 - s_n)}, \quad (32)$$

$$TR = \frac{T_{max} - T_{min}}{T_{avg}}, \quad (33)$$

where $T_{(max)}$ is the maximum torque, $T_{(min)}$ is the minimum torque, and T_{avg} is the average torque.

While the motion equation is given as:

$$T_n - T_L = J\frac{d\omega_r}{dt}, \quad (34)$$

where T_L is the load torque and J is the inertia constant.

Based on the classical equation, efficiency of the electrical machine is defined as:

$$\eta = \frac{P_{out}}{P_{in}} = \frac{P_{out}}{P_{out} + P_{loss}}, \quad (35)$$

with the P_{out} and total loss (P_{loss}) calculated as

$$P_{out} = 3I_r^2 R_r \left(\frac{1-s}{s}\right) - P_{mv} - P_{stray}, \tag{36}$$

$$P_{loss} = P_{cu} + P_{Al} + P_{iron} + P_{mv} + P_{stray}, \tag{37}$$

where P_{cu} is the stator winding losses and P_{iron} is the core loss.

The core loss is computed based on the formula proposed by [13] as:

$$P_{iron} = 7.16 \left(\frac{f}{50}\right)^{1.3} G_{ts} + 5.12 \left(\frac{f}{50}\right)^{1.3} G_{yl} + 31.03 \times 10^{-7} \left[\left(N_r \frac{f}{P_n}\right)^2 G_{ts} + \left(N_s \frac{f}{P_n}\right)^2 G_{tr}\right], \tag{38}$$

where G_{ts}, G_{yl}, and G_{tr} stand for stator tooth, yoke, and rotor tooth masses, respectively. The other loss components are defined as:

$$P_{cu} = 3R_s I_{in}^2, \ P_{Al} = 3R_r I_{in}^2, \ P_{mv} = 0.01 P_n, \ P_{stray} = 0.01 P_n. \tag{39}$$

In the study, P_{mv} represents mechanical losses that are assumed to be 1% of the nominal power P_n. This value is chosen based on standards referenced in [13], which suggest using $0.01\,P_n$. P_{stray} represents stray load losses, which account for additional losses due to harmonic fields and leakage fluxes not accounted for in the analytical model. Based on [13], we conservatively assume these to be 1% of P_n. The reason for choosing these specific values is the challenge of accurately measuring losses in high-efficiency machines. Even small measurement errors can cause significant uncertainty in calculated loss percentages. Given the four-pole design of our SCIM, the 1% contributions for both P_{mv} and P_{stray} follow recommended practice to ensure the overall losses and efficiency are within acceptable bounds.

The calculations in this section provide valuable insights into the performance of SCIM, enabling informed decisions for motor selection, design optimization, and system integration. Furthermore, a comprehensive understanding of the performance characteristics empowers researchers to explore innovative approaches and strategies for improving the efficiency, reliability, and overall performance of SCIMs.

2.2. Design Process of WFFSM

WFFSM is a double-salient pole motor with a robust rotor and yields sinusoidal flux linkages. The operating theory of the WFFSM, whereby the flux generated in the stator by the field coils is switched between adjacent armature coils and then linked through the rotor, is clearly elucidated in ref. [14].

2.2.1. Design Sizing for WFFSM

Figure 3 shows the flowchart for setting up the design process of the WFFSM. The design process starts at I with a selection of appropriate topology of the machine (slots/pole combination) which normally should consider the rotor mechanical phase and phase number of the proposed motor. Thereafter, the complete specification of the technical data, such as rated torque, power factor, target efficiency, and supply voltage, is made in II. In III, the machine is then sized based on the generic torque per rotor volume (TRV) equation as follows [15]:

$$D_{out} = \sqrt[3]{\frac{4T_e P_s}{\pi^2 P_r k_e k_L \Lambda_0^3 A_\Sigma B_g \eta c_s}}, \tag{40}$$

where D_{out} is the stator outer diameter, T_e is electromagnetic torque, Λ_0 is the split ratio, P_s is the number of slots for the phase windings, P_r is the number of rotor slots, k_e is a factor that accounts for leakage, A_s is the electrical loading of the phase windings, B_g is the peak airgap flux density, η is the efficiency, k_L is the aspect ratio, c_s is the stator tooth arc factor, and A_- is the summation of the preselected electrical loading for both the phase and field windings.

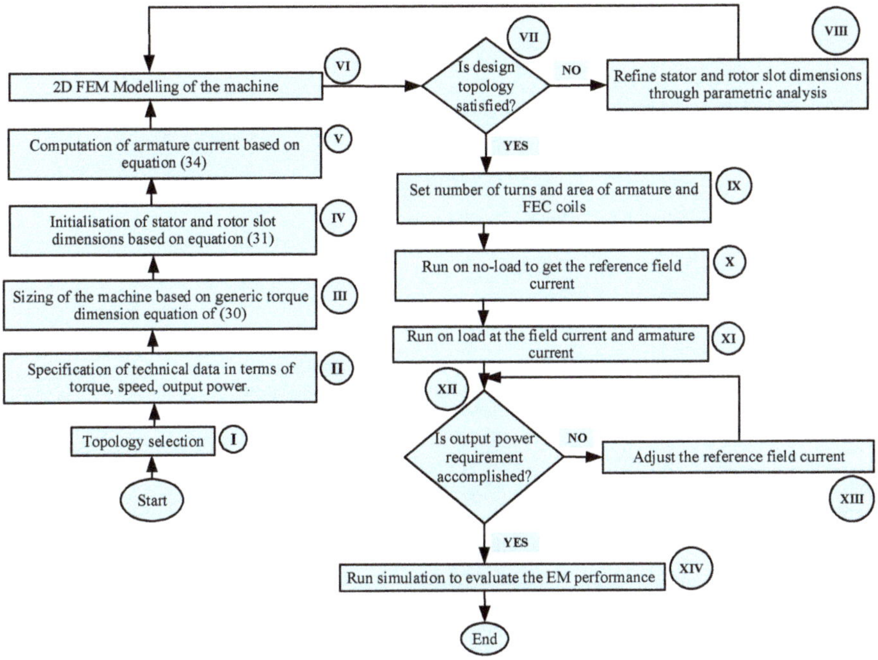

Figure 3. Flowchart for WFFSM semi-analytical design process.

Thereafter, parameters such as outer and inner diameter of stator and rotor and stack length are obtained. In IV, the stator and rotor dimensions such as stator pole width (b_{ps}), stator yoke height (h_{ys}), rotor pole width (b_{pr}), and rotor yoke height (h_{yr}) are scaled to the same value as follows:

$$b_{ps} = h_{ys} = b_{pr} = h_{yr} = \frac{\pi D_{in(WFFSM)}}{P_s}, \qquad (41)$$

where $D_{in(WFFSM)}$ is the stator inner diameter and P_s is the stator slot number.

If the design topology is not satisfied in VII, parametric analysis will be carried out in VIII to fine tune the stator and rotor dimensions. These new optimal values are subsequently integrated into the 2D FEM modeling phase of step VI and the iterative cycle continues until the design topology is achieved.

2.2.2. Performance Calculations for WFFSM

The performance characteristics of a WFFSM are crucial for optimizing its design and understanding its capabilities. By analyzing parameters such as power, electromotive force (EMF), current, torque, efficiency, and core loss, engineers gain valuable insights into the machine's performance, aiding in the selection of optimal designs. In this section, we establish and relate these performance characteristics with those of the SCIM. Additionally, the equivalent circuit and phasor diagram of the WFFSM are presented in Figure 4 to provide model representation of the motor.

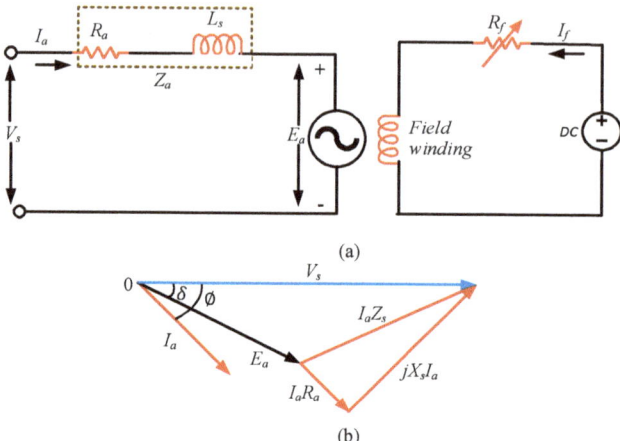

Figure 4. Model representation of WFFSM: (**a**) per phase equivalent circuit and (**b**) phasor diagram.

The equivalent circuit and phasor diagram in Figure 4 provide valuable insights into the motor behavior of the WFFSM. These representations promote the analysis of the flow of current, voltage distribution, and power transfer within the motor. By examining the equivalent circuit and phasor diagram, a deeper understanding of the interplay between the field winding, armature winding, and the magnetic field is achieved, providing a foundation for studying the motor's performance characteristics and optimizing its design. In Figure 4, V_s, I_a, R_a, L_s, X_s, Z_a, E_a, R_f, I_f, $\varnothing$, and δ represent phase voltage, armature current, armature resistance, WFFSM inductance, reactance, impedance, back-EMF, field resistance, field current, phasor angle, and load angle, respectively.

Assuming that sinusoidal armature current is injected into phase winding of the WFFSM in phase with back-EMF to produce unidirectional power, the input power, considering negligible resistance, is expressed as follows:

$$P_{in} = \frac{1}{T}\int_0^T e_a(t)\cdot i_a(t)dt, \tag{42}$$

where $e_a(t) = E_m sin\left(\frac{2\pi}{T}t\right)$ and $i_a(t) = I_m sin\left(\frac{2\pi}{T}t\right)$, m is the phase number, and E_m and I_m are the magnitude of sinusoidal phase back-EMF and current, respectively. Also, $e_a(t)$ can be calculated as:

$$e_a(t) = \frac{d\psi_m}{dt} = n_{ph}\frac{d\phi_m}{d\theta}\frac{d\theta}{dt} = n_{ph}\frac{d\phi_m}{d\theta}\omega_r = \omega_r i_f \frac{dM_{fa}}{d\theta}, \tag{43}$$

where ψ_m, n_{ph}, θ, ϕ_m, ω_r, i_f, and M_{fa} are total flux linkage, winding turns per phase, rotor position, flux per turn, rotor electrical angle velocity, field current, and mutual inductance between the field and armature windings, respectively.

The peak and field currents for the WFFSM can be calculated as:

$$I_m = \frac{2P_{in}}{3E_m PF}, \tag{44}$$

$$i_f = \frac{J_f \alpha S_f}{N_f}, \tag{45}$$

where PF, J_f, α, S_f, and N_f are power factor, current density, slot filling factor, slot area, and number of turns of the field windings, respectively. Moreover, by applying the co-energy concept to the basic geometry of WFFSM, it has been shown in [16] that the electromagnetic torque (T_e) is given as:

$$T_e = i_a i_f \frac{dM_{fa}}{d\theta}. \qquad (46)$$

The importance of Equation (46) is that, as in the DC machine, torque can be independently controlled using either the armature or field currents.

Since the WFFSM is doubly excited, unlike the SCIM, the following equation is devised based on Figure 4 to numerically approximate the rated field current from the back-EMF (E_m):

$$E_a = \frac{V_s - I_a R_a}{1 - j}. \qquad (47)$$

The phase resistance (R_a) and the field winding (R_f) resistance are evaluated using cross-sections of the slot phase winding information as in [15]:

$$R_a = \frac{2Z_a N_{ph}^2 \rho_{cu} L_{st}}{A_{ph}}, \qquad (48)$$

$$R_f = \frac{2Z_f N_f^2 \rho_{cu} L_{st}}{A_f}, \qquad (49)$$

where N_{ph} and N_f are the turns number per coil for the phase and field windings, ρ_{cu} is the resistivity of copper at room temperature, A_{ph} and A_f are the area per coil for the phase and field windings, and Z_a and Z_f are the number of coils for the armature and field windings. The phase and field winding copper losses at rated conditions are calculated respectively as:

$$P_{cu} = 3I_a^2 R_a, \qquad (50)$$

$$P_f = I_f^2 R_f. \qquad (51)$$

The core losses are approximated from the sum of the hysteresis and eddy current losses and calculated by means of the modified Steinmetz empirical equation given in [15] as:

$$P_{core} = C_m B_{gmax}^{\sigma} f_e^{\beta} G, \qquad (52)$$

where C_m, G, β, σ, and B_{gmax} represent the material coefficient, the mass of the corresponding iron part, the exponent of the machine's electrical frequency, the exponent of the peak airgap flux density, and the maximum flux density of the airgap, respectively.

The output power is evaluated as:

$$P_{out} = \frac{2\pi n_s}{60} T_{avg}. \qquad (53)$$

The output power as a function of the load-angle (δ) is given as:

$$P_{out} = \frac{3V_s E_a}{X_s} \sin \delta, \qquad (54)$$

where n_s is the motor speed in rpm and T_{avg} is the average torque.

The power factor and efficiency are calculated as the same equations of (31) and (35), respectively. Lastly, the torque per volume (T_d) and torque per mass (T_m) of the motors can be determined using:

$$T_d = \frac{4T_{avg}}{\pi D_{out}^2 L_{st}}, \qquad (55)$$

$$T_m = \frac{T_{avg}}{mass}. \qquad (56)$$

In summary, this section outlines the sizing, design algorithms, and analytical performance calculations for both the SCIM and WFFSM. The step-by-step design processes are presented for each machine, highlighting key equations for determining dimensions, winding parameters, and performance metrics.

For the SCIM, the design process focuses on establishing specifications, calculating flux densities and current densities, determining stator and rotor dimensions, verifying loadings, and iteratively adjusting parameters until performance criteria are met. Critical sizing equations defines the stator bore diameter, stack length, slot pitch, and outer diameter. Performance calculations enable analysis of the equivalent circuit, losses, rated slip, efficiency, and other characteristics.

The design process for the WFFSM begins with selecting an appropriate slots/pole combination and specifying technical data such as rated torque and power factor. The machine is then sized based on the torque per rotor volume equation. Stator and rotor dimensions are scaled and parametric analysis fine-tunes the geometry. Back EMF, torque, power, losses, and other performance metrics are analytically derived.

In summary, the theoretical design algorithms and performance calculations provide valuable insights into the electrical and mechanical characteristics of the SCIM and WFFSM. The next section focuses on FEA modeling of both machines. By combining theoretical foundations with computational simulations, more accurate and efficient models can be developed to conclusively compare the two motors. This will elucidate their strengths and weaknesses, guiding future optimization for high-speed brushless industrial AC motors.

3. FEA Design and Modeling

The selected SCIM is a three-phase four-pole motor with 48 stator slots and 44 rotor bars. Figure 5 is the preliminary design of the SCIM in FEA software, as well as the stator and rotor slot dimensions. The specifications and sizing parameters, which are based on equations in Section 2.1.1., are presented in Table 1.

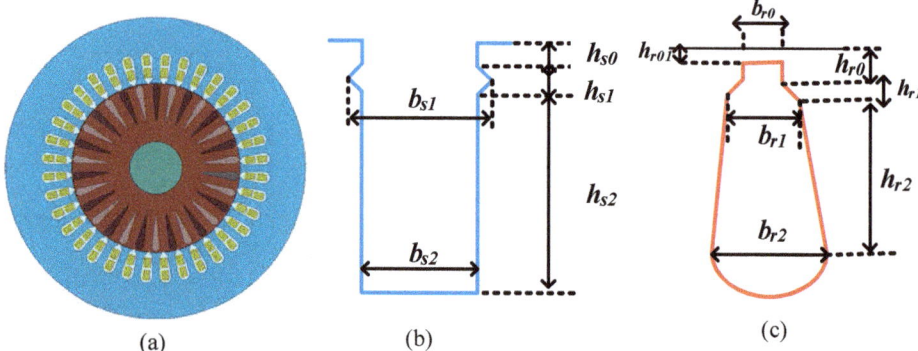

Figure 5. SCIM cross-sectional and slot parameters: (**a**) FEA model, (**b**) stator slot, and (**c**) rotor slot.

Table 1. Specifications and calculated parameters of the SCIM.

Description	Parameter	Value
Rated speed (rpm)	ω_s	1450
Number of poles	P_1	4
Number of stator slots	N_s	36
Split ratio	K_D	0.62
Stack length (mm)	$L_{st(SCIM)}$	154
Stator outer diameter (mm)	$D_{os(SCIM)}$	210.8
Stator inner diameter (mm)	$D_{is(SCIM)}$	130.7
Stator slot height (mm)	h_{s0}	0.8
Number of turns per phase	W_1	48
Number of rotor slots	N_r	30
Airgap length (mm)	$g_{(SCIM)}$	0.33
Rotor outer diameter (mm)	$D_{or(SCIM)}$	130.37
Rotor slot height (mm)	h_{r0}	0.5

For the WFFSM, a 24-stator slot 10-rotor pole design is selected. In the preliminary design, the split ratio, Λ_0, is maintained, same as for the SCIM. For fair comparison, other parameters of the WFFSM that are kept similar to those of the SCIM are output power (7.5 kW), supply frequency (50 Hz), number of phases (3), fill factor (0.45), target power factor (0.83), and target efficiency (0.895 p.u.). Also, the core material used in both SCIM and WFFSM is M400-50A. The FEA model and slot dimensions of the WFFSM are shown in Figure 6. The design parameters of the WFFSM evinced from Section 2.2.1 are presented in Table 2.

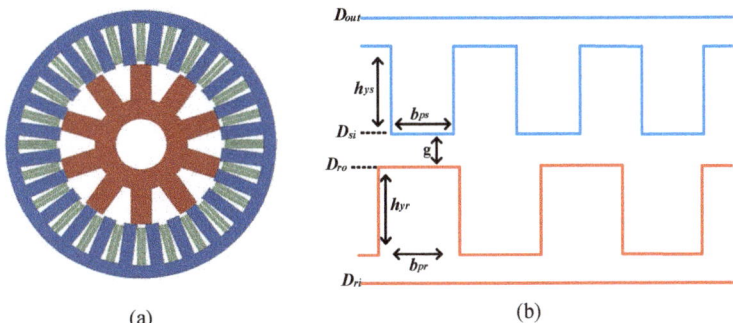

Figure 6. WFFSM cross-sectional and parameters: (**a**) FEA model and (**b**) cross-sectional design.

Table 2. Design parameters of the WF-FSM model.

Description	Parameter	Values
Rated speed (rpm)	ω_r	1500
Stack length (mm)	L_{st}	107
Aspect ratio	K_L	0.7
Split ratio	Λ_0	0.6
Number of stator slots	P_s	24
Stator outer diameter (mm)	D_{out}	254
Stator inner diameter (mm)	D_{in}	152
Stator tooth arc factor	C_s	0.25
Number of rotor slots	P_r	10
Airgap length (mm)	g	0.5
Rotor outer diameter (mm)	D_{or}	151.5
Slot filling factor	a_f	0.45
Area of field coils (mm^2)	S_f	124

Based on Tables 1 and 2, it can be observed that both SCIM and WFFSM have different stator outer diameters of 210.8 mm and 254 mm, respectively. These variations can be attributed to their structural differences and operating principles. The WFFSM incorporates a wound-field configuration, which requires additional space for the field winding and magnetic components, leading to a larger outer diameter. Meanwhile, the SCIM utilizes squirrel-cage windings for its magnetic-field generation, resulting in a comparatively smaller outer diameter.

The back-EMF of the WFFSM, as evaluated using the technique proposed in Equation (46), is 371.8 V. The field current, when varied from 0 to 4.55 A, yields the no-load characteristic curve shown in Figure 7. For rated conditions, the field current value was later extrapolated to 7.25 A to achieve the target output power.

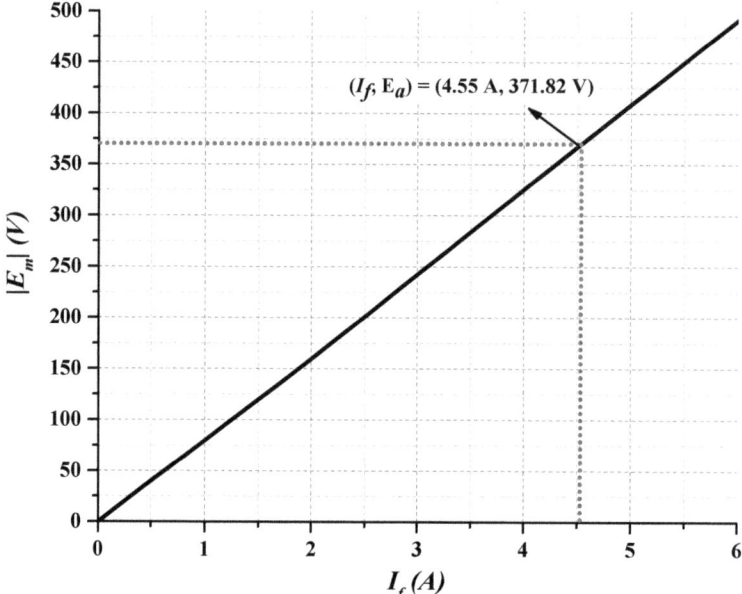

Figure 7. The relationship between no-load voltage and field current.

The calculated flux density maps for the SCIM and WFFSM are compared under rated conditions in a static 2D-FEA transient simulation, as shown in Figure 8. The flux density plots of the WFFSM and SCIM show distinct differences in airgap flux density. The WFFSM exhibits a higher peak value of 2.5 T compared with the SCIM's peak value of 2.1 T. These variations stem from the different design and operational principles of the machines, with the WFFSM's flux modulating configuration contributing to its higher airgap flux density, with potential to enhance torque production and overall machine performance.

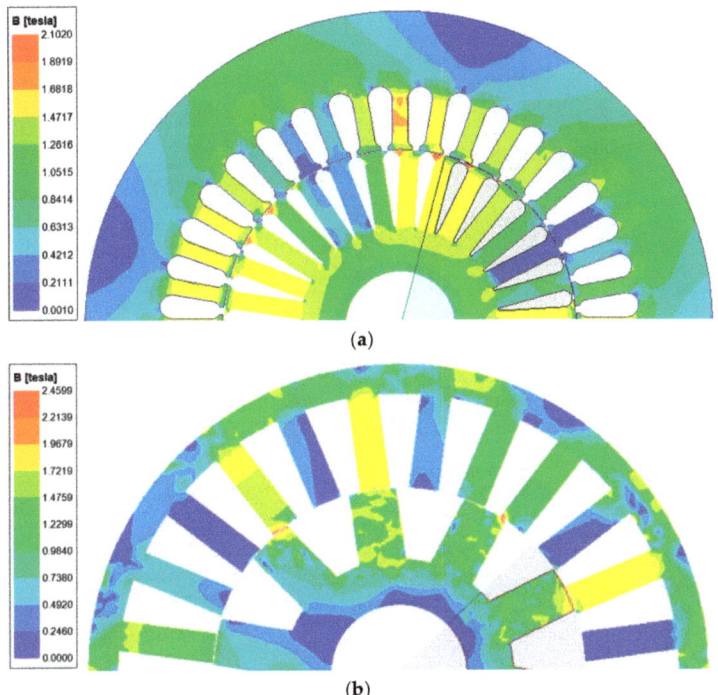

Figure 8. Static 2D-FEA flux density plots: (**a**) SCIM (rotor position of 255.15°) and (**b**) WFFSM (rotor position of 315.64°).

4. Results and Discussion

This section is used to present results on the electromagnetic evaluation and performance comparison of the SCIM and WFFSM under transient steady-state and dynamic on-load conditions. The motor inertia and damping constants are set the same during the simulation of both motors to accurately predict their response under varying load inputs. The highlight of incorporating the motor inertia is to determine how quickly the sampled motors will respond to changes in the load torque, as well as the ability to maintain stable speed under different loads. Damping, on the other hand, is a measure of the motors' resistance to oscillation or vibration. For this study, the motors' inertia and damping constant are both set at 0.0901 kg·m^2 and 0.00932 Nm·s/rad, respectively.

4.1. Steady-State Transient Motor Characteristics

Figures 9–11 show the respective transient steady-state FEA results of supply currents, mechanical speeds, and electromagnetic torque profiles for both motors at the rated condition. These figures provide valuable insights into the transient steady-state behavior and motor characteristics for comparative performance analysis.

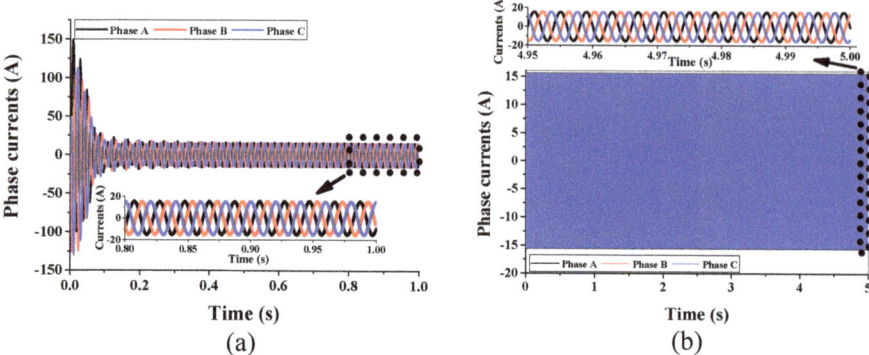

Figure 9. Transient phase current results at rated condition: (**a**) SCIM and (**b**) WFFSM.

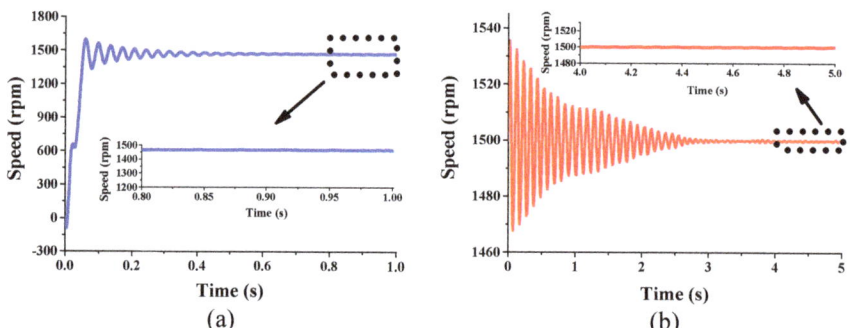

Figure 10. Transient speed results at rated condition: (**a**) SCIM and (**b**) WFFSM.

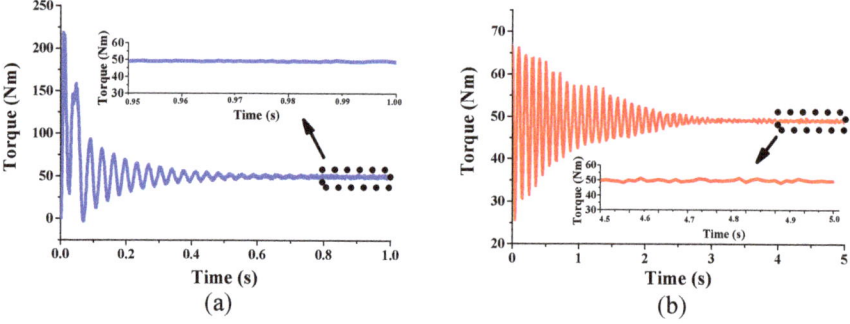

Figure 11. Transient torque results at rated condition: (**a**) SCIM and (**b**) WFFSM.

The discrepancy in the starting behavior of the two motors in Figure 9 is primarily due to their different starting mechanisms. In the SCIM, the rotor consists of conductive bars that are short-circuited at each end by two end rings. During startup, as the rotor is at rest, there is no induced voltage in the rotor bars. Consequently, the rotor current is limited only by the resistance of the bars, which is relatively low. This results in a high initial surge in current, observed as approximately 6.2 times its full load current in Figure 9a. On the other hand, the WFFSM has its windings connected to an external circuit, which limits the current flow to the winding. This results in a lower starting current, as seen in Figure 9b. While the high starting current of the SCIM offers benefits in certain applications, it comes with potential drawbacks, such as increased electrical stress on the motor and power supply system. Additionally, over time, it may lead to reduced efficiency and increased energy costs.

Meanwhile, in Figure 10a, it is observed that the SCIM reaches steady-state time faster compared with the WFFSM in Figure 10b. The speed plot of the SCIM accelerates before meeting the steady-state position at 0.8 s, as indicated in the zoomed plot of Figure 10a, while the WFFSM takes longer to reach its steady-state phase, attaining this at 4 s, which is evident in the zoomed plot of Figure 10b. Under the same inertia, the WFFSM, acting as a synchronous motor, requires more time to synchronize its speed with the frequency of the power supply, resulting in a slower acceleration. On the other hand, the SCIM does not require synchronized field winding excitation when starting up. It can draw high starting current and get up to speed quickly. The SCIM also slips and has a lower synchronous speed compared with the WFFSM operating at the supply frequency. This allows it to accelerate to the working speed more quickly. Lastly, the SCIM has a wider stable operating range and no issues with pull-out during transient conditions, but the WFFSM needs to avoid loss of synchronism during acceleration.

It is important to highlight that, while the high starting current of the SCIM may have its drawbacks, it also results in a higher starting torque, as evident in Figure 11a. This characteristic can be advantageous in applications where quick starts or overcoming high inertia loads are necessary. However, the WFFSM delivers approximately 15.47% more average torque than the SCIM, albeit exhibiting higher torque ripple. This higher torque ripple may introduce mechanical vibrations and other undesirable effects that should be addressed, possibly through future optimization studies to optimize torque performance.

In general, the selection of the appropriate motor type should carefully consider the specific application requirements, balancing the advantages of higher starting torque and quicker starts with potential drawbacks, such as torque ripple. By doing so, the optimal motor can be chosen to meet the specific needs while ensuring efficient and effective performance.

4.2. Dynamic On-Load Transient Motor Characteristics

In this section, SCIM and WFFSM are analyzed using transient FEA with the same inertia and damping constant. We then increase the load torque to investigate the dynamic performance of both motors in terms of mechanical speed, developed torque, power factor, output power, and efficiency. Our results show that the SCIM and WFFSM exhibit different behaviors under dynamic load conditions.

For the SCIM, the speed of the rotor decreases as the load torque increases, as shown in Figure 12, which is characteristic of the negative slope of its speed-torque curve. In contrast, the WFFSM maintains a constant and higher speed throughout, as is characteristic for synchronous motors, which means that the speed of the rotor is determined by the frequency of the AC voltage applied to the phase winding, making it more suitable for applications requiring constant speed.

With an increase in load torque, the developed torque of both the SCIM and WFFSM increases, as shown in Figure 13. For the SCIM, it is because, as the load torque increases, it causes the rotor to slow down slightly; this increases the slip between the rotor speed and the synchronous speed. A higher slip results in increased induced rotor currents, viz. higher rotor torque and overall developed torque. On the other hand, the developed torque of the WFFSM increases with load, which is consistent with the theory in Equation (46).

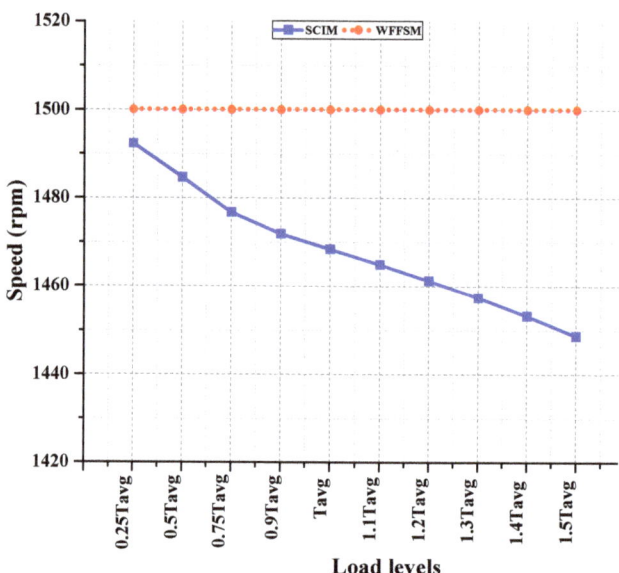

Figure 12. Speed comparison of SCIM and WFFSM at different load levels.

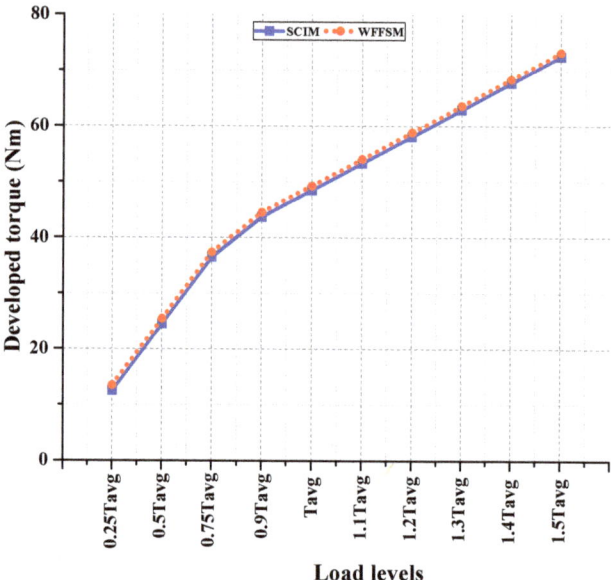

Figure 13. Developed torque comparison of SCIM and WFFSM at different load levels.

The power factor of both motors varies with the load torque, as shown in Figure 14. The power factor of the WFFSM is higher than the SCIM across the load regime. For the SCIM, it appears to increase initially but peaks at rated load torque and then begins to reduce at load torque above 1.1 p.u. For the WFFSM, the power factor is fairly constant throughout the load regime. This is perhaps due to good balance between both MMFs of the excitation of the field and of armature windings to produce the given reference torque of the WFFSM in this study [17]. The reduction in the power factor for the SCIM at higher loads is due to the increased reactive power demand.

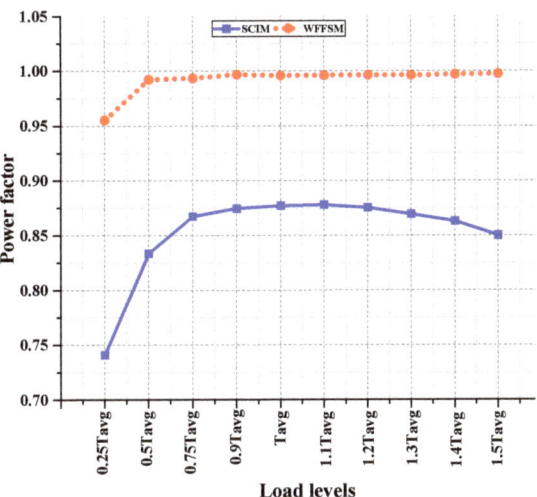

Figure 14. Power factor comparison of SCIM and WFFSM at different load levels.

As shown in Figure 15, the efficiency of the SCIM and WFFSM are varied, with increasing load torque accordingly. For the SCIM, it is observed that the efficiency peaks at 0.75 p.u. of the load torque and then begins to decrease due to the increased losses associated with increased slip viz. reduced speed of the rotor. The efficiency is fairly constant throughout the load regime for the WFFSM and higher throughout compared with the SCIM. This is because DC excitation losses in the field winding are very low since they do not vary significantly with load, unlike the SCIM. Moreover, it is already noticed that the WFFSM operates at near unity power factor, which translates to minimal reactive power, thus lower ohmic losses. Comparing the dynamic performance of both machines so far, it is found that the WFFSM presents a higher but constant speed, slightly higher developed torque, higher efficiency and power factor, and thus higher output power, as shown in Figure 16. Also, based on overload capability studies, the service factor of the SCIM is lower than that of the WFFSM [18].

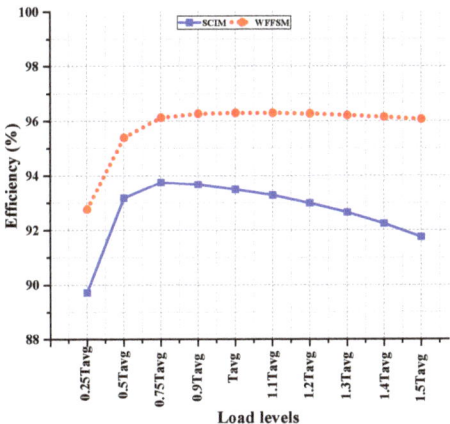

Figure 15. Efficiency comparison of SCIM and WFFSM at different load levels.

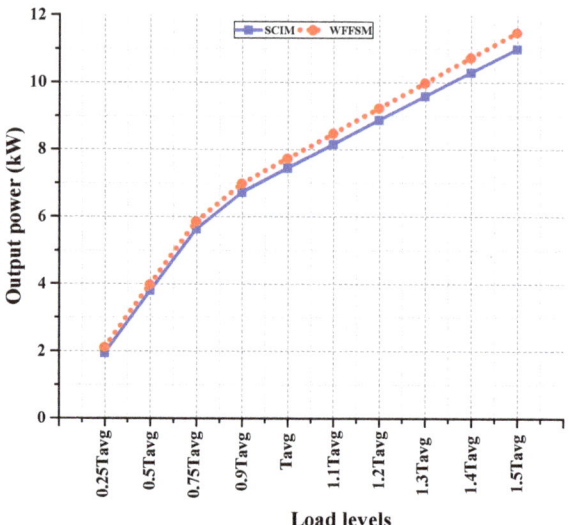

Figure 16. Output power comparison of SCIM and WFFSM at different load levels.

To present a summary of the dynamic performance of both motors, the motor performance characteristics curves are generated and presented, as shown in Figures 17 and 18, for the SCIM and WFFSM, respectively. The starting torque capability of the SCIM is clearly elucidated in Figure 17 compared with the WFFSM, which displays no starting torque in Figure 18. It is observed that the starting torque of the SCIM is about three times (142 Nm) its rated torque, while the pull-out torque occurs at four times (200 Nm), as indicated in Figure 17. The high starting torque of the SCIM allows for direct on-line starting of loads, while the large pull-out torque is an indication of wide stable operating range.

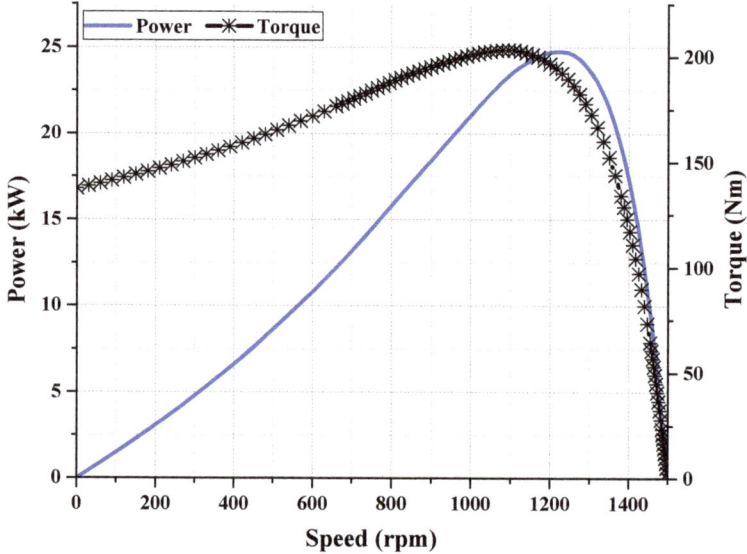

Figure 17. SCIM power/torque-speed characteristic curves.

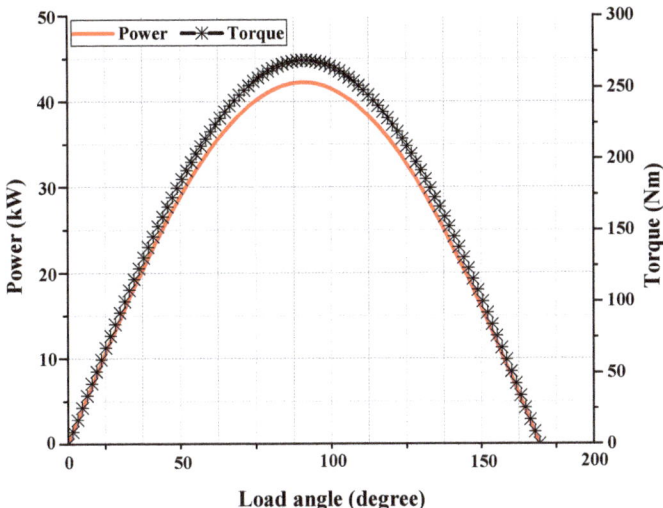

Figure 18. WFFSM power/torque-speed characteristic curves.

On the other hand, Figure 18 shows the power/torque-speed characteristic curves of the WFFSM based on Equation (54). Since the WFFSM is not self-starting, it requires variable frequency drives or other external means for starting. The load angle of the WFFSM at the rated power is 11.5°, which is an indication of a narrow stability range, but it can potentially achieve a higher peak power output considering that the maximum power (where the pull-out torque occurs) is seen to be more than five times (42 kW) the nominal power of the motor.

Overall, the use of field and armature coils allows for precise control of the magnetic field and torque production in WFFSMs, making them suitable for applications where precise control of torque and speed are required. In contrast, the SCIM has a less precise control of torque and speed due to the inherent slip between the rotor and the stator magnetic field. Additionally, WFFSMs can operate at higher speeds without the risk of overheating or damaging the motor compared with the SCIM due to the absence of slip between the rotor and the stator magnetic field, as well as the presence of a robust rotor. This makes the WFFSM suitable for applications where high speeds are required, such as in high-speed machinery; whereas, with a higher starting and pull-out torque, the SCIM is suitable for applications requiring variable speed operation.

4.3. Validation and Cost Analysis Studies

Table 3 presents a summary of the comparative cost analysis and electromagnetic performance of both motors based on rated analytical and steady-state FEA results. The WFFSM has a higher average torque but higher torque ripple than the SCIM, an indication of its flux modulating operation and double-salient topology, respectively [19]. It is observed that the WFFSM has a higher torque-to-mass ratio (1.57 Nm/kg) compared with the SCIM (1.34 Nm/kg), although the torque ripple is ~40% lower in the latter. The output power of the WFFSM is slightly higher than that of the SCIM, although it does not translate to higher torque density (T_d) as indicated, since both machines are yet to be optimized. Moreover, the WFFSM, as a stator-mounted double-excited motor with both its field and armature windings co-located on the stator, has a characteristic high split ratio [20].

Table 3. Comparison of SCIM and WFFSM at rated condition.

Performance Quantity	SCIM		WFFSM	
	Analytical Results	FEA Results	Analytical Results	FEA Results
T_{avg} (Nm)	49.39	48.47	47.75	49.22
TR (%)	-	2.58	-	3.60
P_{out} (kW)	7.50	7.45	7.50	7.73
Slip (%)	3.10	2.93	-	-
P_{cu} (W)	181.23	176.38	135.45	107.39
P_{core} (W)	119.41	104.23	-	191.18
η (%)	89.50	93.48	89.50	96.28
PF	0.8300	0.8767	0.8300	0.9954
T_m (Nm/Kg)	1.34	-	1.57	-
T_d (kNm/m^3)	9.17	9.15	8.81	9.08
Cost (€)	53.50	-	65.45	-

Unlike the WFFSM, the SCIM exploits its squirrel-cage rotor windings to create a rotor magnetic field that interacts with the stator magnetic field to produce torque. Therefore, the copper losses of the WFFSM are observed to be significantly lower than the SCIM, because copper is used for the field windings in the former compared with the latter, whose short-circuited rotor bars are designed using aluminum. Furthermore, the presence of slip in the SCIM results in reduced copper losses. However, the core losses of the WFFSM are higher than the SCIM due to the high number of poles for the WFFSM, which translates to higher electrical frequencies viz. higher core losses [21,22]. Notwithstanding, the efficiency of the WFFSM is seen to be higher than that of the SCIM because of its zero slip, which means that the rotor speed is always synchronous with the stator field, hence with higher output power and efficiency.

In addition, the power factor of the WFFSM is observed to be significantly higher than the SCIM due to reduced reactive power demand. As already indicated, the field and armature windings in WFFSMs are separately excited, making it possible for the field current to be controlled to optimize the power factor. In SCIMs, the magnetizing current is part of the stator current, which reduces the power factor.

The cost analysis is also presented in Table 3, showing the material costs of the WFFSM to be 27% higher than the SCIM. Since the WFFSM is not optimized, this cost differential is expected to be reduced with improved performance benefits. Furthermore, the SCIM rotor bars are designed using aluminum, which costs half the price of copper, as shown in Table 4. But, as already indicated, the higher torque-to-mass ratio of the WFFSM partially offsets the cost differences in applications where torque output is critical.

Table 4. Material cost of SCIM and WFFSM.

Motor Parts	SCIM		WFFSM	
	Mass (Kg)	Cost (€/Kg)	Mass (Kg)	Cost (€/Kg)
Stator	17.60	1	17.30	1
Rotor	15.20	1	8.36	1
Copper coils	2.45	7	5.68	7
Aluminum bars	1.01	3.5	-	-

In summary, the results in Table 3 validate the sizing and analytical modeling techniques developed in this study, with good agreement between the analytical and FEA calculations for both motors, which successfully achieve the prescribed design targets with only marginal differences. The comparative performance tradeoffs between the SCIM and WFFSM highlight that electrical machine design inherently requires balancing strengths and weaknesses based on the intended application. This interim study shows that the SCIM provides lower material cost, torque ripple, and core losses compared with higher torque-per-mass ratio, efficiency, and power factor recorded in the WFFSM. While this initial comparative study reveals distinct performance advantages of both the SCIM and the WFFSM, more extensive and robust multi-objective optimization is required before definitive conclusions can be made regarding the ideal machine design for a given application based on factors such as cost, efficiency, power density, and power quality.

5. Conclusions

In this paper, we present the design process and analytical performance calculations for two different types of brushless motors: SCIM and WFFSM. The design process for both motors involves determining the design specifications, calculating the flux densities and current densities, sizing the stator and rotor components, verifying electric and magnetic loadings, and computing various performance metrics such as torque, efficiency, and power factor. A comparative analysis of the performance of both motors under transient FEA steady-state and dynamic on-load conditions is then undertaken after the initial sizing study. The study reveals key differences between the SCIM and WFFSM.

The transient analysis shows the SCIM reaches steady state in 0.8 s, while the WFFSM takes 4 s to synchronize and stabilize. The SCIM exhibits a starting torque of 142 Nm, close to three times its 48.5 Nm rated torque. In contrast, the WFFSM has no inherent starting torque but can rise to four times the nominal torque to potentially achieve a higher peak power output.

The SCIM's simple squirrel-cage design enables quick starts but requires slip for induction torque production. At a nominal rated load, the SCIM operates with a 5% steady-state slip. In comparison, the WFFSM achieves synchronous speed but needs external methods for starting. The WFFSM also demonstrates superior efficiency across the load range due to stable and higher operating speed, exceeding 96% at below (0.75 p.u.) and above (1.5 p.u.) rated load.

With a higher torque-to-mass ratio of 1.57 Nm/kg versus 1.34 Nm/kg for the SCIM, the WFFSM provides superior power density, offsetting its 27% higher material costs. For budget-driven applications, comprehensive cost analysis beyond initial material costs, evaluating maintenance, operational expenses, and motor lifecycle costs is imperative.

In conclusion, the SCIM and WFFSM offer complementary strengths and weaknesses. The selection of the appropriate motor type should carefully consider the specific application requirements, balancing the advantages of each motor type with potential drawbacks. Therefore, this initial comparative study of the SCIM and WFFSM reveals distinct advantages and tradeoffs for each machine type to inform more extensive multi-objective optimization and design decisions.

Author Contributions: Conceptualization, U.B.A. and O.I.O.; formal analysis, U.B.A. and C.E.A.; funding acquisition, O.I.O., U.B.A. and C.C.A.; investigation, U.B.A., O.I.O. and C.C.A.; methodology, U.B.A., C.E.A. and C.C.A.; project administration, O.I.O.; resources, C.E.A., U.B.A. and O.I.O.; software, U.B.A., C.E.A., C.C.A. and O.I.O.; supervision, U.B.A. and O.I.O.; validation, U.B.A., C.E.A. and C.C.A.; visualization, C.E.A., U.B.A. and C.C.A.; writing—original draft, C.E.A. and U.B.A.; writing—review and editing, C.E.A., U.B.A., O.I.O. and C.C.A. All authors have read and agreed to the published version of the manuscript.

Funding: This research is funded by the 2020 TETFUND National Research Fund (NRF) Intervention (grant number NRF/SET1/III/00055).

Data Availability Statement: Not applicable.

Acknowledgments: The authors wish to thank TETFUND National Research Fund Intervention for their support in carrying out this research.

Conflicts of Interest: The authors declare no conflict of interest.

References

1. Trianni, A.; Cagno, E.; Accordini, D. Energy efficiency measures in electric motors systems: A novel classification highlighting specific implications in their adoption. *Appl. Energy* **2019**, *252*, 113481. [CrossRef]
2. Almeida, A.T.; Ferreira, F.J.; Baoming, G. Beyond Induction Motors—Technology Trends to Move Up Efficiency. *IEEE Trans. Ind. Appl.* **2014**, *50*, 2103–2114. [CrossRef]
3. De Souza, D.F.; Salotti, F.A.M.; Sauer, I.L.; Tatizawa, H.; de Almeida, A.T.; Kanashiro, A.G. A Performance Evaluation of Three-Phase Induction Electric Motors between 1945 and 2020. *Energies* **2022**, *15*, 2002. [CrossRef]
4. Khan, F.; Sulaiman, E.; Ahmad Md, Z. Review of switched flux wound-field machines technology. *IETE Tech. Rev.* **2017**, *34*, 343–352. [CrossRef]
5. Melfi, M.J.; Umans, S.D. Squirrel-Cage Induction Motors: Understanding Starting Transients. *IEEE Ind. Appl. Mag.* **2012**, *18*, 28–36. [CrossRef]
6. Wu, Z.; Zhu, Z.Q.; Hua, W.; Akehurst, S.; Zhu, X.; Zhang, W.; Hu, J.; Li, H.; Zhu, J. Analysis and suppression of induced voltage pulsation in DC winding of five-phase wound-field switched flux machines. *IEEE Trans. Energy Convers.* **2019**, *34*, 1890–1905. [CrossRef]
7. Akuru, U.B.; Kamper, M.J. Potentials of Locally Manufactured Wound-Field Flux Switching Wind Generator in South Africa. *J. Energy S. Afr.* **2019**, *30*, 110–117. [CrossRef]
8. Akuru, U.B.; Kamper, M.J. Performance Comparison of Optimum Wound-field and Ferrite PM Flux Switching Machines for Wind Energy Applications. In Proceedings of the 2016 XXII International Conference on Electrical Machines (ICEM), Lausanne, Switzerland, 4–7 September 2016; pp. 2478–2485. [CrossRef]
9. Zhao, G.; Hua, W.; Qi, J. Comparative Study of Wound-Field Flux-Switching Machines and Switched Reluctance Machines. *IEEE Trans. Ind. Appl.* **2019**, *55*, 2581–2591. [CrossRef]
10. Sethupathi, P.; Senthilnathan, N. Comparative analysis of line-start permanent magnet synchronous motor and squirrel cage induction motor under customary power quality indices. *Electr. Eng.* **2020**, *102*, 1339–1349. [CrossRef]
11. Muteba, M. Comparison of Dynamic Behaviors between a Synchronous Reluctance Motor with Brass Rotor Bars and a Squirrel Cage Induction Motor. In Proceedings of the 2019 IEEE PES/IAS PowerAfrica, Abuja, Nigeria, 20–23 August 2019; pp. 374–378. [CrossRef]
12. Ozcelik, N.G.; Dogru, U.E.; Imeryuz, M.; Ergene, L.T. Synchronous Reluctance Motor vs. Induction Motor at Low-Power Industrial Applications: Design and Comparison. *Energies* **2019**, *12*, 2190. [CrossRef]
13. Boldea, I. *Induction Machines Handbook*; CRC Press: Boca Raton, FL, USA, 2020.
14. Cao, R.; Yuan, X.; Jin, Y.; Zhang, Z. MW-Class Stator Wound Field Flux-Switching Motor for Semidirect Drive Wind Power Generation System. *IEEE Trans. Ind. Electron.* **2019**, *66*, 795–805. [CrossRef]
15. Akuru, U.B.; Kamper, M.J. Formulation and multiobjective design optimization of wound-field flux switching machines for wind energy drives. *IEEE Trans. Ind. Electron.* **2018**, *65*, 1828–1836. [CrossRef]
16. Zulu, A.; Mecrow, B.C.; Armstrong, M. A wound-field three-phase flux-switching synchronous motor with all excitation sources on the stator. *IEEE Trans. Ind. Appl.* **2010**, *46*, 2363–2371. [CrossRef]
17. Sulaiman, E.B.; Kosaka, T.; Matsui, N. Design study and experimental analysis of wound field flux switching motor for HEV applications. In Proceedings of the XXth International Conference on Electrical Machines, Marseille, France, 2–5 September 2012; pp. 1269–1275.
18. Campbell, B.; Galleno, J. Motor life: The effects of loading, service factor and temperature rise on insulation life. In Proceedings of the IEEE Industry Applications Society 45th Annual Petroleum and Chemical Industry Conference, Indianapolis, Indiana, 28–30 September 1998; pp. 303–310.
19. Udosen, D.; Kalengo, K.; Akuru, U.B.; Popoola, O.; Munda, J.L. Non-Conventional, Non-Permanent Magnet Wind Generator Candidates. *Wind* **2022**, *2*, 429–450. [CrossRef]
20. Akuru, U.B.; Kamper, M.J. Intriguing Behavioral Characteristics of Rare-Earth-Free Flux Switching Wind Generators at Small- and Large-Scale Power Levels. *IEEE Trans. Ind. Appl.* **2018**, *54*, 5772–5782. [CrossRef]
21. Yu, W.; Hua, W.; Zhang, Z. High-Frequency Core Loss Analysis of High-Speed Flux-Switching Permanent Magnet Machines. *Electronics* **2021**, *10*, 1076. [CrossRef]
22. Fukami, T.; Aoki, B.; Shima, K.; Momiyama, M.; Kawamura, M. Assessment of Core Losses in a Flux-Modulating Synchronous Machine. *IEEE Trans. Ind. Appl.* **2012**, *48*, 603–611. [CrossRef]

Disclaimer/Publisher's Note: The statements, opinions and data contained in all publications are solely those of the individual author(s) and contributor(s) and not of MDPI and/or the editor(s). MDPI and/or the editor(s) disclaim responsibility for any injury to people or property resulting from any ideas, methods, instructions or products referred to in the content.

Article

A Study on Electric Potential and Electric Field Distribution for Optimal Design of Lightning Rod Using Finite Element Method

Kyung-Hoon Jang [1], Sang-Won Seo [2] and Dong-Jin Kim [2,*]

[1] Korea Conformity Laboratories, Components & Material Division Department, Seoul 08503, Republic of Korea; khjang@kcl.re.kr
[2] Research & Development, SUNKWANG LTI, Seoul 06030, Republic of Korea
* Correspondence: rnd@i-sk.com

Abstract: In this paper, we present an electric field analysis for the optimal structural design of lightning rods for high performance with a charge transfer system (CTS). In the case of a conventional rod that is produced with an empirical design and structure without quantitative data because the design is structurally very simple, only the materials and radius of curvature of the lightning rod to concentrate the electric field at the tip part of the rod are considered. Recently, the development of new types of lightning rods, such as early streamer emission (ESE) and charge transfer system (CTS), has been introduced through simulation analysis and experiments, but detailed specifications and information about the optimal design and structure have not been fully reported. In this paper, we performed an electric field analysis of the structures and materials for the optimal structural design of lightning rods with a function of CTS through computer software analysis with consideration for the radius of curvature, the size of corona ring, and optimal position (X-axis and Y-axis) of the floating electrode. For optimal structural design of lightning rods based on electric field analysis, we used a source of lightning voltage with 1.2/50 μs based on a double exponential equation. The results revealed that the electric field on the relaxation part decreases as the radius of curvature and corona ring increases. For the radius of curvature, the electric field first decreases and then increases with increasing radius of curvature and reaches a minimum at 7 mm and a maximum above 8 mm. For the case of the corona ring, the electric field decreases with increasing corona ring, and the optimal size of the corona ring was selected as 4 mm; the size of the 4 mm corona ring uniformly formed the electric field both at the tip part of the ground current collector and the corona ring. For the electric field concentration part, we found that the optimal X-axis position of the floating electrode and the Y-axis position between the ionizer conductor and floating electrode are 7 mm and 0.1 mm, respectively. These simulation results in this paper are expected to provide useful information for the design of optimized CTS-type lightning rods.

Keywords: lightning rod protection; charge transfer system (CTS); electric field analysis; numerical analysis; optimal design; finite element method; corona discharge; air insulation

MSC: 65K10

1. Introduction

Direct or indirect lightning strikes cause power outages, forest fires, and other damages to various electronic systems as well as infrastructure, with associated costs of billions of euros each year [1,2]. Despite technological advances, lightning protection rods still rely on the nearly 300-year-old basic concept of the lightning rod as invented by Benjamin Franklin [3].

The purpose of a lightning rod is to induce lightning strikes at a particular point with a very sharp tip and low-impedance paths for concentrating the local electric field from direct or indirect lightning strikes. In the case of conventional Franklin lightning rods, the structure and design are very simple; in other words, only the materials, radius of curvature, installation position, height and quantity needed to induce a high local electric field at the tip part of the Franklin rod after lightning strikes need to be considered. However, recently the structure and design of lightning rods have been advanced and diversified, and various types of lightning rods have been developed and introduced, such as early streamer emission (ESE) and charge transfer system (CTS), to effectively protect facilities from indirect or direct lightning strikes [4–6]. In other words, to ensure optimal performance, quantitative analysis of the structures and materials of various types of lightning rods is required. However, studies do not report detailed specifications and information on the structures and materials utilized.

For this reason, a lot of work has recently been devoted to improving lightning rods based on the results of simulation analysis and experimentation methods that quantitatively optimize the performance of lightning rods. Dong-Jin Kim et al. [7] suggested the HEC (Hybrid ESE Conductor) method, which mixes a horizontal conductor and an ESE lightning rod. Their results revealed that the starting point of a corona discharge current is low, and HEC is efficient for lightning protection compared with other methods based on experiment and simulation analysis. Vernon Cooray and Marley Becerra et al. [8,9] investigated the early streamer emission-enhanced ionizing air terminal and multi-points discharge system and conceptual future methods of lightning protection; the ESE lightning rod proposed in NF C 17-102 was compared to the conventional Franklin rods proposed in IEC 62305 using simulation analysis under laboratory conditions. Myung-Ki Baek et al. [10] presented a numerical analysis method and experimental results for the discharge characteristic associated with the presence of a floating conductor with consideration of space charge on the surface of the floating electrode. The results revealed that the floating conductor can generate more electrons than other discharge systems without the floating electrode and that the corona discharge can be controlled using a floating conductor. Bok-hee Lee et al. [11,12] have introduced simulation analysis and experimentation for the validation of the CTS-type lightning rods. They introduced a new type of CTS lightning rod consisting of a floating electrode, 250 brushes, and a cylindrical conducting body. Their results obtained by experiment and simulation analysis showed the CTS-type lightning rods significantly decreased electric field intensification, which reduced the probability of direct lightning strikes. Thomas Produit et al. [13] suggested a new approach for diverting lightning strikes: the laser lightning rod. For experiment and validation of the laser's ability to initiate upward discharge from the tip of a lightning rod, the laser experiment at Säntis Tower was conducted. The results revealed that lasers would allow the protection of industrial sites, including chemical and nuclear power plants, and that creating protected corridors along runways would reduce breaks in airport operations during thunderstorms. However, in the aforementioned studies, a lot of them have only focused on the technology as well as the modification of the structure and shape. In other words, the detailed specifications and information about the optimal design and structure of lightning rods for high performance were not fully reported quantitatively.

From this perspective, the scientific aim of our research work is to investigate the electric potential and electric field distribution for the optimal design of lightning rods that have a charge transfer system (CTS) using a finite element method (FEM). In CTS-type lightning rods, the most important factors for optimal design are divided into two parts: the electric field concentration part and the electric field relaxation part. The optimal design of CTS-type lightning rods in two parts could be to create a non-uniform electric field with a large electric field enhancement, which can delay the upward streamer by generating a corona discharge in the air insulation system [14,15].

In this paper, there are two important parts needed to obtain an optimal design of CTS-type lightning rods. For the electric field relaxation part, we considered the variables of radius of curvature and corona ring size. To maximize the electric field concentration between the ionizer conductor and floating electrode, the positions of the X-axis and Y-axis were varied.

In this paper, Section 1 describes the introduction of the research work, Section 2 presents a model for electric field analysis and a detailed description of the analysis approach, Section 3 describes the results and discussion, and Section 4 presents the final conclusions and future directions of the research in this field. These simulation results are expected to offer useful information for optimizing the design of CTS-type lightning rods as well as lightning rods of other types.

2. Modeling for Electric Field Analysis

The optimized design for CTS-type lightning rods has focused on electric field distribution under lightning impulse voltage using an FEA software package. The basic equations used to calculate the electric field are Maxwell's equations [16], as follows:

$$div\ D = \rho_{charge\ density} \tag{1}$$

$$D = \varepsilon E \tag{2}$$

$$E = \nabla V \tag{3}$$

The general expression of Poisson's equation for electric field analysis from the above three Equations (1)–(3) is as follows [17]:

$$div\ \varepsilon_r \cdot (-\nabla E) = \rho_{space\ charge} \tag{4}$$

where ε_r is the relative permittivity of each material and E is the electric field. In this paper, space charge ρ is negligible such that $\rho = 0$, i.e.,

Calculation of current density depending on time variation can be expressed as a current continuity equation, and the relationship between current density and bulk charge density satisfies the current continuity equation, as follows:

$$\nabla \cdot J = \frac{\delta \rho_{charge\ density}}{\delta t} \tag{5}$$

by substituting Formula (4) into Equation (5), the relationship between current density and field strength can be established.

$$\nabla \cdot \left(J + \varepsilon \frac{\delta E}{\delta t} \right) = 0, \tag{6}$$

if the current density conforms to Ohm's Law:

$$J = \sigma E \tag{7}$$

where J is the current density, and it can be calculated following Equation (6) from Equations (5)–(7) using Ohm's law as follows:

$$\nabla \cdot (\sigma E) = \nabla \cdot J = 0 \tag{8}$$

where σ is the electrical conductivity of each material, and the electric field is determined by the conductivity from Equation (8).

By substituting Equation (8) into Equation (5), the current continuity equation can be expressed as Equations (9) and (10) as follows:

$$\nabla \cdot (\sigma E) + \frac{\delta \rho}{\delta t} = 0 \tag{9}$$

$$\nabla \cdot \left(\sigma E + \frac{\delta E}{\delta t} \right) = 0 \tag{10}$$

Since the electric field is the result of the spatial differentiation of the electric potential, it can be expressed as follows:

$$\nabla \cdot (-\sigma \nabla V) = 0 \tag{11}$$

Table 1 describes the significance of various abbreviations and acronyms used throughout the paper.

Table 1. Abbreviations and acronyms used in this paper.

Symbol	Meaning
D	Electric flux density
V	Electric potential
ε_r	Relative permittivity
ρ	Charge density
J	Current density
σ	Electric conductivity

Figure 1 shows computer-aided numerical model geometry in COMSOL Multiphysics. In order to perform the electric field analysis of the CTS-type lightning rod under lightning impulse voltage, it is necessary to understand the structure of the model and the type of materials. The CTS-type lightning rod consists of not only copper and aluminum conductors but also ionizer conductors and insulating support structures such as epoxy resin, as shown in Figure 1. Equations (5), (8) and (11) are applied to all areas of the simulation model as shown in Figure 1. In addition, the zero charge node adds the condition that there is zero charge on the boundary $n \cdot D = 0$. This is the default boundary condition for exterior boundaries. At interior boundaries, it means that no displacement field can penetrate the boundary and that the electric potential is discontinuous across the boundary. The boundary condition for V in Figure 1 is applied to the ground and CTS-type lightning rod as $V = 0$, while the electric potential considers that the potential voltage is 300 kV (standard lightning impulse voltage is applied until 1.2 μs. The air space between the tip part of the ground current collector and the opposite conductor is 300 mm.

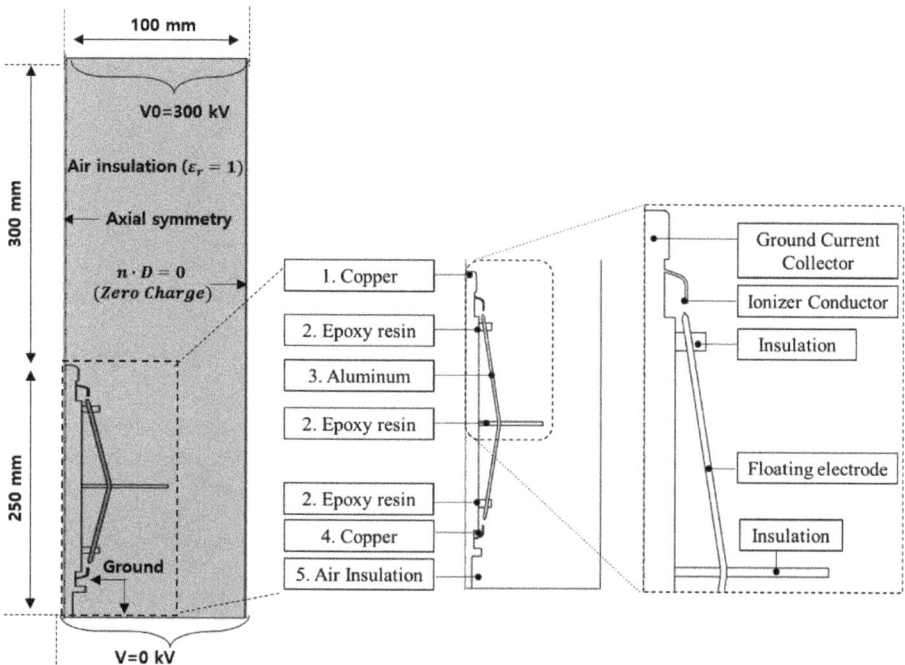

Figure 1. Computer-aided design representation of the numerical model geometry.

The following Table 2 describes the name and function of the main components used throughout the paper [18,19].

Table 2. Name and function of the main components of a CTS-type lightning rod.

	Name	Material	Functions of the Main Component	Relative Permittivity (ε_r)	Conductivity (S/m)
1	Tip part of the ground current collector	Copper	Electric field relaxation part	1	1.6×10^7
2	Insulation support	Epoxy resin	Insulation support for fixing a floating electrode	4.2	3.8×10^{15}
3	Floating electrode	Aluminum	Electric field concentration between the ionizer conductor and floating electrode	1	3.5×10^7
4	Ionizer conductor	Copper	Electric field concentration between the ionizer conductor and floating electrode	1	1.6×10^7
5	Air insulation	Air	-	1	2.6×10^{-17}

Figure 2 shows the main two parts of this simulation: one part is electric field relaxation, and another part is electric field concentration. The main concept of the CTS lightning rod is to form a non-uniform electric field with a large electric field enhancement at the point of the electric field concentration part to delay the upward streamer. However, the tip part of the ground current collector suppresses the upward streamer through electric field relaxation.

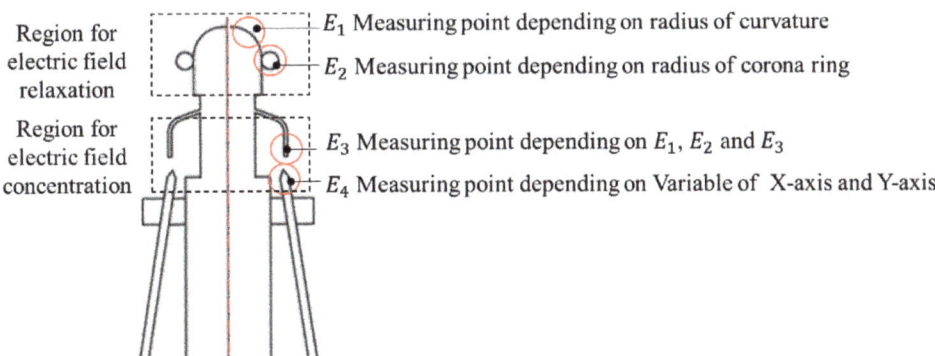

Figure 2. Main measuring points (E_1, E_2: part for electric field relaxation; E_3, E_4: part for electric field concentration)

3. Results and Discussion

3.1. Electric Field Relaxation Part

Figure 3 shows the results of the electric field distribution (contour) and electric potential (solid line) depending on the radius of curvature of the tip part of the grounding current collector in air insulation under 300 kV impulse voltage. The E_1 is a maximum electric field at the tip part of the grounding current collector, $rate_{dec.}$ is the decrease rate (%) of the electric field compared with the reference model (2 mm). To reduce the electric field on the relaxation part, the radius of curvature was changed from 2 mm to 8 mm. The electric field decreases with increasing radius of curvature and reaches a minimum at 7 mm (E_1: 4.88 kV/mm, $rate_{dec.}$: −23%), and then the electric field shows an increasing trend at 8 mm (5.21 kV/mm, $rate_{dec.}$: −17%), as shown in Figure 4. The electric potential also widens the contour gap with increasing radius of curvature; the electric potential distribution at 8 mm curvature radius is formed in the vicinity of the tip part of the ground current collector as dense contour lines.

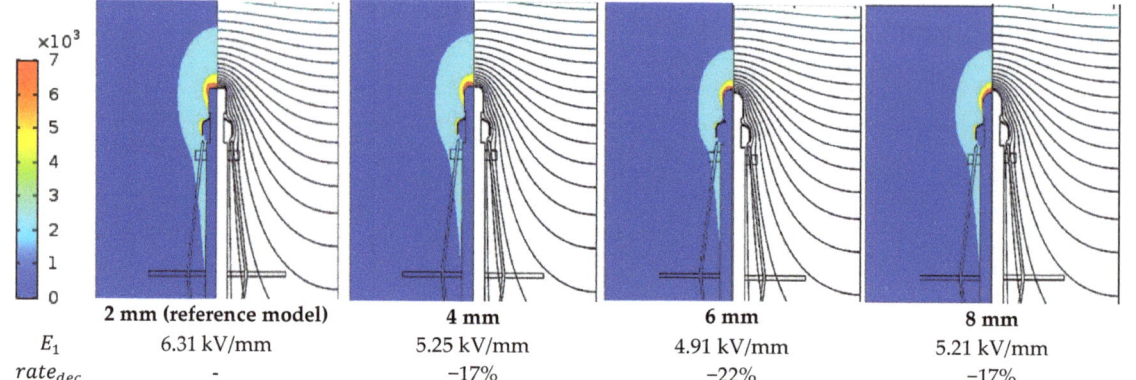

Figure 3. Electric field (contour) and electric potential distribution (solid line) based on the radius of curvature [mm] under lightning impulse voltage.

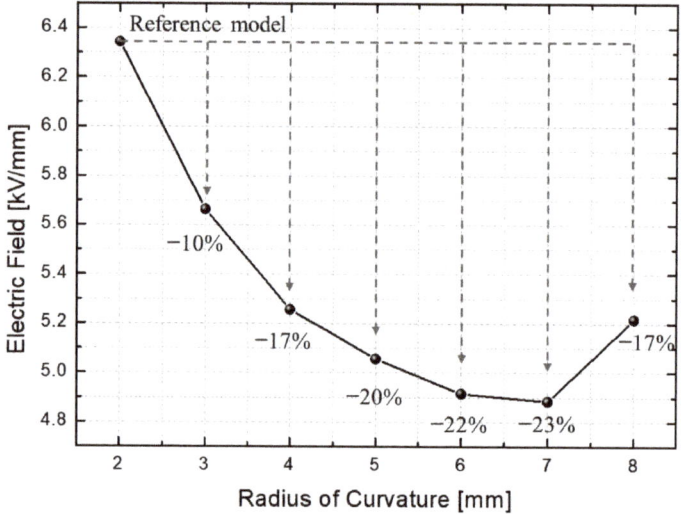

Figure 4. Electric field level based on the radius of curvature.

The reason for the increased electric field at 8 mm radius of curvature is that the tip part of the grounding current collector formed a complete spherical shape at 8 mm radius of curvature, and the electric field was concentrated in the vicinity of the sphere tip part. From the above results, we selected the optimal curvature radius as 7 mm in this simulation model.

Figure 5 shows the electric field and electric potential distribution depending on the size of the corona ring. To reduce the electric field concentration, the size of the corona ring was changed from 2 mm to 15 mm. Electric field and electric potential were measured at three points (E_1, E_2 and E_3) to select the optimal design criteria. The $E_1 rate_{dec.}$ And $E_3 rate_{dec.}$ In Figure 5 are the electric field decrease rates at points E_1 and E_3, respectively, compared to the model without the corona ring.

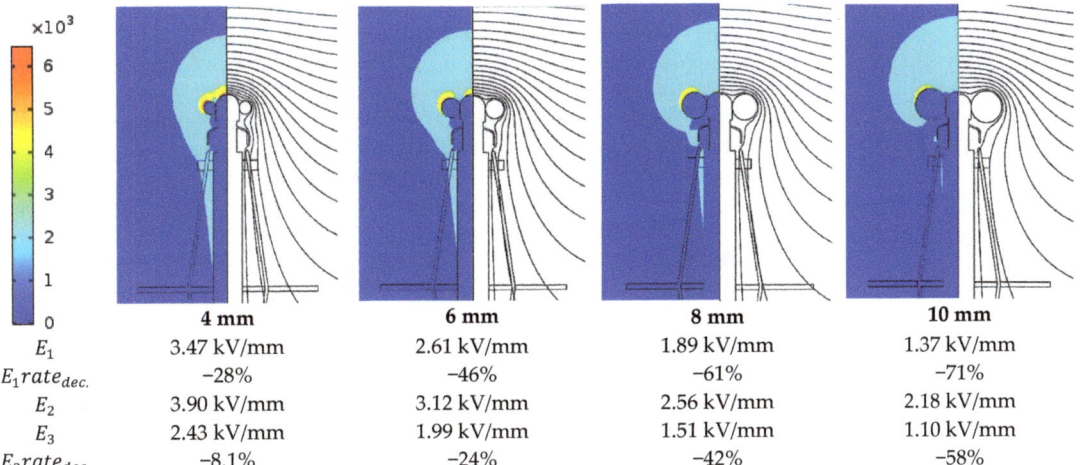

	4 mm	6 mm	8 mm	10 mm
E_1	3.47 kV/mm	2.61 kV/mm	1.89 kV/mm	1.37 kV/mm
$E_1 rate_{dec.}$	−28%	−46%	−61%	−71%
E_2	3.90 kV/mm	3.12 kV/mm	2.56 kV/mm	2.18 kV/mm
E_3	2.43 kV/mm	1.99 kV/mm	1.51 kV/mm	1.10 kV/mm
$E_3 rate_{dec.}$	−8.1%	−24%	−42%	−58%

Figure 5. Electric field (contour) and electric potential distribution (solid line) depending on the size of the corona ring [mm] under lightning impulse voltage.

The results revealed that the electric field at both the tip part of the ground current collector and the corona ring part decreases with increasing the size of the corona ring. In other words, it is found that the corona ring with a large radius of curvature has a definite effect on the electric field relaxation of the tip part compared with the model without a corona ring. However, in order to design a CTS-type lightning rod, the following two points should be considered.

(1) There should be uniform electric field formation in both the corona ring (E_2) and the tip part of the grounding current collector (E_1).

(2) There should be no reduction in the electric field of the ionizer conductor (E_3) due to the increase in the corona ring size.

As shown in Figure 6, the electric field of the E_1 part is relaxed, and the electric field of the E_3 part is also relaxed at the same time as the size of the corona ring increases.

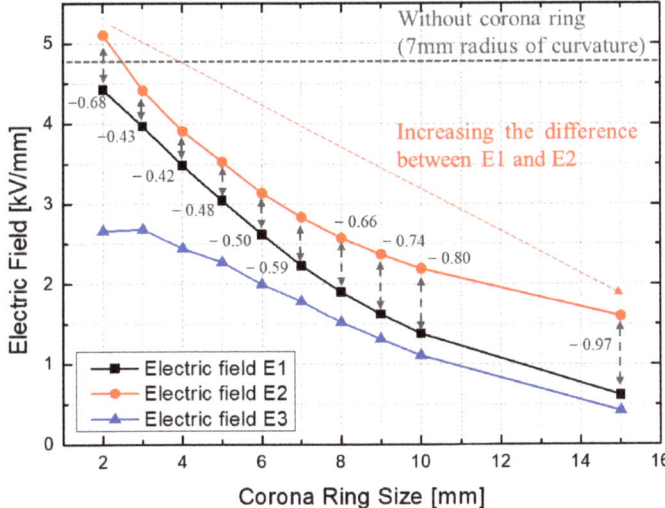

Figure 6. Electric field level depending on the size of the corona ring compared with that without a corona ring.

Firstly, the difference between E_1 and E_2 was calculated and compared according to the corona ring size to select the corona ring size where the electric field difference between the corona ring part and the tip part of the grounding current collector was the minimum. The minimum difference between E_1 and E_2 depending on corona ring size was 4 mm, as shown in Figure 6. On the other hand, as the corona ring size increases, it might be possible that the electric field concentration occurs at the corona ring part, and it could be the starting point of early corona discharge inception under a high electric field.

Secondly, the electric field of E_3 is higher than that of others at the 3 mm corona ring, compared to the 3 mm corona ring, the size with the lowest $E_3 rate_{dec}$ were 2 mm and 4 mm. From the above results, we selected an optimal design that satisfies two conditions; the most reasonable design was the 4 mm corona ring.

3.2. Local Electric Field Concentration Part

Figure 7 shows the locations of the variable Y-axis and X-axis for simulation. In this Section 3.2, we have focused on electric field concentration, which is different from the results of Section 3.1's electric field relaxation. The reference X-axis and Y-axis are 3 mm and 6 mm, respectively. In the case of the Y-axis, the simulation analysis was conducted by changing from 0.05 mm to 5 mm based on the reference model. The X-axis was also changed from 4 mm to 9 mm. The results revealed that the electric field increases with a decreasing air space by adjusting the Y-axis position of the floating electrode, as shown in Figure 8. The electric field strength at 0.1 mm of air space was higher than that of the other results for air space (mm). In other words, when the air space is above 1 mm, the electric field decreases by 30% compared to the 0.1 mm air space. The discussion on electric field distribution according to variable air space has been discussed by introducing the electron mean free path (MFP) theory in our previous research work [19]. If the air space is too far, the generated free electron could not excite another free electron for the electron avalanche between the ionizer conductor and grounding current collector. In other words, the first free electron cannot play a role as the seed of the secondary and tertiary electron avalanches [20]. In contrast, if the floating electrode is located near the ionizer conductor, the positive and negative charges are equally induced on the ionizer conductor and the floating electrode, and then the local electric field strength near them increases. This electric field concentration can cause the ionization process of the surrounding air insulation, and the corona discharge could have a different starting point at the electric field concentration part than at other parts [21].

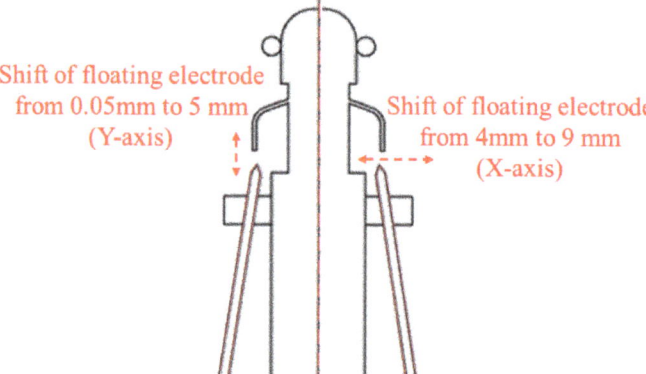

Figure 7. Locations of X-axis and Y-axis for simulation analysis.

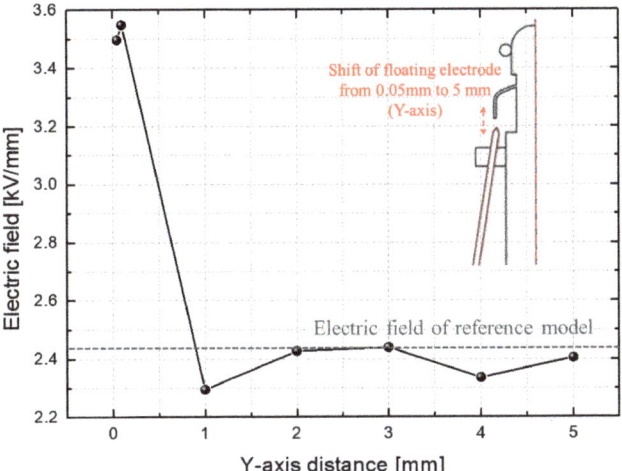

Figure 8. Electric field results according to different Y-axis distances.

For the results of the variable X-axis as shown in Figure 9, element design was performed to increase the maximum electric field value by adjusting the X-axis position of the floating electrode. As a result, it is found that the high electric field distribution is formed at positions where the ionizer conductor and the floating electrode tip part coincide with each other. In addition, the distance of the electric potential line was densely formed at 7 mm on the X-axis model. In the case of a 7 mm X-axis, the maximum electric field value was 4.72 kV/mm, and the electric field value gradually decreased when moved away by ±1 mm from the 7 mm X-axis. In our work, it is possible to reduce the electric field strength of the tip part of the grounding current collector by introducing the optimum design of curvature radius and corona ring. By introducing the optimal distance of X-axis and Y-axis, a high local electric field can be formed between the ionizer conductor and the floating electrode.

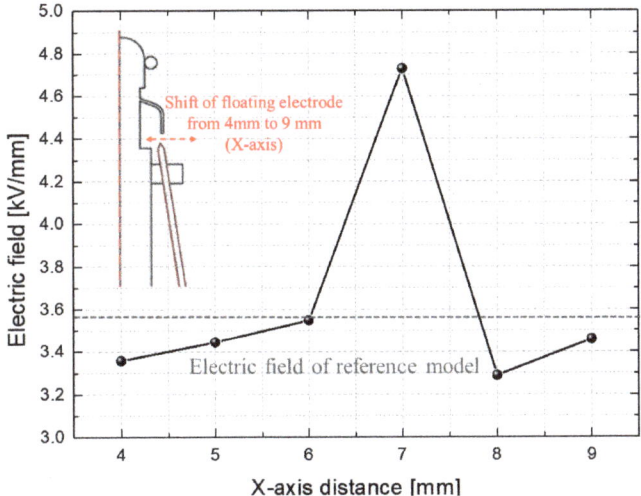

Figure 9. Electric field results according to different X-axis distances.

From the above results, if the maximum electric field distribution is formed at the electric field concentration part rather than the tip part of the ground current collector, the concentration part could be causing the very first ionization and the starting point of early corona discharge inception under a high electric field as a possibility.

However, further experimental research is needed for validation of the above simulation research on the optimal design of CTS-type lightning rods in terms of electric potential and electric field distribution. Also, consideration of the temperature parameter is needed. The characteristics of conductivity materials are dependent on temperature variation by concentrated local electric field distribution.

4. Conclusions

Some researchers have focused more on the performance improvement of lightning rods according to modifications of the structure and shape than the optimal design and performance improvement based on quantitative data. For the aforementioned reason, it is not possible to obtain a detailed rationale for the location and size of each component. This research work has focused on the optimal design of CTS-type lightning rods for high performance. To investigate the optimal structure design, the following four factors are considered:

(1) Radius of curvature (electric field relaxation)
(2) Corona ring size (electric field relaxation)
(3) The position of the floating electrode in the direction of the X-axis (electric field concentration)
(4) The position of the floating electrode in the direction of the Y-axis (electric field concentration)

The results for electric field relaxation revealed that the electric field decreases with increasing radius of curvature and reaches a minimum at 7 mm, with an increasing trend above 8 mm. In addition, electric potential distribution also widens the contour gap with increasing radius of curvature at 7 mm. From the above results, we selected an optimized radius of curvature of 7 mm.

For the optimized size of the corona ring, the electric field of both parts (the tip part of the ground current collector and the corona ring part) decreases with increasing the size of the corona ring. We found that the optimized size of the corona ring was 4 mm due to the following two factors:

(1) Uniform electric field formation, both at the corona ring and tip part of the grounding current collector.
(2) The reduction in the electric field of the ionizer conductor due to the increase in corona ring size.

The optimized position of the floating electrode in the direction of the X-axis and Y-axis was 7 mm and 0.1 mm, respectively. If it is possible to locate the floating electrode near the ionizer conductor, the positive and negative charges are equally induced on the ionizer conductor and a floating electrode, and then the local electric field strength near them increases. In addition, this electric field concentration can cause the ionization of the surrounding air insulation, and the corona discharge could have a different starting point at the electric field concentration part than other parts.

For further work, experiments on the optimized design of a CTS-type lightning rod under standard lightning impulse voltage should be performed while considering neighboring grounded objects. Besides, a comparison with experimental and simulation results should be performed. For further computational work, some parameters such as electron mobility, positive and negative ion mobility under high electric field distribution, temperature, and melting of the electrode as a result of a local heat source from a lightning strike should be considered. A comparison between a conventional lighting rod and a CTS-type lightning rod should be performed to verify the performance of the CTS-type lightning rod.

Author Contributions: K.-H.J., contributed to conceptualization, methodology, analysis, and writing. S.-W.S., contributed to validation, formal analysis, data curation, and visualization. D.-J.K., contributed to reviewing and editing. All authors have read and agreed to the published version of the manuscript.

Funding: This study was supported by the Korea Institute of Energy Technology Evaluation and Planning (KETEP), and we were granted financial resources from the Ministry of Trade, Industry and Energy (MOTIE) (20212020800090, Development and Demonstration of Energy-Efficiency Enhanced Technology for Temperature-Controlled Transportation and Logistics Center).

Institutional Review Board Statement: Not applicable.

Informed Consent Statement: Not applicable.

Data Availability Statement: Not applicable.

Conflicts of Interest: The authors declare no conflict of interest.

References

1. Lightning Costs and Losses from Attributed Sources, Lightning Costs and Losses from Attributed Sources—National Lightning Safety Institute. 2020. Available online: https://lightningelectricity.com/ (accessed on 22 March 2020).
2. Facts + Statistics: Lightning. 2022. Available online: https://www.iii.org/fact-statistic/facts-statistics-lightning (accessed on 22 March 2020).
3. History.com Editors. Benjamin Franklin. A&E Television Networks, 2022 (Last Updated). Available online: https://www.history.com/topics/american-revolution/benjamin-franklin (accessed on 22 March 2022).
4. Kim, H.-G.; Lee, K.-S. The Change of Lightning Air Terminal and Trend of the World. In Proceedings of the International Symposium on High Voltage Engineering, Hannover, Germany, 22–26 August 2011.
5. Cooray, V. The Similarity of the Action of Franklin and ESE Lightning Rods under Natural Conditions. *Atmosphere* **2018**, *9*, 225. [CrossRef]
6. Cooray, V. Non-Conventional Lightning Protection System. In Proceedings of the 30th International Conference on Lightning Protection (ICLP), Cagliari, Italy, 13–17 September 2010.
7. Kim, D.-J. A Study on the HEC (Hybrid ESE Conductor) Method for Lightning Protection of Buildings. *Trans. Korean Inst. Electr. Eng.* **2008**, *57*, 2.
8. Mladen, B. Early Streamer Emission Vs Conventional Lightning Protection Systems. *B&H Electr. Eng.* **2019**, *13*, 24–34.
9. Becerra, M.; Cooray, V. The Early Streamer Emission Principle Does Not Work Under Natural Lightning. In Proceedings of the International Symposium on Lightning Protection, Foz do Iguaçu, Brazil, 26–30 November 2007.
10. Baek, M.K.; Chung, Y.K.; Park, I.H. Experiment and Analysis for Effect of Floating Conductor on Electric Discharge Characteristic. *IEEE Trans. Magn.* **2013**, *49*, 5. [CrossRef]
11. Lee, B.-H. Effect of the Corona Shield of the OMNI Bipolar Conventional Air Terminals. In Proceedings of the 2016 International Conference on ElectroMagnetic Interference & Compatibility (INCEMIC), Bengaluru, India, 8–9 December 2016.
12. Lee, B.-H. Analysis and Test on Electric Field Concentration Effect of Bipolar Conventional Air Terminal. In Proceedings of the International Conference on Lightning Protection (ICLP), Shanghai, China, 11–18 October 2014.
13. Produit, T. The Laser Lightning Rod Project. *Eur. Phys. J. Appl. Phys.* **2020**, *92*, 30501. [CrossRef]
14. Jang, K.-H. Corona Discharge Behaviors with Presence of Floating Electrode in Air Insulation using Numerical Model Analysis. *J. Mater. Sci. Manuf. Res.* **2023**, *4*, 1–4. [CrossRef]
15. Jang, K.-H. Numerical Simulation Analysis on Non-Conventional Lightning Protection System. In Proceedings of the IEEE 4th Eurasia Conference on IOT, Communication and Engineering (ECICE), Yunlin, Taiwan, 28–30 October 2022.
16. Jackson, J.D. *Classical Electrodynamics*, 3rd ed.; Wiley: New York, NY, USA, 1999.
17. Benguesmia, H.; M'ziou, N. Simulation of the Potential and Electric Field Distribution on High Voltage Insulator using the Finite Element Method. *Diagnostyka* **2018**, *19*, 2. [CrossRef]
18. Shi, Y.; Xie, G.; Wang, Q. Simulation Analysis and Calculation of Electric Field Distribution Characteristics of UHV Wall Bushing. In Proceedings of the 4th International Conference on Electrical Engineering and Green Energy CEEGE, Munich, Germany, 10–13 June 2021.
19. Liao, L.; Xue, T.; Xiong, J.; Lu, B. Research on the Influence of the Relative Permittivity of the Reinforced Insulation of the Cable Intermediate Joint on the Electric Field Intensity. *J. Phys. Conf. Ser.* **2021**, *1815*, 012037. [CrossRef]
20. Jang, K.-H.; Seo, S.-W.; Kim, D.-J. Electric Field Analysis on the Corona Discharge Phenomenon according to the Variable Air Space between the Ionizer and Ground Current Collector. *Appl. Syst. Innov.* **2023**, *6*, 10. [CrossRef]
21. Lee, J.-H. Local Electric Field Analysis for Evaluation of Charge Transfer System Using Sequential Subwindow Technique. *IEEE Trans. Magn.* **2004**, *40*, 2. [CrossRef]

Disclaimer/Publisher's Note: The statements, opinions and data contained in all publications are solely those of the individual author(s) and contributor(s) and not of MDPI and/or the editor(s). MDPI and/or the editor(s) disclaim responsibility for any injury to people or property resulting from any ideas, methods, instructions or products referred to in the content.

Article

Investigation of the Effect of the Voltage Drop and Cable Length on the Success of Starting the Line-Start Permanent Magnet Motor in the Drive of a Centrifugal Pump Unit

Aleksey Paramonov, Safarbek Oshurbekov, Vadim Kazakbaev, Vladimir Prakht * and Vladimir Dmitrievskii

Department of Electrical Engineering, Ural Federal University, 620002 Yekaterinburg, Russia
* Correspondence: va.prakht@urfu.ru; Tel.: +7-909-028-49-25

Abstract: The use of Line-Start Permanent Magnet Synchronous Motors (LSPMSM) improves the efficiency of conventional direct-on-line electric motor-driven fluid machinery such as pumps and fans. Such motors have increased efficiency compared to induction motors and do not have an excitation winding compared to classical synchronous motors with an excitation winding. However, LSPMSMs have difficulty in starting mechanisms with a high moment of inertia. This problem can be exacerbated by a reduced supply network voltage and a voltage drop on the cable. This article investigates the transients during the startup of an industrial centrifugal pump with a line-start permanent magnet synchronous motor. The simulation results showed that when the voltage on the motor terminals is reduced by 10%, the synchronization is delayed. The use of the cable also leads to a reduction in the voltage at the motor terminals in a steady state, but the time synchronization delay is more significant than that with a corresponding reduction in the supply voltage. The considered simulation example shows that the line-start permanent magnet synchronous motor has no problems with starting the pumping unit, even with a reduced supply voltage. The conclusions of this paper support a wider use of energy-efficient electric motors and can be used when selecting an electric motor to drive a centrifugal pump.

Keywords: line-start permanent magnet synchronous motor; centrifugal pumps; electric motors; energy efficiency class; energy saving; motor starting

MSC: 00A06

1. Introduction

At present, most typical industrial mechanisms (pumps, fans, blowers, compressors, etc.) of medium and small power with a direct start from the mains are driven by induction motors (IMs) [1–4]. IMs are reliable and affordable [5], but their significant drawback is their relatively low efficiency. For this reason, they usually have a relatively low energy efficiency class, not exceeding IE3, in accordance with the IEC 60034-30-2 standard "Rotating electric machines—Part 30-2: efficiency class of variable speed AC motors (IE code)" [6]. Rising prices for electricity [7], the tightening of the requirements of the European regulator for the energy efficiency of electric motors used in enterprises [8] and requirements to reduce CO_2 consumption [9] force enterprises to use more efficient motors. However, improving the efficiency of induction motors leads to a significant increase in the cost and dimensions of the machine [1].

Linear start permanent magnet motors (LSPMSMs) can be a more energy-efficient alternative in direct-on-line start applications compared to induction motors. LSPMSMs have a stator design similar to IMs, a squirrel cage and permanent magnets on the rotor. LSPMSMs are already commercially available as general-purpose motors [10–12] and as part of a compressor drive [13]. Due to their principle of operation, such motors can have an IE4 energy efficiency class and higher, without going beyond the dimensions of induction motors with the IE3 efficiency class [14] and a constant operating speed, which may be necessary in some applications, such as weaving machines. The use of such motors is limited by their high cost and the technological dependence of their manufacture on the availability of rare earth magnets (China is the main supplier of rare earth magnets).

In addition to this, the LSPMSM has limits on the moment of inertia of the loading mechanism. This is due to the fact that, when starting, the motor works in an asynchronous mode. In this case, the positive torque is characterized as the asynchronous torque of the short-circuited winding, and the magnets create an oscillating torque. If the moment of inertia is large enough, the speed fluctuations are flattened, and, like an induction motor, the LSPMSM does not reach synchronous speed and has a slip at the steady state.

As the inertia decreases, the fluctuations of the speed caused by the fluctuations of the torque increase. The synchronization process is observed when the speed fluctuations are sufficient to achieve synchronous speed. As a rule, in this case, the speed exceeds the synchronous speed, after which calming down follows, i.e., the speed stabilizes and becomes equal to the synchronous speed. The critical moment of inertia of the loading mechanism (the value of the maximum moment of inertia of the loading mechanism with which the LSPMSM synchronization is successful) is indicated in the catalogs of commercially produced LSPMSMs [10–12]. Figure 1 compares the IM and LSPMSM designs. In addition to the use of more energy-efficient motors, an increase in the efficiency of pump units can be achieved through the use of frequency converters [15,16], but such an improvement in efficiency requires a significant increase in capital costs.

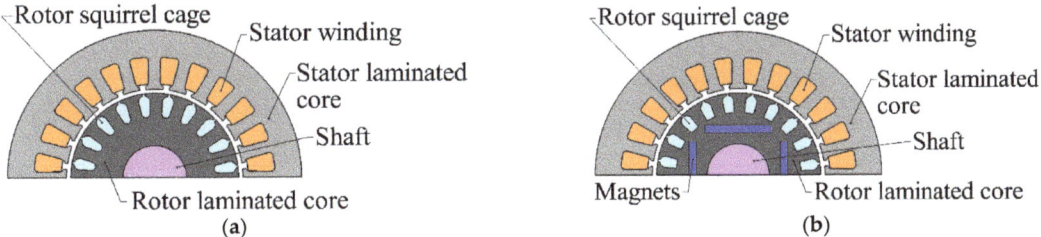

Figure 1. Sketches of electric motors. (**a**) Induction motor (IM); (**b**) Line-start permanent magnet synchronous motor (LSPMSM).

In energy-saving drives, it is also possible to use a converter-fed PMSM without starting winding [17]. However, the use of a frequency converter leads to a significant increase in the cost of the drive. This increase is especially significant for low-power general-purpose drives, for which the cost of the converter can be several times higher than the cost of the motor.

The starting processes of LSPMSM in the drives of a piston pump [18] and a centrifugal fan [19] were studied. In [18], simulation in Matlab Simulink is given, based on the results from which the data of the torque and current of the LSPMSM were obtained during the start-up and operation in a steady state. In [19], dynamic processes in LSPMSM under fan load were studied. It was found that the start of the fan with a direct start from the network can be implemented by regulating the air flow by throttling. However, there is a lack of research into the process of starting LSPMSM as part of a centrifugal pump. Compared to centrifugal fans, in which the working fluid is a gas, in pumping systems, it is a liquid. A distinctive feature of the liquid is its high density and, hence, a significant kinetic energy accumulated in the pipes.

Typically, the LSPMSM start-up process is investigated at the rated value and frequency of the sinusoidal voltage, without taking into account the voltage drop and the drop in the cable in particular. There are also studies that study the start of LSPMSM [20] and the operation of LSPMSM with distorted voltage [21,22], studies that study the start of IM with reduced voltage [23–25] and studies that take into account the effect of the cable [23–25].

However, studies on the assessment of the effect of a decrease in the mains voltage and the effect of a cable on the start of an LSPMSM are not presented in the literature.

The starting current of LSPMSMs is five to seven times greater than the rated one, which can cause an additional voltage drop both on the cable and on the internal impedance of the supply transformer. In addition, in the asynchronous mode, the frequency of the electromotive force (EMF) generated by the LSPMSM magnets does not match the frequency of the mains voltage, that is, the action of magnets is equivalent to the action of a closed-circuit generator on the resistance of the winding and the resistance of the network supplying the motor (total impedance of the cable, transformer and other network elements). Thus, magnets not only create an oscillatory torque. Therefore, with an increase in resistance due to the cable, the power of the "generator" and the braking torque increase, which additionally complicates synchronization. As a result, synchronization may not occur, the motor speed will fluctuate at speeds below the synchronous speed and the motor current will pulsate and reach the starting level. Such conditions during prolonged operation lead to the failure of the LSPMSM.

This article, which fills in the gaps noted above in previous studies, is devoted to assessing the success of the synchronization of an LSPMSM as a part of a centrifugal pump unit, taking into account the reduced supply voltage and the influence of the cable. The success of the synchronization is determined using a lumped parameter model, due to the simplicity of the model and the fewer computational efforts required. The parameters for such a model can be determined as a result of the motor's tests or by calculation, if the details of the motor design are known. To obtain the results of this study, the processes of starting a four-pole LSPMSM in centrifugal pump units of capacities of 0.55 kW and 3 kW are considered. Motors of two power ratings are considered to compare the effect of the cable on the starting process of motors of different powers.

The choice of such a range of rated power of motors for research is due to the fact that, at present, only low-power LSPMSMs are available on the general-purpose motor market [12], which is associated with their increased cost compared to that of IM. Another reason is that, with an increase in the power rating, the available energy efficiency class of mass-produced IMs increases [26]. Additionally, for medium and high powers, it becomes more feasible to use a variable frequency drive.

2. Problem Statement

Figure 2 shows the layout of the pump unit under consideration. The LSPMSM is powered by the three-phase 400 V, 50 Hz voltage via the cable.

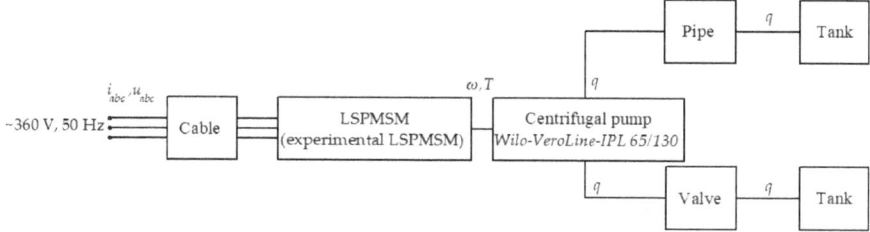

Figure 2. Diagram of the pump unit under consideration. Here, i_{abc} and u_{abc} are the mains phase currents and mains voltage; ω is the mechanical angular frequency; T is the shaft torque; q is the pump volume flow rate.

The IEC 60038 standard [27] regulates the tolerance to $\pm 10\%$ of the rated phase-to-neutral voltage and additionally allows for a 4% voltage drop due to the connecting cable and ancillary equipment in a steady state. Based on this, the motor start is simulated at a reduced voltage equal to 90% of the rated value. The motor shaft is connected directly to the shaft of a centrifugal pump "Wilo-VeroLine-IPL 65/130" [28], which pumps liquid from one reservoir (tank) to another. The fluid flow enters through a pipe and is regulated by a valve. The model was developed in Matlab Simulink. This scheme is typical for separately operating horizontal pumps.

The following assumptions were made to model the system under consideration:
- The fluid is incompressible;
- Only one pump unit operates in the pipeline;
- All hydraulic resistances (such as couplings, filters, etc.), except for the valve resistance, are reduced to one hydraulic diameter and to one pipeline;
- There are no leaks in the hydraulic system.

3. Hydraulic System Model

Similarity laws are used to calculate the pressure drop at different angular speeds. The volume flow is calculated using the following formula [29]:

$$q = q_{ref} \cdot (\omega/\omega_{ref}); \qquad (1)$$

where q_{ref} is the volume flow rate of the pump at the rated speed according to the catalog data; ω—the angular frequency of the rotation of the pump shaft; ω_{ref}—the rated angular frequency of the rotation of the pump shaft.

The differential pressure is calculated using the following formula:

$$p = p_{ref} \cdot \rho/\rho_{ref} \cdot (\omega/\omega_{ref})^2; \qquad (2)$$

where p_{ref} is the differential pressure at the rated speed and the rated density, according to the catalog; ρ is the density of the pumped liquid; ρ_{ref} is the rated density of the pumped liquid.

The mechanical power consumed by the pump is determined according to the following equation:

$$N = N_{ref} \cdot \rho/\rho_{ref} \cdot (\omega/\omega_{ref})^3; \qquad (3)$$

where N_{ref} is the mechanical power at the rated speed of the pump. The mechanical torque of the pump is:

$$T = N/\omega. \qquad (4)$$

The inertia of the fluid results in a pressure drop due to the change in the volume flow rate of the fluid. This pressure drop Δp is determined by the following equation [30]:

$$\Delta p = \rho \cdot L/A \cdot dq/dt, \qquad (5)$$

where L is the length of the pipeline; A is the cross-sectional area of the pipe; t is the time variable. The initial condition for the volume flow rate of the fluid is $q = 0$.

Regulation by means of a valve allows for changing the characteristics of the hydraulic system by means of changing the degree of the valve opening. This method of regulation allows for reducing the hydraulic power during the motor start-up and makes it possible to adjust the operating point of the system throughout the entire life of the pump unit. However, this control method creates additional hydraulic losses [31].

The simulation in this study is carried out in order to determine the conditions under which the LSPMSM successfully enters into synchronism. For this reason, it is especially important to carefully simulate physical processes, including gas flows, occurring at rotation speeds close to the synchronous one.

At sub-synchronous speeds, the phase shift between the motor current and voltage changes slowly, and the LSPMSM alternates between the motor and generator modes until synchronism is achieved. Therefore, for such an analysis, it is sufficient to use a quasi-static model (quasi-stationary), i.e., a model in which, although the hydrodynamic quantities in the elements of a liquid pipeline are assumed to be non-stationary, the relationships connecting them do not contain time derivatives and do not depend on the previous state.

4. LSPMSM Model

When modeling the LSPMSM, it is assumed that:

- The magnetic fields generated by the stator and rotor windings have a sinusoidal spatial distribution;
- The magnetic permeability of the steel is constant;
- The stator and rotor windings are symmetrical;
- Each winding is powered by a separate source;
- The mains voltage phasor is constant throughout the entire starting process, and an increase in the motor current does not cause a decrease in its amplitude;
- Magnetic core losses are not taken into account.

The system of ordinary differential LSPMSM equations to be solved is represented as:

$$\begin{aligned}
& d\lambda_{sd}/dt - Z_p \cdot \lambda_{sq} \cdot d\varphi/dt + R_s \cdot I_{sd} = U_{sd}; \\
& d\lambda_{sq}/dt + Z_p \cdot \lambda_{sd} \cdot d\varphi/dt + R_s \cdot I_{sq} = U_{sq}; \\
& d\lambda'_{rd}/dt + r'_d \cdot I'_{rd} = 0; \\
& d\lambda'_{rq}/dt + r'_q \cdot I'_{rq} = 0; \\
& I'_{rd} = (\lambda'_{rd} - \lambda_{sd})/L_{\sigma d}; \\
& I'_{rq} = (\lambda'_{rq} - [\lambda_{sq} - \lambda'_0])/L_{\sigma q}; \\
& I_{sd} = \lambda_{sd}/L_{sd} - I'_{rd}; \\
& I_{sq} = (\lambda_{sq} - \lambda'_0)/L_{sq} - I'_{rq}; \\
& T = 3/2 \cdot Z_p \cdot (\lambda_{sd} \cdot I_{sq} - \lambda_{sq} \cdot I_{sd}); \\
& J \cdot d^2\varphi/dt^2 = T - T_{load}.
\end{aligned} \qquad (6)$$

where U_{sd} and U_{sq} are the stator voltages along the d and q axes; I_{sd}, I_{sq}, I'_{rd} and I'_{rq} are the stator and rotor currents; L_{sd}, L_{sq}, L_{rd} and L_{rq} are the stator and rotor total inductances; $L_{\sigma d}$ and $L_{\sigma q}$ are the rotor leakage inductances; $\lambda_{sd}, \lambda_{sq}, \lambda'_{rd}$ and λ'_{rq} are the stator and rotor flux linkages; λ'_0 is the permanent magnet flux linkage; ω_r is the rotor electrical angular frequency; R_s is the stator resistance; Z_p is the number of motor pole pairs; r'_{rd} and r'_{rq} are the rotor resistances; φ is the mechanical rotational angle equal to the integral of the motor speed ω; T is the motor torque; T_{load} is the loading torque; J is the total moment of inertia.

All initial conditions are equal to zero, except for the stator flux along the q-axis $\lambda_{sq.0} = \lambda'_0$. Figure 3 shows the implementation of equation system (6) in Simulink.

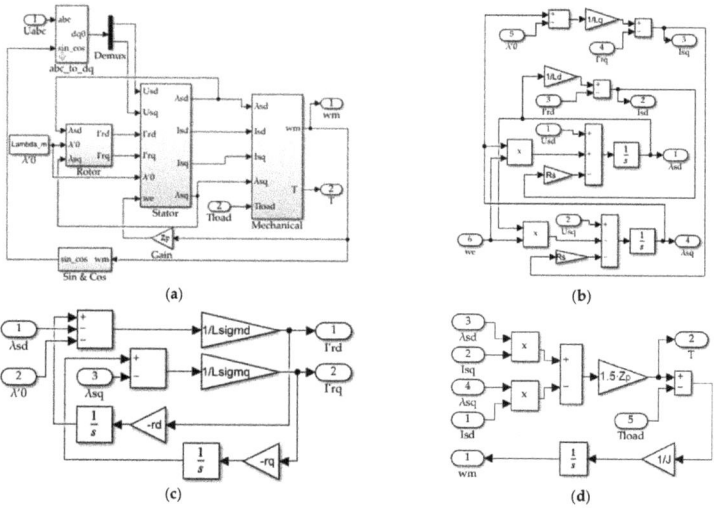

Figure 3. Simulink model of the LSPMSM motor in the d-q axes: (**a**) General view of the model; (**b**) Calculation of stator currents; (**c**) Calculation of rotor currents; (**d**) Torque calculation.

5. Model of the Pump Unit in Matlab Simulink and Parameters of the 0.55 kW Pump Unit

To simulate a pump unit, the Sim-scape/Fluids/Hydraulics library was used in the Matlab Simulink environment, designed to simulate the movement of an ideal fluid. Figure 4 shows a block diagram of the model under study in the Matlab Simulink environment. The model uses the following fluid dynamics modeling blocks: Centrifugal Pump [29], Hydraulic Pipeline [32], Fluid Inertia [30], Variable Area Hydraulic Orifice [33] and Tank [34].

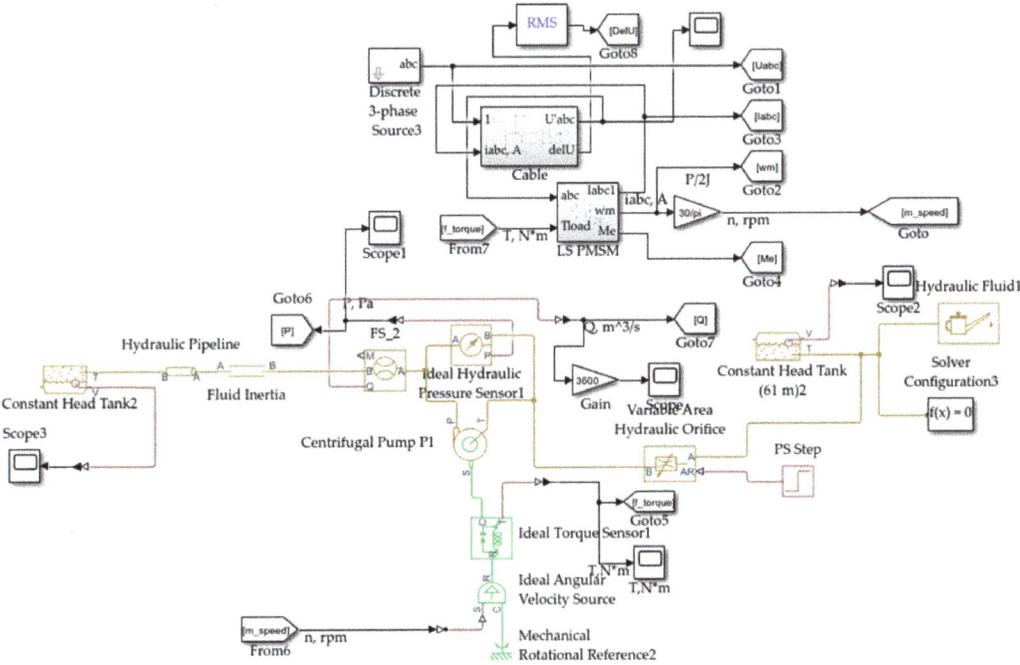

Figure 4. Implementation of the model of the pump unit with the LSPMSM drive in the Matlab Simulink.

The LSPMSM, fed directly from the mains ("Discrete 3-phase Source"), is mechanically connected directly to the shaft of a centrifugal pump ("Centrifugal Pump"). The pump moves the working fluid from one reservoir to another ("Constant Head Tank" and "Constant Head Tank") through a pipe ("Hydraulic Pipeline"). The fluid flow is controlled by a valve ("Variable Area Hydraulic Orifice"). Fluid inertia in the pipe is pointed out with a separate element ("Fluid Inertia"). The load torque is measured between the motor shaft and the pump shaft ("Ideal Torque Sensor") and enters to the input of the motor unit. The motor speed is transmitted to the input ("Ideal Angular Velocity Source"), after which it is connected to the pump shaft. The fluid parameters are set in the block ("Hydraulic Fluid"), and the solver parameters are set in the block ("Solver Configuration"). The parameterization of the model blocks is carried out in such a way that the pump in a steady state operates within the limits of the operating curve given in the catalog. The LSPMSM parameters (rated power 550 W, rated speed 1500 rpm) given in Table 1 were taken as motor parameters.

Table 1. 0.55 kW LSPMSM parameters.

Parameter	Value
Rated power P_{rate}, kW	0.55
Rated line-to-line voltage U_{rate}, V	380
RMS rated stator current, A	1.11
Rated power factor	0.85
Rated frequency f, Hz	50
Pole pair number Z_p	2
Stator phase resistance R_s, Ohm	15.3
Total direct inductance L_d, H	0.26
Total quadrature inductance L_q, H	0.15
Leakage direct inductance $L_{\sigma d}$, H	0.038
Leakage quadrature inductance $L_{\sigma q}$, H	0.051
Rotor direct resistance r'_d, Ohm	9.24
Rotor quadrature resistance r'_q, Ohm	10.1
Permanent magnet flux linkage λ'_0, Wb	0.76
Motor inertia moment J_m, kg·m²	0.003
Pump impeller inertia moment J_i, kg·m²	0.00022

The model uses the parameters of a centrifugal pump with a steel impeller "Wilo-VeroLine-IPL 65/130". The catalog does not specify the moment of inertia, so the moment of inertia of the impeller was assumed to be a tenth of the moment of inertia of the motor, $J_i = 0.1 \cdot J_m$, as proposed for pump drives with an induction motor [35]. Given that the moment of inertia of an M3BP80MA4 induction motor with a rated power of 0.55 kW and a rated speed of 1406 rpm is $J_m = 0.0022$ kg·m² [26], the moment of inertia of the pump impeller is:

$$J_i = 0.1 \cdot J_m = 0.1 \cdot 0.0022 = 0.00022 \text{ kg·m}^2. \tag{7}$$

Table 2 shows the catalog characteristics of the pump. These characteristics correspond to a rotational speed of 1450 rpm.

Table 2. 0.55 kW pump characteristics.

Flow Q, m³/h	Pressure P, Pa	Braking power N_m, W
4.77	54,229	239.15
10.00	53,304	300.71
19.97	52,259	418.44
29.82	44,691	502.98
39.73	31,490	545.82
48.42	16,106	545.82

Figure 5 shows the curves of the pump head-flow characteristics corresponding to speeds of 1450 and 1500 rpm, as well as the characteristics of the hydraulic load with the valve open.

For the 0.55 kW pump unit, a pipe with a length of 94.5 m and a cross-sectional area of 0.0137 m² was chosen. The calculation of the characteristics of the pump for rotation speeds other than the rated one is carried out using the laws of similarity (1)–(3).

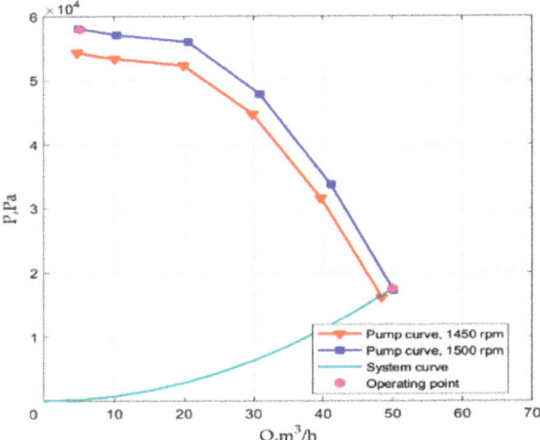

Figure 5. Pump head-flow curves taken at 1450 and 1500 rpm, as well as the system response with the valve open.

6. The Model of the Cable

For modeling, a copper cable with lengths of l = 0, 100, 300 and 500 m and a cross section of 1.5 mm² was adopted. ρ_{rcu} = 0.0225 Ohm·mm²/m is the specific resistance of copper; λ_{lcu} = 0.08 mOhm/m is the specific reactance at 50 Hz of the cable, according to [36]. Below is a calculation of the impedance components of a cable that is 100 m long:

$$R_{c100} = \rho_{rcu} \cdot l/S_c = 1.5 \text{ Ohm;} \tag{8}$$

$$X_{c100} = \lambda_{lcu} \cdot l = 0.008 \text{ Ohm;} \tag{9}$$

Assuming that the inductive reactance was calculated for 50 Hz, the cable inductance is:

$$L_c = X_{c100}/(2\pi \cdot f) = 2.55 \cdot 10^{-5} \text{ H;} \tag{10}$$

Figure 6 shows the block diagram of the cable model in the Matlab Simulink environment.

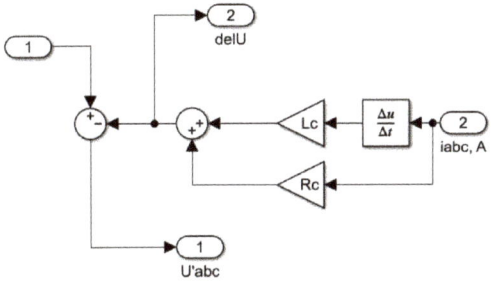

Figure 6. The block diagram of the cable model.

7. Simulation Results for the 0.55 kW Pump Unit

Figures 7–12 show the results of the simulation of the electric drive of the centrifugal pump at a rated voltage of 400 V and at a voltage reduced by 10% (360 V), as well as start-up graphs for different cable lengths and the reduced voltage.

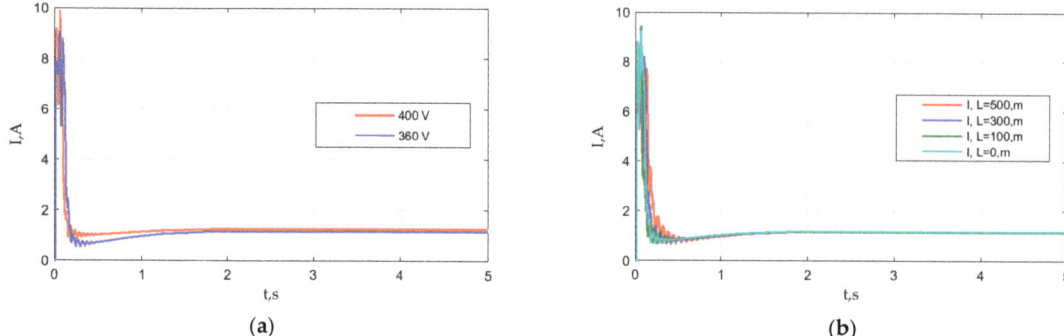

Figure 7. Motor current simulation results (envelopes of its amplitude). (**a**) Start with the rated and reduced voltages; (**b**) with the reduced voltage and different cable length.

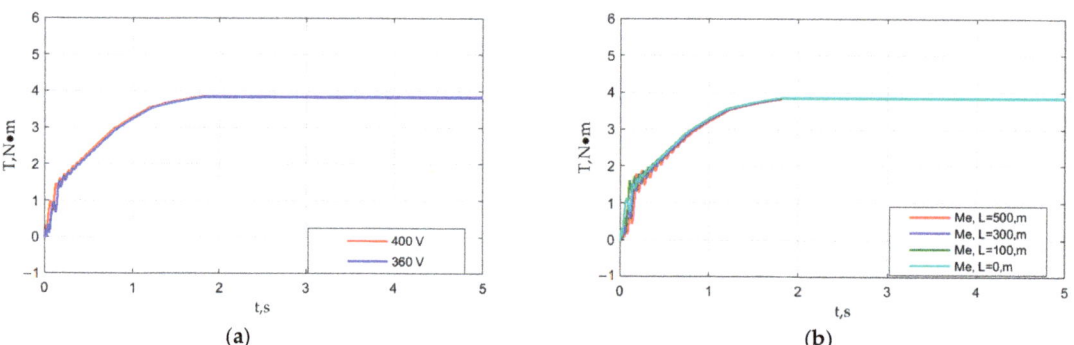

Figure 8. Electromagnetic torque simulation results. (**a**) Start with the rated and reduced voltages; (**b**) Start with the reduced voltage and different cable length.

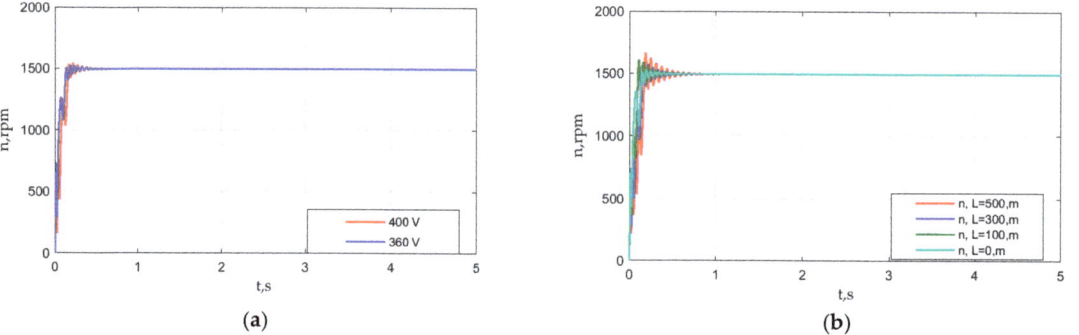

Figure 9. Motor speed simulation results. (**a**) Start with the rated and reduced voltages; (**b**) Start with the reduced voltage and different cable length.

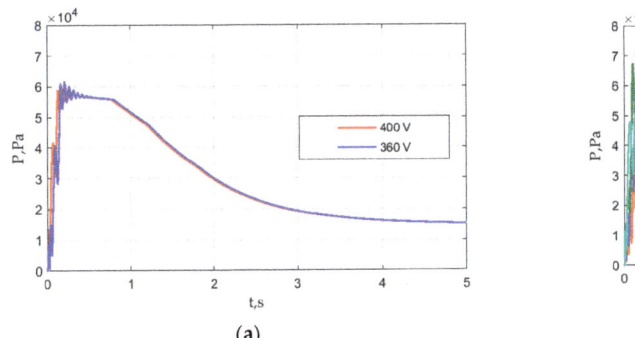
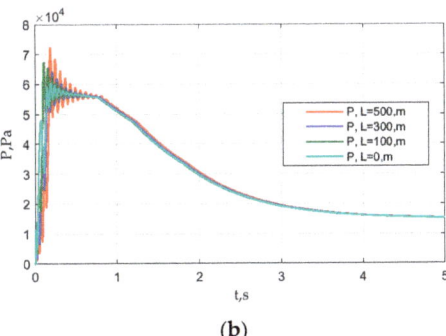

Figure 10. Pump differential pressure simulation results. (**a**) Start with the rated and reduced voltages; (**b**) with the reduced voltage and different cable length.

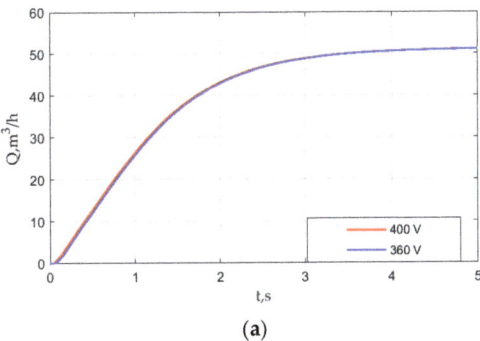
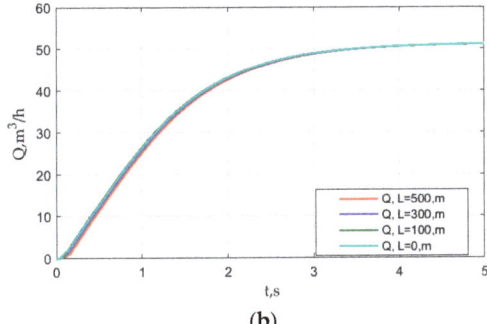

Figure 11. Pump flow simulation results. (**a**) Start with the rated and reduced voltages; (**b**) Start with the reduced voltage and different cable length.

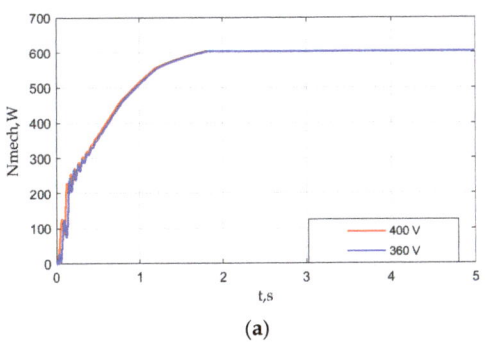
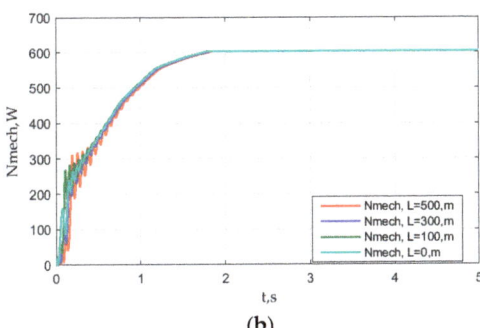

Figure 12. Pump mechanical power simulation results. (**a**) Start with the rated and reduced voltages; (**b**) Start with the reduced voltage and different cable length.

Figures 7–12 show the instantaneous rms values of the current, the electromagnetic torque, the motor speed, the pressure, the volume flow rate and the mechanical power at the rated voltage, at a 10% reduced voltage and at a 10% reduced voltage when powered through a cable of various lengths. It can be seen from the graphs that synchronization occurs in all considered cases. Figure 9 shows that, in all the cases considered, the first achievement of the synchronous speed occurs within the time τsynch1, no more than

0.2 s, which is followed by relaxation to the synchronous speed. The characteristic time dependences of the volume flow rate on time, shown in Figure 11, are approximately τ_{hydr} = 2 s, i.e., much larger than τ_{synch1}. The dependences of the flow rate on the time in the considered cases are close to each other, since, after the synchronous speed is reached, the speed fluctuations are not large, the motor rotates at an almost synchronous speed and damped speed fluctuations have practically no effect on the flow due to hydraulic inertia. The mechanical power increases as the volumetric flow rate of the fluid increases, as can be seen in Figure 12, i.e., synchronization occurs in "soft" conditions.

As seen in Figure 7, during the starting process, the current reaches a value seven times higher than the rated one. Despite this, there are no torque surges (Figure 8). On the contrary, as the volume flow increases, the torque gradually increases, making damped oscillations, which ensures a high mechanical reliability. This behaviour of the torque is explained by a soft start and the rather small moments of inertia of the rotor and the impeller.

In some applications, such as blowers, the moment of inertia is much greater. At the same time, to facilitate starting the motor, the number of turns of the stator winding is reduced, which is equivalent to raising the voltage. Figure 7a shows that the current at the start at the reduced voltage is less than that at the rated one, which leads to the losses reduction in the winding and the increase in efficiency. So, due to the facilitated starting conditions, it is possible to recommend the manufacture of LSPMSMs for pumping applications in order to produce LSPMSMs with an increased number of turns, which is equivalent to a decrease in the supply voltage and therefore to increased efficiency.

The decrease in the current (and, therefore, the voltage) on the motor terminals in a steady state due to the presence of the cable is so small that it is not noticeable in Figure 7b. At the same time, the presence of a cable leads to a significant delay in synchronization. Thus, the effect of the cable on delaying synchronization does not come down to an equivalent decrease in the voltage/current on the motor, which is additionally explained by the generation of power by the motor into the grid at a frequency different from the frequency of the grid.

8. Parameters of the 3 kW Pump Unit

The parameters of the experimental four-pole 3 kW synchronous motor with the starting winding and permanent magnets were taken for the calculation. Table 3 shows the motor parameters. The centrifugal pump "Wilo-VeroLine-IPL 100/175-3/4 28" was chosen for modeling. Table 4 shows the pump characteristics. The pipe length l is 450 m. Its diameter D is 0.168 m.

Table 3. 3 kW LSPMSM parameters.

Parameter	Value
Rated power P_{rate}, kW	3
Rated line-to-line voltage U_{rate}, V	380
RMS rated stator current, A	5.84
Rated power factor	0.82
Rated frequency f, Hz	50
Pole pair number Z_p	2
Stator phase resistance R_s, Ohm	1.281
Total direct inductance L_d, H	0.110
Total quadrature inductance L_q, H	0.0567
Leakage direct inductance $L_{\sigma d}$, H	0.0186
Leakage quadrature inductance $L_{\sigma q}$, H	0.0108
Rotor direct resistance r'_d, Ohm	2.943
Rotor quadrature resistance r'_q, Ohm	2.426
Permanent magnet flux linkage λ'_0, Wb	0.6095
Motor inertia moment J_m, kg·m^2	0.010
Pump impeller inertia moment J_i, kg·m^2	0.0011

Table 4. 3 kW pump characteristics.

Flow Q, m³/h	Pressure P, Pa	Power N_m, W
20.67	92,337	1613.5
40.00	91,669	1968.5
79.51	84,825	2475.3
118.96	70,390	2819.2
158.92	49,404	3011.0
188.19	30,138	3052.1

Figure 13 shows the curves of the pressure-flow characteristics of the pump, taken at 1450 and 1500 rpm, as well as the system characteristic with an open valve.

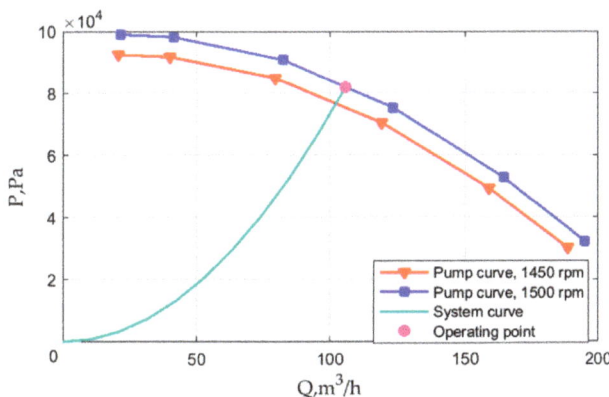

Figure 13. Pump head-flow curves taken at 1450 and 1500 rpm, as well as the system response with the valve open.

A voltage drop on the cable is defined as a decrease in the voltage on the motor. It is possible to estimate the rms value of the voltage drop on the cable in the rated mode by the rated current and power factor [37]:

$$u = b \cdot I \cdot (\rho_{rcu} \cdot L/S \cdot \cos\varphi + \lambda \cdot \sin\varphi); \tag{11}$$

where u is the voltage drop on the cable; b is the coefficient (1 if the circuit is three-phase and 2 if the circuit is single-phase); I is the rms value of the current; $\cos\varphi$ is the motor power factor. The relative voltage drop on the cable is determined as follows:

$$\Delta u = 100 \, u/U_0; \tag{12}$$

where U_0 is the phase voltage. The dependence of the relative voltage drop on the cable on its length is given in Table 5. In accordance with [37], since the voltage drop with cable lengths of 300, 500 and 700 m exceeds 4%, such cable lengths cannot be used.

Table 5. Relative cable voltage drop versus cable length.

l_c, m	Δu, %
0	0
100	3.12
300	9.37
500	15.61
700	21.85

9. Simulation Results for the 3 kW Pump Unit

Figures 14–20 show the results of the simulation of the pump electric drive at the rated voltage and at the voltage reduced by 10%, as well as the start-up graphs at different cable lengths and at the reduced voltage.

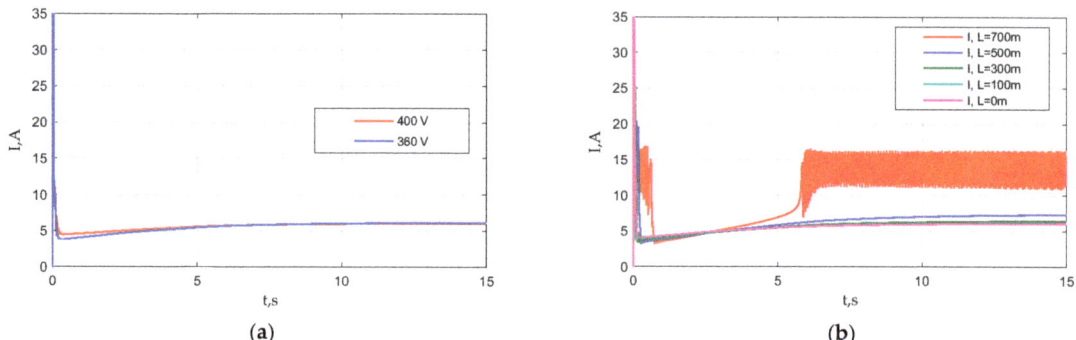

Figure 14. Motor current simulation results (envelopes of its amplitude): (**a**) Start with the rated and reduced voltages; (**b**) Start with the reduced voltage and different cable length.

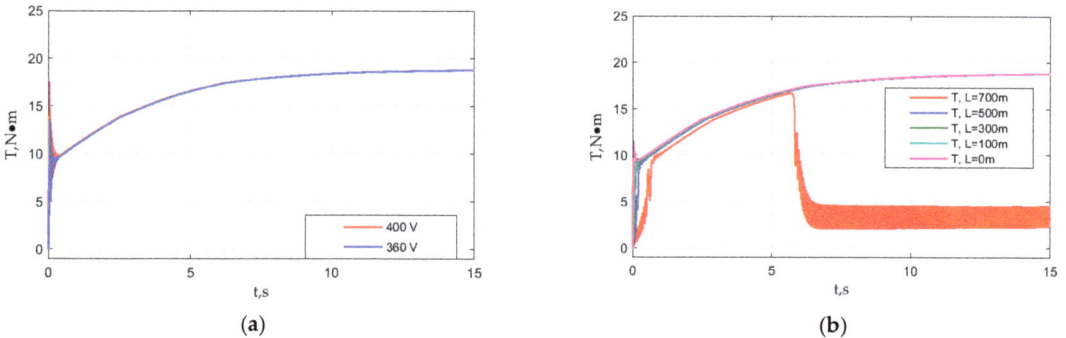

Figure 15. Electromagnetic torque simulation results: (**a**) Start with the rated and reduced voltages; (**b**) Start with the reduced voltage and different cable length.

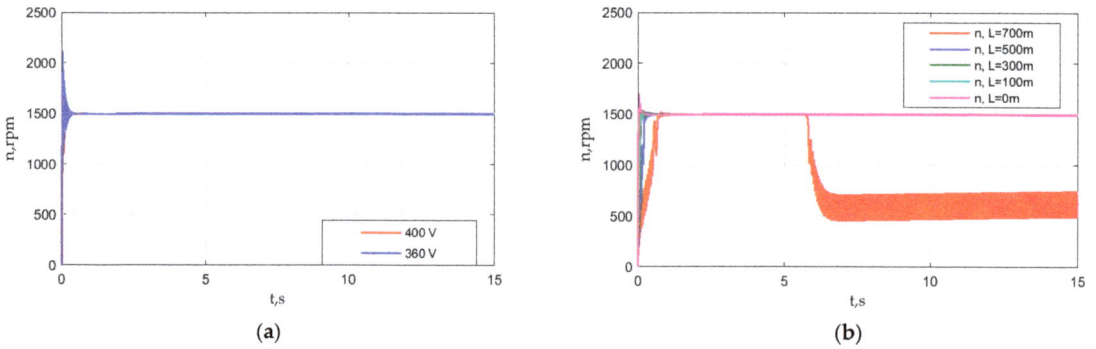

Figure 16. Motor speed simulation results: (**a**) Start with the rated and reduced voltages; (**b**) Start with the reduced voltage and different cable length.

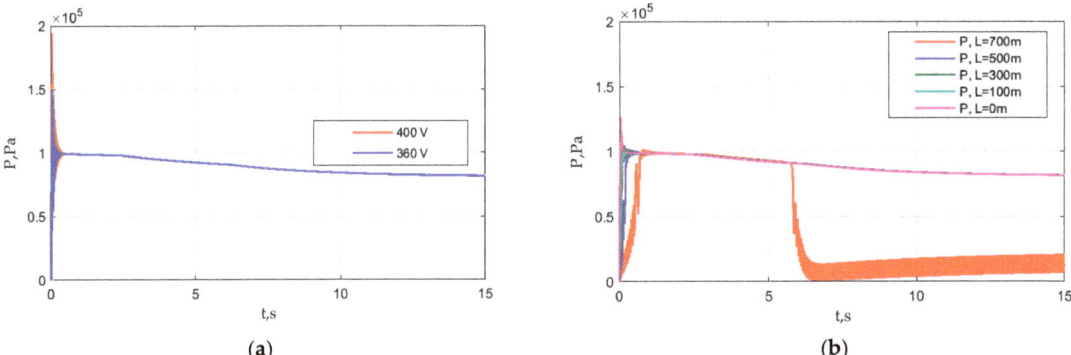

Figure 17. Pump differential pressure simulation results: (**a**) Start with the rated and reduced voltages; (**b**) Start with the reduced voltage and different cable length.

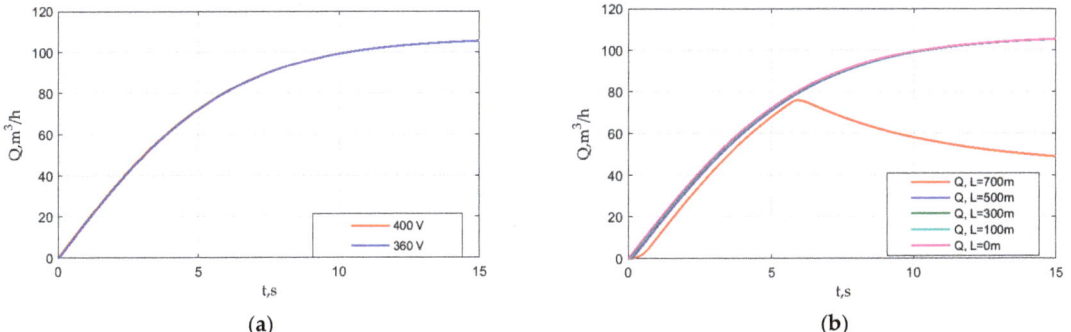

Figure 18. Pump flow simulation results: (**a**) Start with the rated and reduced voltages; (**b**) Start with the reduced voltage and different cable length.

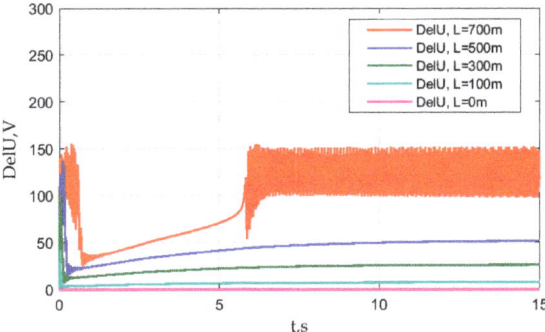

Figure 19. Rms value of the voltage drop on the cable.

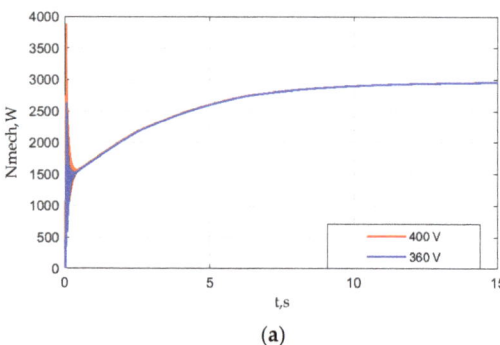
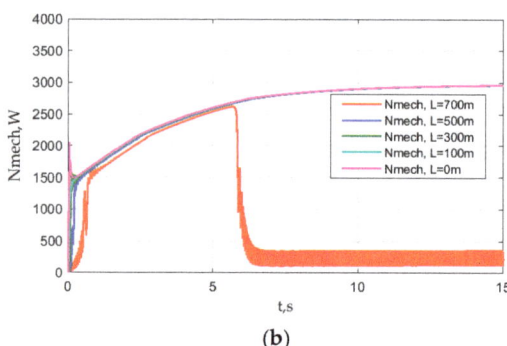

Figure 20. Pump mechanical power simulation results: (**a**) Start with the rated and reduced voltages; (**b**) Start with the reduced voltage and different cable length.

The results of this simulation are mainly similar to those for the 0.55 kW LSPMSM. The characteristic time of the hydrodynamic process in the pipe and in the pump is much greater than that of the mechanical and electric process in the motor. The impeller inertia is much less than that of the rotor. These result in a "soft start". Thus, there are no torque surges exceeding the torque rated value.

Figures 14–20 present the simulation results at the reduced voltage and at cable lengths of 0, 100, 300, 500 and 700 m. The standard [37] allows for a voltage drop on the cable of not more than 4%, which is satisfied only in the cases of the cable lengths of 0 and 100 m. The rest simulations are carried out to demonstrate the behavior long cables. The synchronization problems are observed only with the cable length of 700 m, which is much more than the allowed length. Namely, the synchronization occurs due to a "soft start". However, in increasing the mechanical power, the synchronization cannot be sustained at some point.

10. Conclusions

This paper analyzes the effect of the reduced voltage and an increase in the cable length on the success of starting LSPMSM in the drive of centrifugal pump units with a power of 0.55 kW and 3 kW.

A feature of the pumps, in comparison, for example, with fans with large centrifugal masses, is that the moment of inertia of the pump impeller is several times less than the moment of inertia of the rotor itself, which makes it possible to hope that there will be no problem of unsuccessful starts of the LSPMSM in the pump drive.

The simulation examples show that, for pump drives of the considered type and power rating, the stable synchronization occurs at the rated voltage, at the voltage of 90% of the rated one and, additionally, with the supply via the cable, except for the cases in which the voltage drop in the cable significantly exceeds the level of 4% recommended in IEC 60364-5-52.

The characteristic time of the hydrodynamic process in the pipe and in the pump is much greater than that of the mechanical and electrical process in the motor. The impeller inertia is much less than that of the rotor. These result in a "soft start". Therefore, there are no torque surges exceeding the torque rated value, which prolongs the system lifetime. Additionally, due to a "soft start", it is expected that the number of turns of the stator winding to ensure the successful synchronization can be greater than that in other applications, which can decrease the current and increase the LSPMSM efficiency in the pump applications.

Author Contributions: Conceptual approach, A.P., V.K. and V.P.; data duration, A.P. and S.O.; software, A.P., S.O. and V.K.; calculations and modeling, A.P., S.O., V.D., V.K. and V.P.; writing—original draft, A.P., S.O., V.D., V.K. and V.P.; visualization, A.P. and V.K.; review and editing, A.P., S.O., V.D., V.K. and V.P. All authors have read and agreed to the published version of the manuscript.

Funding: The work was partially supported by the Ministry of Science and Higher Education of the Russian Federation (through the basic part of the government mandate, Project No. FEUZ 2020-0060).

Institutional Review Board Statement: Not applicable.

Informed Consent Statement: Not applicable.

Data Availability Statement: Data are contained within the article.

Acknowledgments: The authors thank the editors and reviewers for the careful reading and constructive comments.

Conflicts of Interest: The authors declare no conflict of interest.

Glossary

List of Abbreviations

EMF	Electromotive force
IM	Induction motor
LSPMSM	Linear start permanent magnet motor

List of Mathematical Symbols

A	Cross-sectional area of the pipe, m^2
i_{abc}	Mains phase currents, A
I_{sd}, I_{sq}	Stator currents, A
I'_{rd}, I'_{rq}	Rotor currents, A
f	Mains voltage frequency, Hz
J	Total moment of inertia, kg · m^2
J_i	Moment of inertia of the impeller, kg · m^2
l	Cable length, m
L	Length of the pipeline, m
L_{sd}, L_{sq}	Stator total inductances, H
L_c	Cable length, m
$L_{\sigma d}, L_{\sigma q}$	Rotor leakage inductances, H
N	Mechanical power consumed by the pump, W
N_{ref}	Mechanical power at the rated speed of the pump, W
p	Differential pump pressure, Pa
p_{ref}	Differential pressure at the rated speed and the rated density, Pa
q	Flow volume, m^3/s
q_{ref}	Volume flow rate of the pump at the rated speed according to the catalog data, m^3/s
R_{c100}	Resistance of a cable 100 m long, Ohm
R_s	Stator resistance, Ohm
t	Time variable
T	Motor shaft torque, N · m
T_{pump}	Mechanical torque of the pump, N · m
u_{abc}	Mains phase voltage, V
U_{sd}, U_{sq}	Stator voltages along d and q axes, V
X_{c100}	Reactance of a cable that is 100 m long, Ohm
Z_p	Number of motor poles
Δp	Pressure drop due to fluid inertia, Pa
λ_{lcu}	Specific cable reactance, mOhm/m
$\lambda'_{rd}, \lambda'_{rq}$	Rotor flux linkages, Wb
$\lambda_{sd}, \lambda_{sq}$	Stator flux linkages, Wb
λ'_0	Permanent magnet flux linkage, Wb

ρ		Density of the pumped liquid, m^3/s
ρ$_{rcu}$		Specific resistance of copper, Ohm·mm^2/m
ρ$_{ref}$		Rated density of the pumped liquid, m^3/s
τ$_{synch1}$		Motor synchronization time, s
τ$_{hydr}$		Hydraulic system transient time, s
φ		Mechanical rotational angle, rad
ω		Angular frequency of the rotation of the pump shaft, rad/s
ω$_{ref}$		Rated angular frequency of the rotation of the pump shaft, rad/s

References

1. Goman, V.; Oshurbekov, S.; Kazakbaev, V.; Prakht, V.; Dmitrievskii, V. Energy Efficiency Analysis of Fixed-Speed Pump Drives with Various Types of Motors. *Appl. Sci.* **2019**, *9*, 5295. [CrossRef]
2. Baka, S.; Sashidhar, S.; Fernandes, B.G. Design of an Energy Efficient Line-Start Two-Pole Ferrite Assisted Synchronous Reluctance Motor for Water Pumps. *IEEE Trans. Energy Conv.* **2021**, *36*, 961–970. [CrossRef]
3. Ismagilov, F.; Vavilov, V.; Gusakov, D. Line-Start Permanent Magnet Synchronous Motor for Aerospace Application. In Proceedings of the IEEE International Conference on Electrical Systems for Aircraft, Railway, Ship Propulsion and Road Vehicles and International Transportation Electrification Conference (ESARS-ITEC 2018), Nottingham, UK, 7–9 November 2018; pp. 1–5. [CrossRef]
4. Kurihara, K.; Rahman, M. High-Efficiency Line-Start Interior Permanent-Magnet Synchronous Motors. *IEEE Trans. Ind. Appl.* **2004**, *40*, 789–796. [CrossRef]
5. Kazakbaev, V.; Paramonov, A.; Dmitrievskii, V.; Prakht, V.; Goman, V. Indirect Efficiency Measurement Method for Line-Start Permanent Magnet Synchronous Motors. *Mathematics* **2022**, *10*, 1056. [CrossRef]
6. Rotating Electrical Machines—Part 30-2: Efficiency Classes of Variable Speed AC Motors (IE-Code). IEC 60034-30-2/IEC: 2016-12. Available online: https://webstore.iec.ch/publication/30830 (accessed on 19 December 2022).
7. Quarterly Report On European Electricity Markets, Market Observatory for Energy DG Energy, 15(1), Covering First Quarter of 2022, European Commission. 2022. Available online: https://energy.ec.europa.eu/system/files/2022-07/Quarterly%20Report%20on%20European%20Electricity%20markets%20Q1%202022.pdf (accessed on 19 December 2022).
8. European Commission Regulation (EC). No. 640/2009 Implementing Directive 2005/32/EC of the European Parliament and of the Council with Regard to Ecodesign Requirements for Electric Motors, (2009), Amended by Commission Regulation (EU) No 4/2014 of January 6, 2014. Document 32014R0004. Available online: https://eur-lex.europa.eu/legal-content/EN/TXT/?uri=CELEX%3A32014R0004 (accessed on 19 December 2022).
9. European Council Meeting (10 and 11 December 2020)—Conclusions. EUCO 22/20 CO EUR 17 CONCL 8. Available online: https://www.consilium.europa.eu/media/47296/1011-12-20-euco-conclusions-en.pdf (accessed on 23 January 2021).
10. Addendum to the Operating Instructions: AC Motors DR.71.J-DR.100.J with LSPM Technology, 21281793/EN, 09/2014, SEW Eurodrive. Available online: https://download.sew-eurodrive.com/download/pdf/21343799.pdf (accessed on 19 December 2022).
11. Catalogue of Super Premium Efficiency SynchroVERT LSPM Motors. Available online: https://www.bharatbijlee.com/media/14228/synchrovert_catalogue.pdf (accessed on 19 December 2022).
12. WQuattro, Super Premium Efficiency Motor, Product Catalogue, WEG Group-Motors Business Unit, Cod: 50025713, Rev: 03, Date (m/y): 07/2017. Available online: https://static.weg.net/medias/downloadcenter/h01/hfc/WEG-w22-quattro-european-market-50025713-brochure-english-web.pdf (accessed on 19 December 2022).
13. KT-420-5, Operation of Bitzer Reciprocating Compressors with External Frequency Inverters, Bitzer, 01.2022. Available online: https://www.bitzer.de/shared_media/html/kt-420/Resources/pdf/279503819.pdf (accessed on 19 December 2022).
14. Kazakbaev, V.; Prakht, V.; Dmitrievskii, V.; Golovanov, D. Feasibility Study of Pump Units with Various Direct-On-Line Electric Motors Considering Cable and Transformer Losses. *Appl. Sci.* **2020**, *10*, 8120. [CrossRef]
15. Vasilev, B. Analysis and improvement of the efficiency of frequency converters with pulse width modulation. *Int. J. Electr. Comput. Eng.* **2019**, *9*, 2314–2320. [CrossRef]
16. Churkin, A.; Churkin, B.; Savelieva, I. Improving the energy efficiency of frequency converters in networks of limited power. *IOP Conf. Ser. Mater. Sci. Eng.* **2019**, *643*, 012056. [CrossRef]
17. Wang, S.-C.; Nien, Y.-C.; Huang, S.-M. Multi-Objective Optimization Design and Analysis of V-Shape Permanent Magnet Synchronous Motor. *Energies* **2022**, *15*, 3496. [CrossRef]
18. Wang, D.; Wang, X.; Chen, H.; Zhang, R. Matlab/Simulink-Based Simulation of Line-start PMSM Used in Pump Jacks. In Proceedings of the 2nd IEEE Conference on Industrial Electronics and Applications, Harbin, China, 23–25 May 2007; pp. 1179–1181. [CrossRef]
19. Paramonov, A.; Oshurbekov, S.; Kazakbaev, V.; Prakht, V.; Dmitrievskii, V. Study of the Effect of Throttling on the Success of Starting a Line-Start Permanent Magnet Motor Driving a Centrifugal Fan. *Mathematics* **2022**, *10*, 4324. [CrossRef]
20. Sethupathi, P.; Senthilnathan, N. Comparative analysis of line-start permanent magnet synchronous motor and squirrel cage induction motor under customary power quality indices. *Electr. Eng.* **2020**, *102*, 1339–1349. [CrossRef]
21. Faiz, J.; Ebrahimpour, H.; Pillay, P. Influence of unbalanced voltage on the steady-state performance of a three-phase squirrel-cage induction motor. *IEEE Trans. Energy Convers.* **2004**, *19*, 657–662. [CrossRef]

22. Gnaciński, P.; Muc, A.; Pepliński, M. Influence of Voltage Subharmonics on Line Start Permanent Magnet Synchronous Motor. *IEEE Access* **2021**, *9*, 164275–164281. [CrossRef]
23. Blair, T.H. 3 phase AC motor starting methods: An analysis of reduced voltage starting characteristics. In Proceedings of the IEEE SoutheastCon 2002 (Cat. No.02CH37283), Columbia, SC, USA, 5–7 April 2002; pp. 181–186. [CrossRef]
24. Hsu, C.-T.; Chuang, H.-J. Power quality assessment of large motor starting and loading for the integrated steel-making cogeneration facility. In Proceedings of the Fortieth IAS Annual Meeting. Conference Record of the 2005 Industry Applications Conference, Hong Kong, China, 2–6 October 2005; pp. 59–66. [CrossRef]
25. Patil, P.; Porate, K. Starting Analysis of Induction Motor: A Computer Simulation by Etap Power Station. In Proceedings of the Second International Conference on Emerging Trends in Engineering & Technology, Nagpur, India, 16–18 December 2009; pp. 494–499. [CrossRef]
26. Low Voltage Process Performance Motors 400 V 50 Hz, 460 V 60 Hz., ABB, Catalog, February 2022. Available online: https://search.abb.com/library/Download.aspx?DocumentID=9AKK105944&LanguageCode=en&DocumentPartId=&Action=Launch (accessed on 19 December 2022).
27. International Standard IEC 60038, IEC Standard Voltages, Edition 7.0 2009-06. Available online: https://webstore.iec.ch/publication/153 (accessed on 19 December 2022).
28. Wilo, Circulation Pumps, Catalog. 2009. Available online: http://sirkulasyonpompasi.org/sirkulasyonpompasi/wilo-katalog.pdf (accessed on 19 December 2022).
29. Centrifugal Pump, Pumps and Motors. Model Description. MathWorks. Available online: https://www.mathworks.com/help/hydro/ref/centrifugalpump.html;jsessionid=9398cce96bf548d19dfefaa598e3 (accessed on 19 December 2022).
30. Fluid Inertia. Model description. MathWorks. Available online: https://www.mathworks.com/help/simscape/ref/fluidinertia.html?s_tid=doc_ta (accessed on 19 December 2022).
31. Improving Pumping System Performance, a Sourcebook for Industry. Prepared for the United States Department of Energy Office of Energy Efficiency and Renewable Energy Industrial Technologies Program, Second Edition, May 2006. Available online: https://www.energy.gov/sites/prod/files/2014/05/f16/pump.pdf (accessed on 19 December 2022).
32. Hydraulic Pipeline. Model Description. MathWorks. Available online: https://www.mathworks.com/help/hydro/ref/hydraulicpipeline.html (accessed on 19 December 2022).
33. Variable Area Hydraulic Orifice. Model Description. MathWorks. Available online: https://www.mathworks.com/help/simscape/ref/variableareahydraulicorifice.html?searchHighlight=%20hydraulic%20orifice&s_tid=srchtitle_%20hydraulic%20orifice_2 (accessed on 19 December 2022).
34. Tank. Low-Pressure Blocks. Model Description. MathWorks. Available online: https://www.mathworks.com/help/hydro/ref/tank.html?s_tid=doc_ta (accessed on 19 December 2022).
35. Estimation of Pump Moment of Inertia. Article. Native Dynamics. 10 September 2013. Available online: https://neutrium.net/equipment/estimation-of-pump-moment-of-inertia (accessed on 19 December 2022).
36. Kirar, M.; Aginhotri, G. Cable sizing and effects of cable length on dynamic performance of induction motor. Proceedings of IEEE Fifth Power India Conference, Murthal, India, 19–22 December 2012; pp. 1–6. [CrossRef]
37. *IEC 60364-5-52*; Electrical Installations in Buildings—Part. 5-52: Selection and Erection of Electrical Equipment—Wiring Systems, Is the IEC Standard Governing Cable Sizing. IEC: Geneva, Switzerland, 2009. Available online: https://webstore.iec.ch/publication/1878 (accessed on 19 December 2022).

Disclaimer/Publisher's Note: The statements, opinions and data contained in all publications are solely those of the individual author(s) and contributor(s) and not of MDPI and/or the editor(s). MDPI and/or the editor(s) disclaim responsibility for any injury to people or property resulting from any ideas, methods, instructions or products referred to in the content.

Design Optimization of a Synchronous Homopolar Motor with Ferrite Magnets for Subway Train

Vladimir Dmitrievskii, Vladimir Prakht * and Vadim Kazakbaev

Department of Electrical Engineering, Ural Federal University, 620002 Yekaterinburg, Russia
* Correspondence: va.prakht@urfu.ru; Tel.: +7-909-028-49-25

Abstract: Brushless synchronous homopolar machines (SHM) have long been used as highly reliable motors and generators with an excitation winding on the stator. However, a significant disadvantage that limits their use in traction applications is the reduced specific torque due to the incomplete use of the rotor surface. One possible way to improve the torque density of SHMs is to add inexpensive ferrite magnets in the rotor slots. This paper presents the results of optimizing the performances of an SHM with ferrite magnets for a subway train, considering the timing diagram of train movement. A comparison of its characteristics with an SHM without permanent magnets is also presented. When using the SHM with ferrite magnets, a significant reduction in the dimensions and weight of the motor, as well as power loss, is shown.

Keywords: ferrite magnets; synchronous homopolar motor; electrically excited synchronous motor; Nelder–Mead method; optimal design of electric machines; subway train; constant power speed range; traction drive

MSC: 00A06

1. Introduction

Interior rare-earth permanent magnet synchronous motors (PMSM) are a popular choice in electric vehicle drives of various power ratings. As they have no excitation winding and, consequently, excitation loss, these motors have high power density and efficiency [1,2]. However, their drawbacks are: (1) Rare-earth magnets required in the PMSMs production are expensive and their cost can vary by several times in a year or two because of a limited number of manufacturers; it also makes it inevitable to rely on a limited number of magnet suppliers in the world market [3–5]; (2) The rare-earth elements extraction is not environmentally friendly [6]; (3) Strong magnetic field and high temperature in PMSMs with high power density can result in rare-earth magnets demagnetization; (4) If a wide constant power speed region is required, achieving high efficiency at speeds close to maximum is not so easy because of increased winding losses in field weakening mode [7–9]; (5) In addition, in electric drives such as drives of subway and railway trains, a large value of uncontrolled electromotive force (EMF) in the windings during rotation of the PMSM creates a fire hazard in the event of an emergency short circuit. Since trains have high inertia and cannot stop quickly in the event of an emergency short circuit, the use of PMSM in this application should be avoided.

Wound rotor synchronous motors (WRSM) have no disadvantages such as this and are used in traction applications by BMW (BMW iX3 crossover) and Renault (Renault Zoe supermini electric car, Fluence sedan, Megane E-TECH small family car) (BMW [10,11], Renault [12]). WRSMs have no magnets. Excitation current as well as demagnetizing part of stator current can be reduced at high speed, which makes it possible to achieve high efficiency throughout a wide constant power speed region [13,14]. However, slip rings in the WRSM design limit the motor speed and reduce its reliability [14].

Homopolar machines (SHM) combine the advantages of PMSMs and WRSMs: they have no slip rings such as PMSMs. Similar to WRSMs, they have no magnets and can have a wide constant power speed range due to controlling the excitation current. An additional advantage of SHM over WRSMs is that the number of the excitation coils does not depend on the number of poles, while it grows with the number of poles in WRSMs, reducing excitation magnetomotive force (MMF).

In the SHM design considered in this paper, there is only one excitation coil. This results in a decrease in the mass and the loss in excitation winding in SHMs compared with those of WRSMs. Also, since there are no losses in the SHM rotor, no rotor cooling is required. There are many applications of SHMs as high-reliable generators: in passenger railway cars, in ships and aircraft [15,16], in welding units [17], and as automotive generators [18].

Applications of SHMs as a traction motor of a mining truck are discussed in [7,19–23]. In [19], the computation method of the traction SHM, based on the set of 2D magnetostatic was described and verified in the experiment in [19]. The control strategy for traction SHMs was described in [20]. The examples of the optimization of the SHM for traction applications based on Nelder–Mead algorithm and the model described in [19] are provided in [21]. Paper [7] provide a comparison of the SHM and PMSM characteristics in the mining truck application.

However, as shown in [23], traction SHMs have the following disadvantages compared to WRSMs: (1) The mass and dimensions of SHMs are greater than those of WRSMs, since each rotor tooth covers approximately one pole pitch, and about half of the pole pitches of the SHMs is not used; (2) SHMs require a higher inverter power rating than WRSMs.

There are multi-pole SHMs with an excitation winding on the stator and rare-earth magnets in the rotor slots [24–26]. Such SHMs with rare-earth magnets in the rotor slots are superior to SHMs without permanent magnets due to the better use of the rotor surface, and their weight and dimensions are close to conventional PMSM. The main advantage of SHMs with rare earth magnets, compared to conventional PMSM in traction drives with a wide constant power speed range (CPSR): the inverter utilization is higher, and the cost and rated power of the inverter are lower, since due to the excitation winding, it is possible to set the optimal excitation flux in a wide range of speeds. Thus, the excitation winding current is an additional control signal that expands the opportunities for optimizing the operation of the SHM with magnets. The main disadvantage of the SHM with rare earth magnets, which limits their use, is the high cost of rare earth magnets and raw material dependence on a limited number of suppliers of rare earth elements. In addition, the depth of the rotor slots of the SHM with rare earth magnets is much smaller than that of the SHM without magnets, which worsens the saliency of the rotor of the SHM with rare earth magnets. Therefore, although the use of rare earth magnets creates a significant additional torque, the main torque generated by the interaction of the field of the excitation winding modulated by rotor stacks and the field of the stator winding is reduced.

Further improvement of the characteristics of SHMs with magnets is possible by using inexpensive ferrite magnets in their design. Ferrite magnets are much cheaper than rare-earth ones and produced in many countries throughout the world [5].

An SHM with ferrite magnets is presented in [15,27] as an undercar generator for railway passenger cars. It is shown that the use of SHMs with ferrite magnets has the following main advantages compared to SHMs without magnets: (1) The reduction of the weight and dimensions of the machine; (2) Power loss reduction. However, the review of the literature shows that traction SHMs with ferrite magnets in the rotor slots have not been studied in detail. This article presents a novel design of the traction SHM with ferrite magnets. An example of the optimal design of the SHM with ferrite magnets for the subway train is provided. A comparison between the SHM with ferrite magnets and the SHM without magnets in the target application is also presented. The characteristics of the SHM without magnets for comparison are adopted from our previous study [28].

This study provides the optimization of the SHM with ferrite magnets for subway trains by means of the Nelder–Mead method and the mathematical model described in [19]. The Nelder–Mead method is a one-criterion, unconstrained, local optimum search method. The optimization is based on several criteria such as the minimization of losses, and the armature winding current. The merits of these criteria are chosen and taken into account at building up the cost function. Using the Nelder–Mead method significantly reduces computational efforts compared with multicriteria or global search methods [29–33].

2. Main Design Parameters of the SHM

Figure 1 illustrates the design of the traction motor in question. Its specifications are shown in Table 1. Two stator lamination stacks with 60 teeth and a nonmagnetic supporting core are installed in the housing made of ferromagnetic non-laminated structural steel. Common 8-pole 3-phase armature winding is installed between the teeth of the stator stacks. The rotor stacks are installed opposite the stator stacks, on the shaft using an intermediate sleeve made of ferromagnetic non-laminated structural steel.

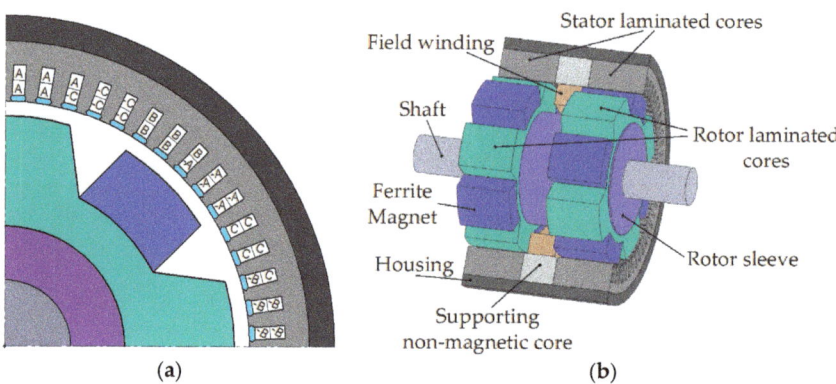

Figure 1. SHM feature representation: (**a**) 1/4 cross-section and stator armature winding layout; (**b**) 3D cutout view. A 1/2 stator cutout is shown. The rotor is shown without cutout. The stator winding is not shown to avoid cluttering up the figure.

Table 1. Technical specifications of the SHM.

Parameter	Value
Rated power, kW	370
Peak torque, N·m	1240
CPSR (motor mode), rpm	1427–4280
CPSR (braking mode), rpm	2854–4280
Phase number	3
Pole number	8
Number of pairs of stator and rotor stacks	2
Stator slot number	60
Rotor slot number	4

A supporting core fixes an excitation winding between the stator and rotor stack pairs. The rotor stacks are rotated with respect to each other by 180 electrical degrees (45 mechanical degrees).

As can be seen in Figure 2, the power supply circuit for the SHM consists of an ordinary three phase invertor to supply the multiphase armature winding and a chopper to supply the excitation winding.

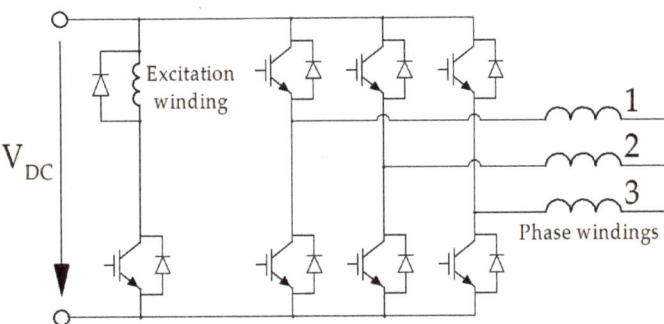

Figure 2. Diagram of a three-phase inverter with a DC breaker for the excitation winding, where '1,2,3' are numbers of the phases of the SHM armature winding.

3. Representation of the Train Flow Pattern in the Motor Optimization Routine

There are following stages in the motion of the subway train between stations from one station to another, as shown in Figure 3 [34]:

1. It accelerates with the constant torque T_0 = 1240 N·m, achieving speed n_m = 1427 rpm;
2. Continues acceleration with the constant mechanical power until the maximum speed n_{max} = 4280 rpm is achieved;
3. Then it sustains the constant speed or decelerates slowly;
4. Brakes with the constant power to the speed n_g = 2854 rpm;
5. Braking with the constant torque T_0 to the stop.

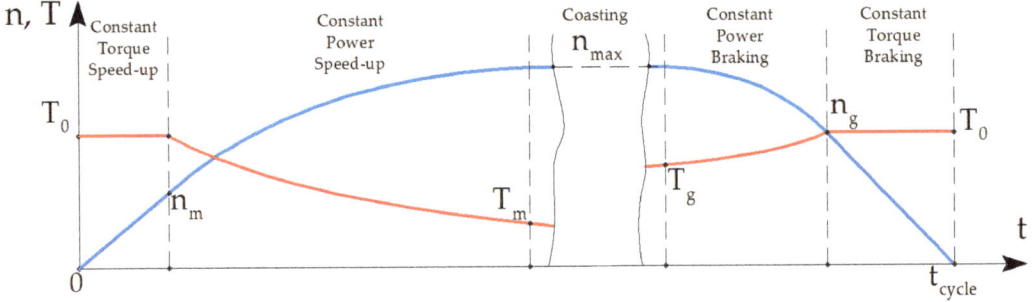

Figure 3. Sketch of the speed (blue) and torque (red) profiles in time of the traction motor of the subway train.

The required torque dependence on the rotational speed is shown in Figure 4. To depict this dependence in the motor and generator modes, there are two abscissa axes. The first abscissa axis directed to the right is for the motor mode, and the second one directed to the left is for the generator mode. Let's list the specific modes used in the optimization procedure, in order of increasing torque:

1. Driving mode at the maximum speed n_{max};
2. Braking mode at the maximum speed n_{max};
3. Braking mode joining the constant power and constant torque modes;
4. Zero speed mode with the maximum torque T_0;
5. Driving mode joining constant torque and constant power at the speed n_m = 1427 rpm.

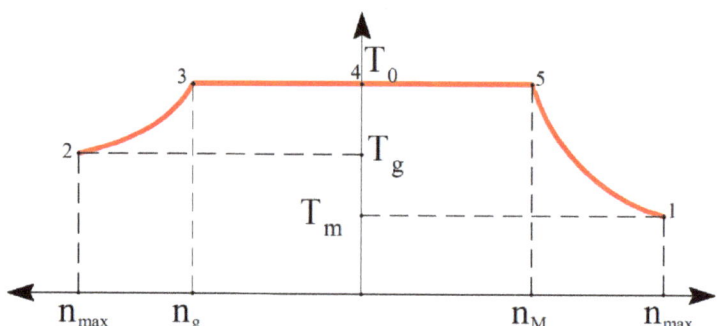

Figure 4. Demanded speed-torque curve of the subway traction drive.

According to the technical assignment for the motor, the braking mode torque is equal to the motor torque, and n_g is twice as large as n_m. So, maximum (constant) power in the generator mode is twice as large as in motor mode.

DC catenary voltage is $V_{DC\,rated}$ = 750 V. Supercapacitors are installed in the train invertor. They charge at braking up to the voltage $V_{DC\,max}$ = 1050 V. The energy used in supercapacitors is used for the acceleration in the next cycle. Also, during the braking, supercapacitors provide a DC voltage higher than $V_{DC\,rated}$, which facilitates implementing brake modes.

4. Average Electrical Loss Calculation

The average of electrical loss over the cycle, that is over the trip from station to station, is estimated with the following assumptions:

- The subway stations are close to each other. The cruising time is excluded from the average loss calculation routine;
- The train accelerates and decelerates only due to the torque produced by the motor. Slopes, windage friction, the friction in the gearbox, etc., are neglected;
- Linear dependence of the losses on speed is assumed at each stage of the cycle.

The average electrical losses <$P_{loss\,el}$> can be calculated as the weighted average of electrical losses $P_{loss\,el\_i}$ in the operating points of the cycle:

$$< P_{loss\_el} > \approx \sum_{i=1,2,3,4,5} w_i P_{loss\_el\_i} \qquad (1)$$

where w_i is a normalized weight coefficient defined by the cycle parameters as follows:

$$w_i = \frac{W_i}{\sum_{i=1}^{5} W_i}, \qquad (2)$$

$$W_1 = \frac{n_{max}(n_{max} - n_m)}{n_m};\ W_2 = \frac{n_{max}(n_{max} - n_g)}{n_g};\ W_3 = n_{max};\ W_4 = (n_g + n_m);\ W_5 = n_{max} \qquad (3)$$

The detailed derivation of (1) is given in [28]. The average of any other value can be found in this way if a linear approximation of its dependance on stages of the cycle is assumed.

5. Voltage Limit

The maximum allowed line-to-line voltage in i-th operation point V_i is approximately equal to the DC voltage but not exactly. Due to voltage drop in the switches, V_i is slightly higher in the generator modes and slightly lower in the other modes than DC voltage. Let's introduce the ratio of V_i and DC voltage in catenary $k_i = V_i/V_{DC\,rated}$. It is assumed that in modes 1,2,4,5, the DC voltage in modes is equal to that in a catenary, and k_i is close to 1. In generator mode 2, k_2 can be chosen greater than in motor modes 1,4,5. With some margin, the following values are chosen $k_2 = 0.99$; $k_1 = k_4 = k_5 = 0.97$.

On braking, supercapacitors charge, and DC voltage increases up to $V_{DC\ max}$. It is claimed in [28] that DC voltage $V_{DC\ 3}$ in the operation point 3 can be calculated knowing the ratio n_g/n_{max} and assuming that the charging goes with constant efficiency:

$$V_{DC\ 3} = \sqrt{V_{DC\ rated}^2 + (V_{DC\ max}^2 - V_{DC\ rated}^2)(1 - \frac{n_g^2}{n_{max}^2})} = 929\ V \quad (4)$$

Since $V_{DC\ 3}/V_{DC\ rated} = 1.24$, $k_3 = 1.1$ is chosen with a good margin.

The parameters of the operating points including w_i, calculated according to (2) and k_i are summarized in Table 2.

Table 2. Operating points of the traction motor considered during the optimization.

Operating Point, i	Operating Point Name	Speed, rpm	Torque, N·m	w_i	k_i
1	Driving mode; maximum speed	4380	413.4	0.363	0.97
2	Braking mode; maximum speed	4380	826.9	0.091	0.99
3	Braking mode; changing from constant power to constant torque operation	2854	1240	0.182	1.1
4	Zero speed	0	1240	0.182	0.97
5	Driving mode; changing from maximum torque to constant power operation modes	1427	1240	0.182	0.97

6. Objectives and Parameters of the Optimization

To use Nelder–Mead method, the cost function is constructed from four targets:

1. Minimization of the estimated average losses $< P_{loss\ el} >$ obtained as the weighted average (1);
2. Minimization of the maximum armature winding current max $(I_{arm\ i})$ among 5 considered operating points;
3. Minimization of the maximum symmetrized torque ripple max $(TRsym_i)$ among 5 considered operating points;
4. Minimization of the maximum nonsymmetrized torque ripple max (TR_i) among 5 considered operating points.

A nonsymmetrized torque ripple is produced by a single pair of the stator and rotor stacks. A symmetrized torque ripple is produced by all pairs of the stator and rotor stacks that is by the SHM as a whole. The paper [19] describes the terms TR and $TRsym$ in detail.

$$F = < P_{loss\_el\_i} > \max(I_{arm\ i})^{0.7} \max(TRsym_i)^{0.025} \max(TR_i)^{0.01}, \quad (5)$$

In developing an SHM the discrete nature of some parameters such as the number of turns N_{sec} in the section of armature winding and discrete values of the rectangular wire width and height standardized in [35] must be taken into account. In this study, the discreteness of these values are neglected, and they can have any positive real value. It provides a more objective picture by excluding random factors arising because of a variety of technical assignments for developing an SHM: in one design the optimal real values of these parameters can be closer or further than in another design. In particular, the number of turns in the armature winding N_{sec} is selected so as to maximum value V_i/k_i over 5 operation points to be equal $V_{DC\ rated} = \max(V_i/k_i)$ [23]. Number of parallel branches is assumed to be equal to 4.

Magnet Y30H-2 has a residual flux density of 3.95–4.15 kG and a coercive force of 3.9–4.2 kOe under rated conditions. As the temperature increases, the coercive force of ferrite magnets increases [36].

The Nelder–Mead method being unconstrained optimization method requirean s only initial design to be given. The parameters fixed during the optimization and varied ones are given in Tables 3 and 4. They uniquely specify the SHM design and electromagnetic processes in the considered operating points together with the following relations:

- The cross-sections of the stator housing and the rotor sleeve have equal areas, for the same excitation flux is conducted through them;
- The shaft is made of nonmagnetic material;
- The following relationship between stator slot depth h_p and width b on one hand and the height w_y and w_x width of the rectangular wire is assumed (see Figure 5b):

$$b_p = w_x + a_x;\ h_p = 2\cdot(w_y + \Delta w)\cdot N_{sec} + a_y, \qquad (6)$$

Table 3. Motor parameters unchanged during optimization.

Parameter	Value
Machine length excluding winding end parts L, mm	260
Stator housing radius, mm	267
Axial clearance between excitation winding and rotor, Δ_a, mm	29
Radial clearance between field winding and rotor Δ_r, mm	22
Shaft radius R_{shaft}, mm	40
Stator lamination yoke $h_{s\ yoke}$, mm	21
Rotor lamination yoke $h_{r\ yoke}$, mm	17
Stator wedge thickness, ε_2, mm	2
Stator unfilled area thickness, ε_1, mm	3

Table 4. Variable motor parameters.

Parameter	Initial Design	Optimized Design
Housing thickness h, mm	23.4	15.6
Total stator stacks length L_{stator}, mm	200	219.6
Stator slot depth, h_p, mm	34	35.9
Stator slot width, b_p, mm	7.7	9.0
Air gap width δ, mm	1	4.4
Rotor slot thickness, α_1	$0.4\cdot t_z$ *	$0.423\cdot t_z$ *
Rotor slot thickness, α_2	$0.7\cdot t_z$ *	$0.664\cdot t_z$ *
Current angles at operating points 1,2,3,4 electrical radians	0.638; 0.941; 0.409; 0.311	0.943; 0.921; 0.404; 0.118
Current ratio **	10.96	10.55

Notes: * the rotor tooth pitch $t_z = 360°/4 = 90$ mechanical degrees; ** the current ratio is the ratio of the current in the armature winding layer to the current in the excitation winding.

The space taken by the slot and layer insulations is taken into account through the constant $a_x = 1.51$ mm, $a_y = 1.8$ mm, $\Delta w = 0.31$ mm.

- DC losses in the excitation colis are calculated assuming the net copper fill factor of 0.8. Eddy current losses in the excitation winding are neglected.
- Only the current angles in modes 1,2,3,4 varied during the optimization.
- The following assumption is made to reduce the number of variable parameters during optimization. The current angle in operating point 5 is equal to that in operating point 4. The ratio of currents in the excitation winding section and the armature winding layer is constant among all modes. Additionally, the ratio of the rotor slot widths α_2/α_1 (see Figure 5c) is constant during the optimization.

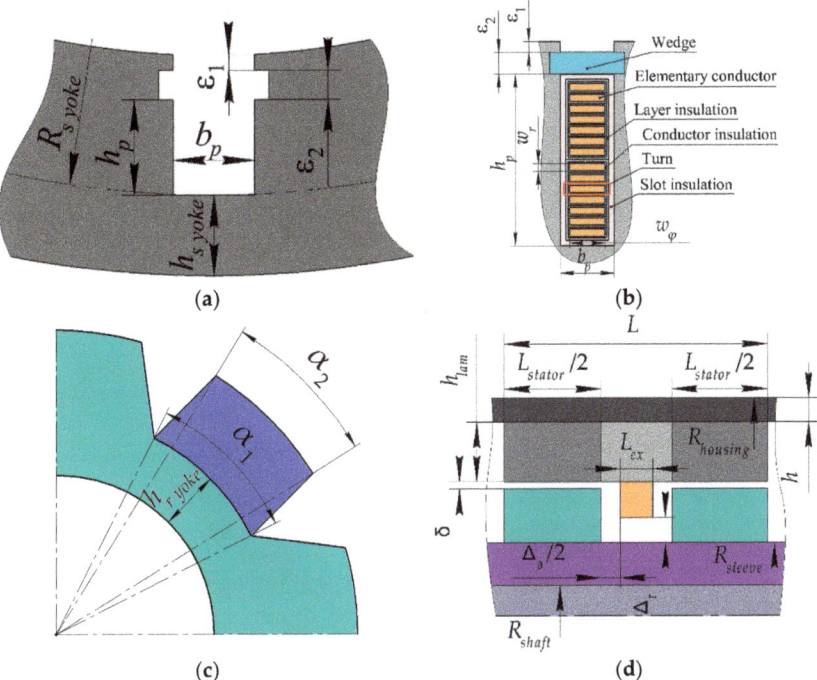

Figure 5. SHM parameters. (**a**) Stator; (**b**) Armature winding; (**c**) Rotor; (**d**) Axial plane.

Figure 5 shows sketches explaining the parameters listed in Tables 3 and 4.

Since the principle of operation of SHM is the interaction of the excitation flux and the current of the armature winding, the width of the air gap does not directly affect the torque as long as both the excitation flux and the armature current vector are fixed. However, too small an air gap increases the leakage flux induced by armature winding and flowing through the rotor teeth, which contributes to the reactive power and saturation of the machine. An increase in the air gap results in an increase in the excitation current required to generate the same excitation flux. Therefore, a too large or small air gap is not optimal. During optimization, the optimal gap width was found.

7. Optimization Results

Figure 6 shows that during optimization, the cost function decreased and became almost constant in the end. Figures 7 and 8 show some optimization targets can increase, and less valuable ones can increase as a compromise to allow the cost function to decrease. Figure 7 shows the change in the electrical losses and current magnitude during the optimization stage. Figure 8 depicts the simultaneous change of the symmetrized and nonsymmetrized torque ripple.

A comparison of Figures 9 and 10. shows that as a result of optimization, the permanent magnet area between the teeth has increased significantly, and the rotor teeth have changed shape and become thinner. As the comparison of the flux density plots for points with the maximum torque (Figures 9c–e and 10c–e) shows, after optimization, the area of regions with a flux density value of more than 2 T is significantly reduced. A comparison of Figure 11a,b shows that as a result of the optimization, the maximum demagnetizing force has been reduced from 3 kOe to about 2.5 kOe. So, no check of the demagnetization is needed in this optimization. Table 5 shows the characteristics of the SHM with ferrite magnets before and after optimization.

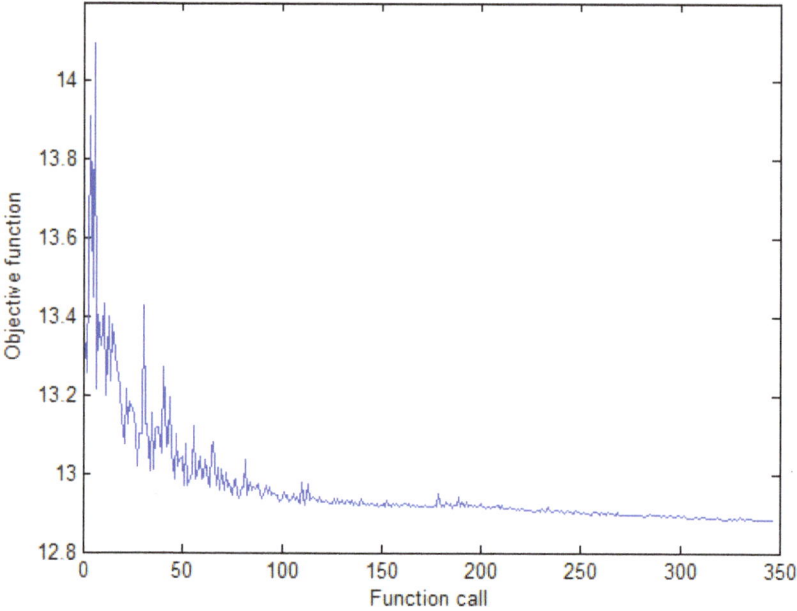

Figure 6. Variation of the cost function value during optimization.

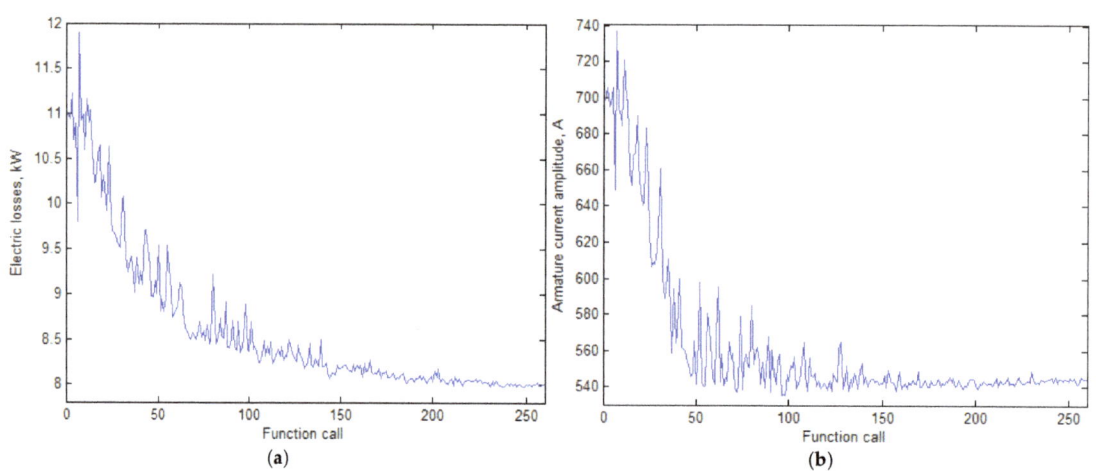

Figure 7. History of change of the target performances. (**a**) Average electrical loss; (**b**) Upper limit of the motor current.

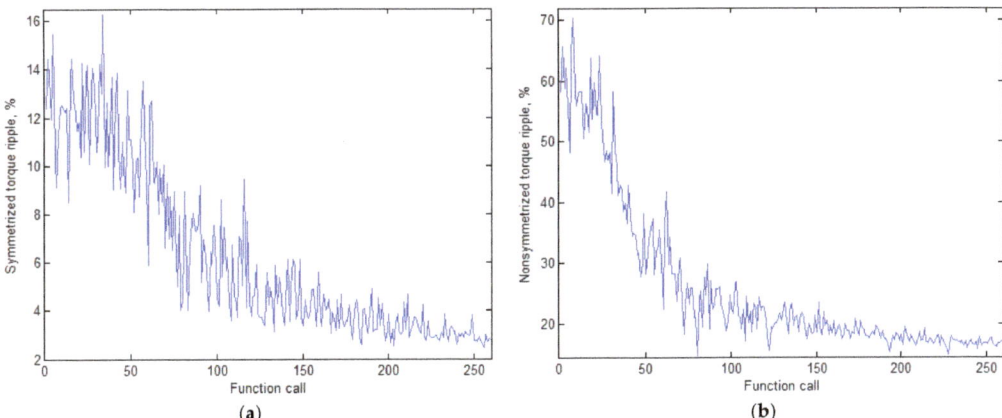

Figure 8. History of change of the target performances. (**a**) Symmetrized torque ripple; (**b**) Nonsymmetrized torque ripple.

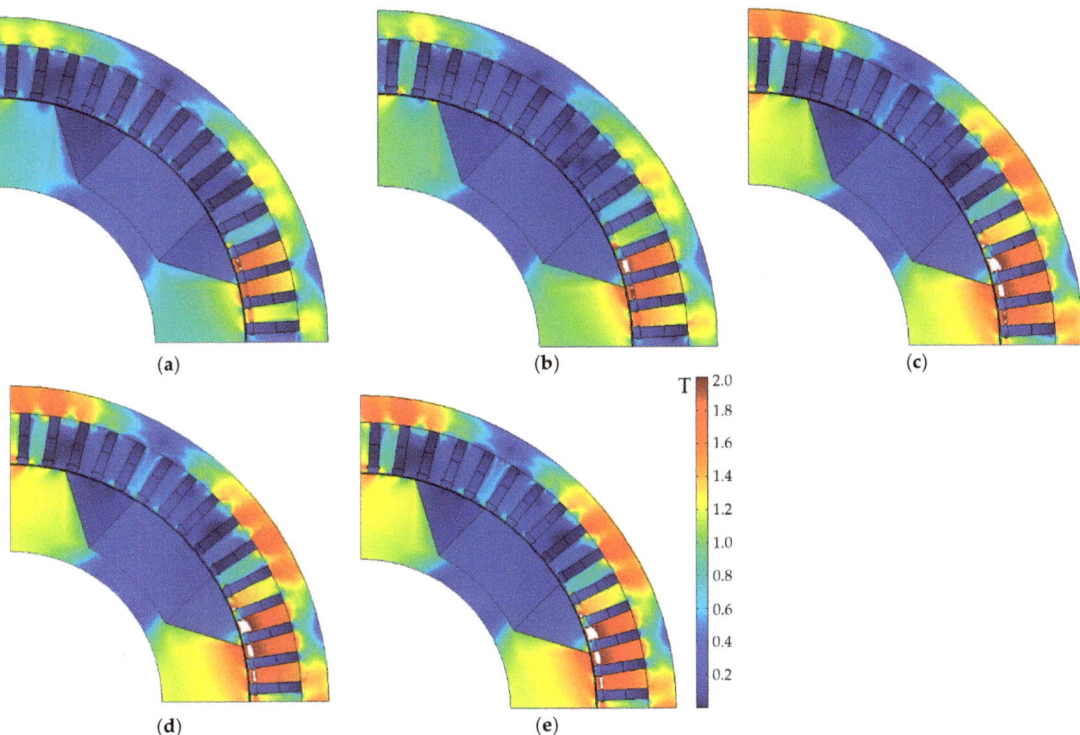

Figure 9. The cross-section of the initial design of the motor and the plot of flux density magnitude; white areas are the extreme saturation areas (>2 T). (**a**) Operating point 1; (**b**) Operating point 2; (**c**) Operating point 3; (**d**) Operating point 4; (**e**) Operating point 5.

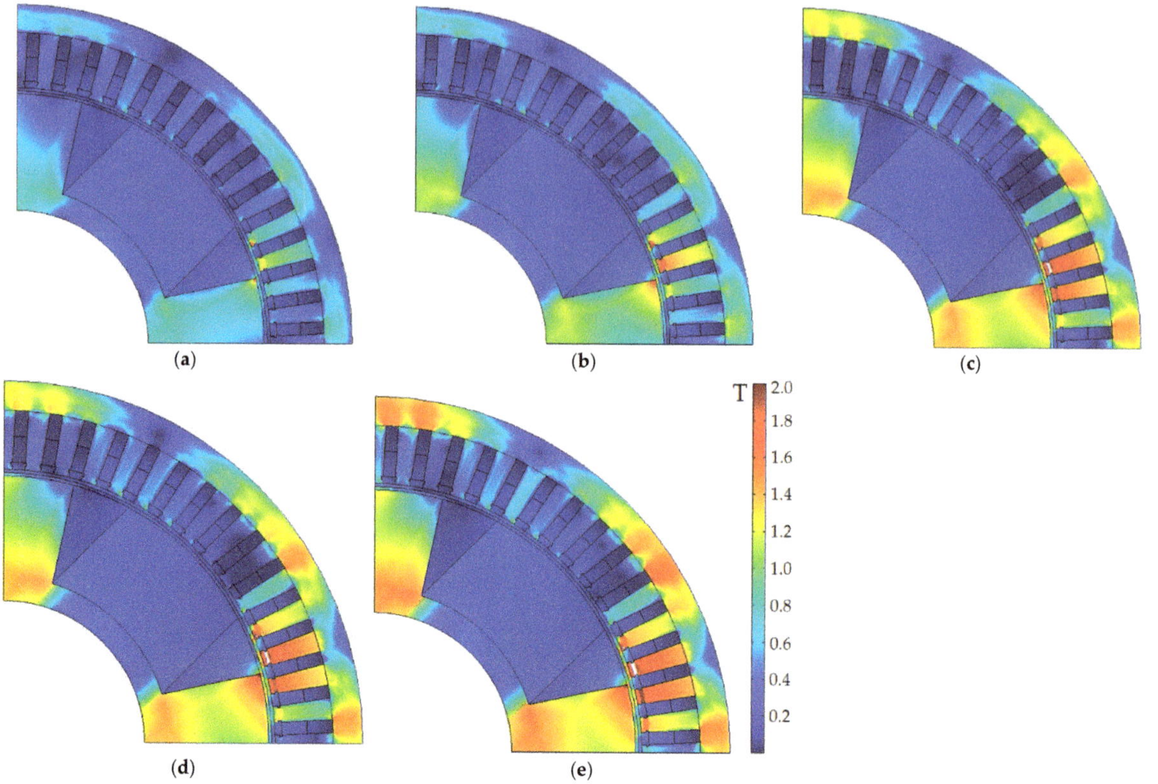

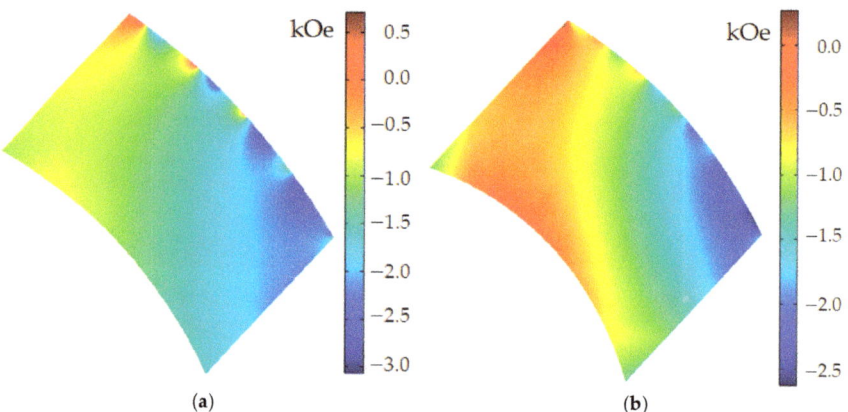

Figure 10. The cross-section of the optimized design of the motor and the plot of flux density magnitude; white areas are the extreme saturation areas (>2 T). (**a**) Operating point 1; (**b**) Operating point 2; (**c**) Operating point 3; (**d**) Operating point 4; (**e**) Operating point 5.

Figure 11. Demagnetizing force (kOe) in the area of the permanent magnet at operating point 4. (**a**) Before optimization; (**b**) After optimization.

Table 5. Optimization results.

Parameter	Initial Design					Optimized Design				
Operating Point, i	1	2	3	4	5	1	2	3	4	5
Rotational speed n, rpm	4280	4280	2854	0	1427	4280	4280	2854	0	1427
Amplitude of the armature phase current I_{arm}, A	330	651	676	686	694	370	541	541	529	531
Efficiency, %	93.2	94.0	95.4	0	95.2	95.0	95.7	96.5	0	95.7
Output mechanical power P_{mech}, kW	185.3	−370.6	−370.6	0	185.3	185.3	−370.6	−370.6	0	185.3
Torque, N·m	413.4	−826.9	−1240	1240	1240	413.4	−826.9	−1240	1240	1240
Input electrical power, kW	198.8	−348.5	−353.6	5.4	194.6	195.0	−354.8	−357.7	5.6	193.6
Mechanical loss, kW *	3.55	3.55	1.06	0	0.14	3.55	3.55	1.06	0	0.14
Armature DC copper loss, kW	0.97	3.76	4.05	4.18	4.27	2.20	4.71	4.71	4.49	4.53
Armature eddy-current copper loss, kW	1.57	5.99	4.91	0	1.30	1.11	2.70	1.99	0	0.55
Stator lamination loss, kW	5.86	5.90	4.72	0	2.01	2.26	3.56	3.81	0	1.90
Rotor lamination loss, kW	1.31	1.70	1.00	0	0.26	0.04	0.09	0.07	0	0.02
Excitation copper loss, kW	0.28	1.19	1.29	1.27	1.28	0.55	1.21	1.21	1.13	1.14
Total loss, kW	13.54	22.09	17.03	5.45	9.26	9.72	15.81	12.85	5.62	8.27
Average losses according to formula (1)			12.70					9.84		
Number of turns in armature winding			4.52					6.58		
Power factor	0.969	−0.878	−0.999	1.0	0.994	0.949	−0.989	−0.963	1.0	0.906
Line-to-line voltage amplitude V_{arm}, V	728	716	626	7	334	640	758	797	10	467
Nonsymmetrized torque ripple, %	61.2	46.9	33.0	32.9	32.8	16.3	15.4	13.5	13.6	13.6
Symmetrized torque ripple, %	10.73	11.10	7.68	7.55	7.54	2.50	2.50	2.42	2.61	2.61
Magnetic flux density in the housing and the sleeve, T	0.62	0.99	1.17	1.20	1.20	0.30	0.71	1.04	1.15	1.15

Note: * The mechanical losses are assumed to be proportional to the speed with a maximum value of 3.55 kW at n_{max}.

The following conclusions can be made from Table 5 on the optimization results:

1. The optimization significantly reduced losses at all operating points, except for operating point 4 at zero speed with maximum torque; it also significantly reduced the average losses and maximum current of the armature winding;
2. Average losses according to (1) were reduced by 100% (12.7 − 9.84)/12.7 = 22.5%;
3. The maximum current magnitude before optimization was 694 A at operation point 5 at the speed n_m, joining the motor modes with the constant power and the constant torque. After optimization, the maximum current magnitude becomes at the operation point 3 at the speed n_g, joining the generator modes with the constant power and the constant torque. Thus, the maximum inverter current magnitude was reduced by (694 − 541)/694 = 22%;
4. In the initial design, the line-to-line voltage limit is reached at operating point 1 in motor mode. In the optimized design, it is reached at operating point 2 in generator mode. Both operating points are at the maximum rotational speed;
5. Before optimization, the maximum torque ripple occurs at operating point 2 at maximum speed in generator mode. After optimization, it occurs at operating point 4 at zero speed at maximum torque. Thus, after optimization, the maximum torque ripple was reduced by (11.1 − 2.61)/11.1 = 76.5%, which is because of a significant increase in the airgap;
6. Since the exact magnetic properties of the structural non-laminated steel are unknown, the drop in MMF in the non-laminated parts of the magnetic circuit can only be estimated approximately. For the error of such an estimate not to strongly affect the results of the evaluation of the characteristics of the machine, the flux density in these parts made of structural steel must be small. It can be seen that for the initial design, the maximum flux density is 1.2 T, which is less significant than the acceptable values of the magnetic flux density in laminated cores of 1.8 T and more, and for the optimized design, this value is further reduced by 100% (1.2 − 1.15)/1.2 = 4.2%. So,

no restriction on the magnetic flux density in non-laminated parts was needed in the optimization.

7. As shown in Figure 11, in the initial design, during optimization, and in the optimized design, the demagnetizing field does not exceed 3.2 kOe, while the coercive force of the Y30H-2 magnet is about 4 kOe [36]. As a result of optimization, the demagnetizing field even weakened. Thus, there is no risk of demagnetization of ferrite magnets in all operating points, due to the well-chosen parameters of the initial design.

8. Comparison of SHM with Ferrite Magnets and SHM without Magnets

This section discusses the comparison between the SHM with ferrite magnets, which characteristics are presented in this article, and the SHM without magnets, which characteristics were calculated in our previous study [28]. To avoid repetition in this article, in Table 6 we present only the final optimized characteristics of the SHM without magnets. Details of the SHM design without magnets and its optimization can be found in [28]. Table 6 shows a comparison of the performances of the optimized SHM designs with ferrite magnets and without magnets. The dimensions, masses, and costs of active materials for the SHMs with and without ferrite magnets are compared in Table 7.

Table 6. Comparison of SHM with ferrite magnets and SHM without magnets.

Parameter	SHM without Magnets					SHM with Ferrite Magnets				
Operating Point, i	1	2	3	4	5	1	2	3	4	5
Rotational speed n, rpm	4280	4280	2854	0	1427	4280	4280	2854	0	1427
Amplitude of the armature phase current I_{arm}, A	311	541	547	542	545	370	541	541	529	531
Efficiency, %	94.8	95.0	95.6	0	94.2	95.0	95.7	96.5	0	95.7
Output mechanical power P_{mech}, kW	185.3	−370.6	−370.6	0	185.3	185.3	−370.6	−370.6	0	185.3
Torque, N·m	413.4	−826.9	−1240	1240	1240	413.4	−826.9	−1240	1240	1240
Input electrical power, kW	195.4	−352.2	−354.4	8.9	196.8	195.0	−354.8	−357.7	5.6	193.6
Mechanical loss, kW	3.55	3.55	1.06	0	0.14	3.55	3.55	1.06	0	0.14
Armature DC copper loss, kW	2.48	7.53	7.67	7.55	7.62	2.20	4.71	4.71	4.49	4.53
Armature eddy-current copper loss, kW	0.59	2.25	2.16	0	0.60	1.11	2.70	1.99	0	0.55
Stator lamination loss, kW	2.87	3.39	3.64	0	1.65	2.26	3.56	3.81	0	1.90
Rotor lamination loss, kW	0.19	0.30	0.22	0	0.06	0.04	0.09	0.07	0	0.02
Excitation copper loss, kW	0.46	1.41	1.44	1.40	1.40	0.55	1.21	1.21	1.13	1.14
Total loss, kW	10.14	18.43	16.19	8.95	11.47	9.72	15.81	12.85	5.62	8.27
Average losses according to formula (1)			12.02					9.84		
Number of turns in armature winding			6.68					6.58		
Power factor	0.994	−1.0	−0.927	1.0	0.905	0.949	−0.989	−0.963	1.0	0.906
Line-to-line voltage amplitude V_{arm}, V	718	755	825	16	468	640	758	797	10	467
Nonsymmetrized torque ripple, %	33.9	25.9	18.8	19.0	19.0	16.3	15.4	13.5	13.6	13.6
Symmetrized torque ripple, %	7.21	5.47	3.77	3.91	3.91	2.50	2.50	2.42	2.61	2.61

Table 7. Comparison of masses, costs, and dimensions of parts of the SHM without magnets and with ferrite magnets.

Parameter	SHM without Magnets	SHM with Ferrite Magnets
Stator lamination mass, kg	103.9	106.6
Rotor lamination mass, kg	53.5	54.5
Armature copper mass, kg	36.0	53.1
Excitation copper mass, kg	17.4	13.9
Magnets mass, kg	-	31.2
Weight of the rotor sleeve and motor housing without bearing shields, kg	195.1	103.8
The total mass of the active materials, rotor sleeve, and motor housing, kg	405.9	363.1
Stator lamination cost, USD	103.9	106.6
Rotor lamination cost, USD	53.5	54.5
Armature copper cost, USD	252	371.7
Excitation copper cost, USD	121.8	97.3
Magnets cost, USD	-	576.0
Rotor sleeve and motor housing cost, USD	195.1	103.8
The total cost of the active materials (electrical steel, copper, permanent magnets) and structural steel of the rotor sleeve and motor housing, USD *	762.3	1309.9
Total length of the stator lamination, mm	228	219.6
Total length of the machine excluding the winding end parts (including spaces for the excitation coils), mm	302.5	260
Stator lamination outer diameter, mm	534	534
Air gap, mm	3.0	4.4

* Note: the following material costs are assumed: copper is 7 USD/kg; laminated electrical steel is 1 USD/kg; non-laminated structural steel for the housing and rotor sleeve is 1 USD/kg; Y30H-2 grade ferrite magnet is 18.46 USD/kg [28,37].

Comparing the characteristics of the SHMs without permanent magnets and with ferrite magnets, shown in Tables 6 and 7, the following findings can be reported:

1. Average losses according to (1) for the SHM with ferrite magnets are less than for the SHM without ferrite magnets by 100% (12.02 − 9.84)/12.02 = 18.1%;
2. The maximum current for the SHM with ferrite magnets is slightly less than for the SHM without ferrite magnets, by 100% (457 − 451)/457 = 1.3%;
3. The maximum output torque ripple of the SHM with ferrite magnets is less than that of the SHM without ferrite magnets by 100% (3.91 − 2.61)/3.91 = 33.2% due to the increased air gap;
4. The total length excluding the armature winding end parts, including the width of the field winding, for the SHM with ferrite magnets is less than for the SHM without ferrite magnets, by 100% (302.5 − 260)/302.5 = 14%;
5. The weight of the rotor sleeve and housing of the SHM with ferrite magnets is two times less than that of the SHM without magnets. This is so, firstly, because the ferrite magnets contribute to the excitation field, and therefore the field of the excitation winding can be reduced. Secondly, the magnetic fluxes created by the ferrite magnets and the excitation winding have opposite directions in the non-laminated parts (the rotor sleeve and housing). Therefore, the saturation of the non-laminated parts is reduced, and the housing of the SHM with ferrites becomes thinner than the SHM case without magnets.
6. The total mass of active materials including the rotor sleeve and the motor housing for the SHM with ferrite magnets is 100% (405.9 − 363.1)/363.1 = 10.5% less than the mass of the SHM without magnets;
7. The total cost of the active materials (electrical steel, copper, permanent magnets) and structural steel of the rotor sleeve and motor housing for the SHM with ferrite

magnets is 1309.9/762.3 = 1.7 times more than that of the SHM without magnets, primarily due to the addition of the cost of ferrite magnets.

9. Conclusions

The optimization procedure based on the single-objective Nelder–Mead algorithm and its results for a synchronous homopolar motor (SHM) with ferrite magnets with a power of 370 kW for driving subway train, taking into account the subway train moving trajectory, namely acceleration and braking stages is described in the article. For one function call, only 5 operating points are to be computed, which makes computational efforts tolerable for computer-aided optimization.

The cost function is constituted from the following optimization objectives: decreasing the average operational cycle, decreasing the upper limit of armature current, and the reduction of the torque ripple. As a result of optimization, the following characteristics of the traction SHM have been improved. Power loss is reduced by 22.5%. The upper limit current of the solid-state inverter is reduced by 22%. The motor torque ripple is reduced by 76.5%.

Based on the optimization results, the obtained characteristics of the SHM with ferrite magnets are compared with the characteristics of the SHM without permanent magnets, optimized by the same method. The comparison of the characteristics of the SHMs with ferrite magnets and without magnets shows that the SHM with ferrite magnets has significant advantages: power loss is reduced by 18.1%, inverter current is reduced by 1.3%, torque ripple is reduced by 33.2%, the total mass of the active materials, rotor sleeve and motor housing is reduced by 10.5%, the overall machine length is reduced by 14%. The advantage of the SHM without magnets is that the cost is 1.7 times less, since it does not use ferrite magnets.

In future work, the comparison between the SHM with ferrite magnets and other types of electrical machines for subway drives and other applications will be carried out.

Author Contributions: Conceptual approach, V.D. and V.P.; data curation, V.D. and V.K.; software, V.D. and V.P.; calculations and modeling, V.D., V.K. and V.P.; writing—original draft, V.D., V.K. and V.P.; visualization, V.D. and V.K.; review and editing, V.D., V.K. and V.P. All authors have read and agreed to the published version of the manuscript.

Funding: The research funding from the Ministry of Science and Higher Education of the Russian Federation (Ural Federal University Program of Development within the Priority-2030 Program) is gratefully acknowledged.

Institutional Review Board Statement: Not applicable.

Informed Consent Statement: Not applicable.

Data Availability Statement: Data are contained within the article.

Acknowledgments: The authors thank the editors and reviewers for careful reading and constructive comments.

Conflicts of Interest: The authors declare no conflict of interest.

References

1. Gundogdu, T.; Zhu, Z.-Q.; Chan, C.C. Comparative Study of Permanent Magnet, Conventional, and Advanced Induction Machines for Traction Applications. *World Electr. Veh. J.* **2022**, *13*, 137. [CrossRef]
2. Dianov, A. Highly Efficient Sensorless Multicontrol Mode Compressor IPMSM Drive with Seamless Transitions. *IEEE Trans. Power Electron.* **2022**, *37*, 9129–9137. [CrossRef]
3. Schulze, R. Reducing Environmental Impacts of the Global Rare Earth Production for Use in Nd-Fe-B Magnets, How Magnetic Technologies Much Can Recycling Contribute? Ph.D. Thesis, Technical University of Darmstadt, Darmstadt, Germany, December 2018. Available online: https://tuprints.ulb.tu-darmstadt.de/8301/7/Diss%20Rita%20Schulze%2015-12-2018.pdf (accessed on 30 November 2022).
4. Dong, S.; Li, W.; Chen, H.; Han, R. The status of Chinese permanent magnet industry and R&D activities. *AIP Adv.* **2017**, *7*, 1–17. [CrossRef]

5. Carraro, E.; Bianchi, N.; Zhang, S.; Koch, M. Design and Performance Comparison of Fractional Slot Concentrated Winding Spoke Type Synchronous Motors with Different Slot-Pole Combinations. *IEEE Trans. Ind. Appl.* **2018**, *54*, 2276–2284. [CrossRef]
6. De Lima, W.I.B. *Rare Earth Industry*; Elsevier: Amsterdam, The Netherlands, 2015.
7. Dmitrievskii, V.; Prakht, V.; Kazakbaev, V.; Anuchin, A. Comparison of Interior Permanent Magnet and Synchronous Homopolar Motors for a Mining Dump Truck Traction Drive Operated in Wide Constant Power Speed Range. *Mathematics* **2022**, *10*, 1581. [CrossRef]
8. Papini, F.; Osama, M. Electromagnetic Design of an Interior Permanent Magnet Motor for Vehicle Traction. In Proceedings of the 2018 XIII International Conference on Electrical Machines (ICEM), Alexandroupoli, Greece, 3–6 September 2018; pp. 205–211. [CrossRef]
9. Dianov, A. Optimized Field-Weakening Strategy for Control of PM Synchronous Motors. In Proceedings of the 2022 29th International Workshop on Electric Drives: Advances in Power Electronics for Electric Drives (IWED), Moscow, Russia, 26–29 January 2022; pp. 1–6. [CrossRef]
10. The First-Ever BMW iX3, PressClub Global, Article. Available online: https://www.press.bmwgroup.com/global/article/detail/T0310696EN/the-first-ever-bmw-ix3?language=enhttps://www.netcarshow.com/bmw/2021-ix3 (accessed on 30 November 2022).
11. Feustel, S.; Huebner, B.; Loos, D.; Merwerth, J.; Tremaudant, Y.; Vollmer, K. Rotor for Separately Excited Inner Rotor Synchronous Machine, Inner Rotor Synchronous Machine, Motor Vehicle and Method. U.S. Patent US20210006105A1, 7 January 2021. Available online: https://patents.google.com/patent/US20210006105A1 (accessed on 30 November 2022).
12. Raia, M.R.; Ruba, M.; Martis, C.; Husar, C.; Sirbu, G.M. Battery electric vehicle (BEV) powertrain modelling and testing for real-time control prototyping platform integration. In Proceedings of the 2021 23rd European Conference on Power Electronics and Applications (EPE'21 ECCE Europe), Ghent, Belgium, 6–10 September 2021; pp. 1–10. Available online: https://ieeexplore.ieee.org/document/9570616 (accessed on 30 November 2022).
13. Chu, W.Q.; Zhu, Z.Q.; Zhang, J.; Ge, X.; Liu, X.; Stone, D.; Foster, M. Comparison of electrically excited and interior permanent magnet machines for hybrid electric vehicle application. In Proceedings of the 2014 17th International Conference on Electrical Machines and Systems (ICEMS), Hangzhou, China, 22–25 October 2014; pp. 401–407. [CrossRef]
14. Illiano, E. Design of a Highly Efficient Brushless Current Excited Synchronous Motor for Automotive Purposes. Ph.D. Thesis, ETH-Zürich, Zürich, Switzerland, 2014. [CrossRef]
15. Orlova, S.; Pugachov, V.; Levin, N. Hybrid Excitation of the Axial Inductor Machine. *Latv. J. Phys. Tech. Sci.* **2012**, *49*, 35–41. [CrossRef]
16. Bindu, G.; Basheer, J.; Venugopal, A. Analysis and control of rotor eccentricity in a train-lighting alternator. In Proceedings of the 2017 IEEE International Conference on Power, Control, Signals and Instrumentation Engineering (ICPCSI), Chennai, India, 21–22 September 2017; pp. 2021–2025. [CrossRef]
17. Bianchini, C.; Immovilli, F.; Bellini, A.; Lorenzani, E.; Concari, C.; Scolari, M. Homopolar generators: An overview. In Proceedings of the 2011 IEEE Energy Conversion Congress and Exposition, Phoenix, AZ, USA, 17–22 September 2011; pp. 1523–1527. [CrossRef]
18. Lorilla, L.; Keim, T.; Lang, J.; Perreault, D. Topologies for future automotive generators. Part I. Modeling and analytics. In Proceedings of the 2005 IEEE Vehicle Power and Propulsion Conference, Chicago, IL, USA, 7 September 2005; pp. 74–85. [CrossRef]
19. Dmitrievskii, V.; Prakht, V.; Anuchin, A.; Kazakbaev, V. Traction Synchronous Homopolar Motor: Simplified Computation Technique and Experimental Validation. *IEEE Access* **2020**, *8*, 185112–185120. [CrossRef]
20. Lashkevich, M.; Anuchin, A.; Aliamkin, D.; Briz, F. Control strategy for synchronous homopolar motor in traction applications. In Proceedings of the 43rd Annual Conference of the IEEE Industrial Electronics Society (IECON), Beijing, China, 29 October–1 November 2017; pp. 6607–6611. [CrossRef]
21. Dmitrievskii, V.; Prakht, V.; Anuchin, A.; Kazakbaev, V. Design Optimization of a Traction Synchronous Homopolar Motor. *Mathematics* **2021**, *9*, 1352. [CrossRef]
22. Prakht, V.; Dmitrievskii, V.; Anuchin, A.; Kazakbaev, V. Inverter Volt-Ampere Capacity Reduction by Optimization of the Traction Synchronous Homopolar Motor. *Mathematics* **2021**, *9*, 2859. [CrossRef]
23. Prakht, V.; Dmitrievskii, V.; Kazakbaev, V.; Anuchin, A. Comparative Study of Electrically Excited Conventional and Homopolar Synchronous Motors for the Traction Drive of a Mining Dump Truck Operating in a Wide Speed Range in Field-Weakening Region. *Mathematics* **2022**, *10*, 3364. [CrossRef]
24. Bekhaled, C.; Hlioui, S.; Vido, L.; Gabsi, M.; Lecrivain, M.; Amara, Y. 3D magnetic equivalent circuit model for homopolar hybrid excitation synchronous machines. In Proceedings of the 2007 International Aegean Conference on Electrical Machines and Power Electronics, Bodrum, Turkey, 10–12 September 2007; pp. 575–580. [CrossRef]
25. Asfirane, S.; Hlioui, S.; Amara, Y.; Gabsi, M. Study of a Hybrid Excitation Synchronous Machine: Modeling and Experimental Validation. *Math. Comput. Appl.* **2019**, *24*, 34. [CrossRef]
26. Hoang, E.; Lécrivain, M.; Hlioui, S.; Gabsi, M. Hybrid excitation synchronous permanent magnets synchronous machines optimally designed for hybrid and full electrical vehicle. In Proceedings of the 8th International Conference on Power Electronics—ECCE Asia, Seogwipo, Republic of Korea, 30 May–3 June 2011.

27. Ketner, K.; Dirba, J.; Levins, N.; Orlova, S.; Pugachev, V. Inductor Machine with Axial Excitation. L.V. Patent LV13971B, 20 November 2009. (In Latvian).
28. Dmitrievskii, V.; Prakht, V.; Kazakbaev, V.; Anuchin, A. Design Optimization of the Magnet-Free Synchronous Homopolar Motor of a Subway Train. *Appl. Sci.* **2022**, *12*, 12647. [CrossRef]
29. Cupertino, F.; Pellegrino, G.; Gerada, C. Design of synchronous reluctance machines with multiobjective optimization algorithms. *IEEE Trans. Ind. Appl.* **2014**, *50*, 3617–3627. [CrossRef]
30. Krasopoulos, C.T.; Beniakar, M.E.; Kladas, A.G. Robust Optimization of High-Speed PM Motor Design. *IEEE Trans. Magn.* **2017**, *53*, 1–4. [CrossRef]
31. Fatemi, A.; Ionel, D.; Popescu, M.; Chong, Y.; Demerdash, N. Design Optimization of a High Torque Density Spoke-Type PM Motor for a Formula E Race Drive Cycle. *IEEE Trans. Ind. Appl.* **2018**, *54*, 4343–4354. [CrossRef]
32. Montonen, J.; Nerg, J.; Gulec, M.; Pyrhönen, J. A New Traction Motor System with Integrated-Gear: A Solution for Off-Road Machinery. *IEEE Access* **2019**, *7*, 113740–113750. [CrossRef]
33. Guo, S.; Zhao, H.; Wang, Y.; Yin, X.; Qi, H.; Li, P.; Lin, Z. A Design Technique of Traction Motor for Efficiency Improvement Based on Multiobjective Optimization. *World Electr. Veh. J.* **2021**, *12*, 260. [CrossRef]
34. Lin, F.; Li, X.; Zhao, Y.; Yang, Z. Control Strategies with Dynamic Threshold Adjustment for Supercapacitor Energy Storage System Considering the Train and Substation Characteristics in Urban Rail Transit. *Energies* **2016**, *9*, 257. [CrossRef]
35. IEC. *Specifications for Particular Types of Winding Wires—Part 0-2: General Requirements—Enamelled Rectangular Copper Wire*; IEC 60317-0-2:2020; IEC: Geneva, Switzerland, 2020; Available online: https://webstore.iec.ch/publication/63495 (accessed on 30 November 2022).
36. Grades of Ferrite. The Original Online Company. Bunting. E-Magnets. Available online: https://e-magnetsuk.com/ferrite-magnets/grades-of-ferrite/ (accessed on 8 January 2022).
37. Hard Ferrite Magnets, Product Information, IBSMagnet. 2020. Available online: https://ibsmagnet.com/products/dauermagnete/hartferrit.php (accessed on 8 January 2022).

Disclaimer/Publisher's Note: The statements, opinions and data contained in all publications are solely those of the individual author(s) and contributor(s) and not of MDPI and/or the editor(s). MDPI and/or the editor(s) disclaim responsibility for any injury to people or property resulting from any ideas, methods, instructions or products referred to in the content.

Article

Mathematical Modeling of the State of the Battery of Cargo Electric Vehicles

Nikita V. Martyushev [1,*], Boris V. Malozyomov [2], Svetlana N. Sorokova [3], Egor A. Efremenkov [3] and Mengxu Qi [3]

1. Department of Advanced Technologies, Tomsk Polytechnic University, 634050 Tomsk, Russia
2. Department of Electrotechnical Complexes, Novosibirsk State Technical University, 20, Karla Marksa Ave., 630073 Novosibirsk, Russia
3. Department of Mechanical Engineering, Tomsk Polytechnic University, 30, Lenin Ave., 634050 Tomsk, Russia
* Correspondence: martjushev@tpu.ru

Abstract: In this paper, a mathematical simulation model of an electric vehicle traction battery has been developed, in which the battery was studied during the dynamic modes of its charge and discharge for heavy electric vehicles in various driving conditions—the conditions of the urban cycle and movement outside the city. The state of a lithium-ion battery is modeled based on operational factors, including changes in battery temperature. The simulation results will be useful for the implementation of real-time systems that take into account the processes of changing the characteristics of traction batteries. The developed mathematical model can be used in battery management systems to monitor the state of charge and battery degradation using the assessment of the state of charge (SOC) and the state of health (SOH). This is especially important when designing and operating a smart battery management system (BMS) in virtually any application of lithium-ion batteries, providing information on how long the device will run before it needs to be charged (SOC value) and when the battery should be replaced due to loss of battery capacity (SOH value). Based on the battery equivalent circuit and the system of equations, a simulation model was created to calculate the electrical and thermal characteristics. The equivalent circuit includes active and reactive elements, each of which imitates the physicochemical parameter of the battery under study or the structural element of the electrochemical battery. The input signals of the mathematical model are the current and ambient temperatures obtained during the tests of the electric vehicle, and the output signals are voltage, electrolyte temperature and degree of charge. The resulting equations make it possible to assign values of internal resistance to a certain temperature value and a certain value of the degree of charge. As a result of simulation modeling, the dependence of battery heating at various ambient temperatures was determined.

Keywords: mathematical model; simulation model; lithium-ion battery; electric car

MSC: 65C20

1. Introduction

In recent years, electric energy storage systems have been considered a key element in the technological development of vehicles and renewable energy. Currently, lithium-ion batteries are the de facto standard in the field of power supplies for electric vehicles, uninterruptible power systems, mobile devices and gadgets [1]. Another example of the use of lithium-ion batteries is storage for renewable energy sources (mainly solar panels and wind turbines). For example, in 2011, China installed a storage device based on lithium-ion batteries with a total capacity of 36 MWh, capable of supplying 6 MW of electrical power to the grid for 6 h [2]. An example from the opposite end of the scale is lithium-ion batteries for implantable pacemakers, the load current of which is on the order of 10 μA [3]. The capacity of a single commercially produced lithium-ion cell has long crossed the mark of 500 Ah [4].

The use of lithium-ion batteries requires compliance with the parameters of the discharge and charge of the battery; otherwise, irreversible degradation of capacity, failure and even fire due to self-heating may occur. Therefore, lithium-ion batteries are always used together with a monitoring and control system—SKU or BMS (battery management system) [5]. The battery management system performs protective functions by monitoring temperature, charge/discharge current and voltage, thus preventing over-discharging, overcharging and overheating. The BMS also monitors the state of the battery by evaluating the degree of charge (state of charge, SOC) and the state of fitness (state of health, SOH). An intelligent battery management system (BMS) is essential in virtually any application of lithium-ion batteries.

A battery is a complex physical-chemical, electrochemical and electrical object, the simulation of which can be carried out at different depth levels and by different methods. Its application on heavy-duty electric vehicles is poorly understood. In modeling batteries, two areas should be distinguished: representation of current parameters during the charge/discharge cycle and simulation of the parameters of the functional state of the battery for a long time of operation. The latest versions of MatLab-Simulink have a built-in model with degradation of accumulator parameters, but it is rather complicated. In particular, when simulating the operation of more than one battery, the duration of the count increases significantly [6]. The issues of battery life and battery degradation come to the fore both for developers of battery management systems and for researchers who operate electric vehicles every day. In Appendix A, we provide a description of the block diagram of our mathematical model in Matlab, as well as the initial data that we used in the model.

2. System-Level Description of the Electric Vehicle Battery

From a schematic point of view, the battery appears to be a two-terminal battery. In this paper, we will use its description in the form of a black box, as a system with one input (current in the circuit $I(t)$) and voltage at the battery terminals $U(t)$. Open-circuit voltage $U_{OCV}(t)$ (OCV) is voltage at the battery terminals U_T in the absence of current selection $I = 0$. The most important parameter is the battery capacity. Q_{max} is defined as the maximum amount of electrical energy (in Ah) that the battery delivers to the load from the moment of full charge to the state of discharge, without leading to premature degradation of the battery. As mentioned earlier, the main function of an intelligent BMS is the estimation of SOC and SOH.

Battery state of charge (SOC) is an indicator characterizing the degree of battery charge: 100%—full charge, 0%—full discharge. Equivalent depth of discharge (DoD) is $DOD = 100\% - SOC$. Usually, SOC is measured as a percentage, but in this paper we will assume that $SOC \in [0,1]$. Formally, SOC is expressed as $SOC = \frac{Q}{Q_{max}}$, where $Q = \int_0^t I(t) \cdot dt$ is the current charge in the battery [7]. Thus, the current strength I is the current strength in the conductor, equal to the ratio of the charge Q that passed through the cross-section of the conductor to the time interval t during which the current $I = \frac{q}{t}$ flowed.

The battery state of health (SOH) is a qualitative indicator that characterizes the current degree of battery capacity degradation. The result of the SOH rating is not a numerical value but the answer to the question: "Does the battery need to be replaced now?" Currently, there is no standard that regulates which battery parameters should be used to calculate SOH. Different BMS manufacturers use different indicators for this—for example, comparing the original and actual battery capacity or internal resistance [8].

2.1. Types of Models for Determining the State of Charge of a Battery

Determination of the SOC is the task of observing the latent states of the system from a given process model and a measured output response from the input. Models intended for use in battery management systems to determine SOC can be classified into two large groups [9]: empirical models that replicate battery behavior from a "black box" position and physical models that simulate the internal electrochemical processes in the battery.

2.2. Empirical Battery Models

The class of empirical models includes a number of different approaches; common features are a significant simplified modeling of physical processes in a battery. Empirical models are the standard for implementing BMS since, on the one hand, they have sufficient simplicity for implementation, and, on the other hand, they have acceptable accuracy for estimating SOC [10]. A quantitative comparison of 28 different empirical models is given in [11]. The main type of empirical model is substitution schemes. The initial prerequisite for empirical modeling is the observation that the dynamics of the battery can be divided into two parts [12]:

- Slow dynamics associated with the charge and discharge of the battery;
- Fast dynamics associated with the internal impedance of the battery—the active resistance of the electrolyte and electrodes—and also with electrochemical capacities.

The units in a typical time scale are tens of hours for slow dynamics (from full charge to discharge) and tens of minutes for fast dynamics (dispersion of charge carriers in parasitic capacitances and transition of the battery to a steady state). The processes of aging and degradation of the capacitance are modeled as nonstationarity of the system parameters.

In this paper, we will focus on building a battery model under dynamic charge and discharge conditions for electric vehicles in urban traffic and then modeling its state depending on operational factors, including changes in battery temperature. The simulation results will be suitable for real-time implementation in battery management systems to monitor the state of the battery using the assessment of the degree of charge (state of charge, SOC) and the state of fitness (state of health, SOH), which is especially important when designing and operating an intelligent BMS in virtually any application of lithium-ion batteries, providing information on how long the device will run before needing to be recharged (SOC value) and when the battery should be replaced due to loss of capacity (SOH value).

3. Obtaining the Initial Data for the Mathematical Model

In the course of experimental studies, to determine the depth of discharge of the battery, six races were made along predetermined routes. Each race assumed different road conditions and traffic intensity as well as vehicle loading.

The route includes both urban and suburban traffic modes. The maximum speed is 65 km/h; the electric car makes three stops for loading and unloading. During the experiment, 2 rides without load (10,000 kg) and 2 rides with partial load (16,000 kg) were carried out.

Measurements of electrical characteristics were carried out using CAN technology. At the same time, the following indicators were measured and recorded:

- The degree of charge of the battery;
- Torques of asynchronous electric motors;
- Battery voltage;
- Frequency of rotation of electric motors;
- Temperature of electric motors and inverters.

The electric car has on board equipment that fixes the following specifications:

- I_{bat}—battery current;
- P_{bat}—battery power;
- U_{bat}—battery voltage;
- Efficiency is the efficiency of the system;
- W_{bat} is the energy given off by the battery;
- SOC is the degree of charge.

Characteristics are obtained using CAN bus communication technology and are recorded throughout the entire cycle.

The experiment was based on two routes that differ in traffic intensity and the amount of cycling of the lithium-ion battery.

On the first route, the electric car at the same time moved partially loaded, and the mass was 16,000 kg. Figure 1 shows a graph of the speed of an electric vehicle along route 1.

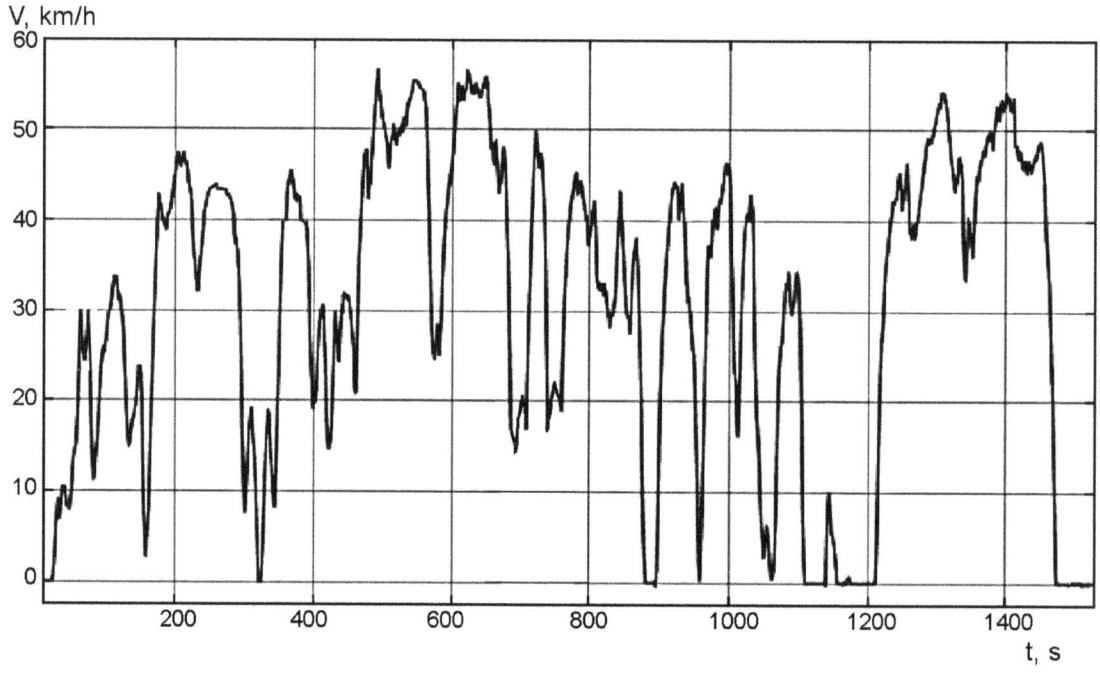

Figure 1. Dependence of speed on time on the first route.

Figure 2 shows the characteristics of the traction electric drive system of an electric vehicle when driving along the first route.

The average battery current in the cycle was 140 A (0.7 C). The degree of charge decreased from 85 to 72 percent.

On the second route, the electric car moved partially loaded, and the mass was 16,000 kg. Figure 3 shows a graph of the speed of the electric vehicle along the second route.

Figure 4 shows the characteristics of the electric vehicle traction drive system. The average battery current in the cycle was 120 A (0.6 C). The degree of charge decreased from 72 to 54 percent. Table 1 shows the cycle parameters and energy characteristics. Unlike the abstract driving cycles often used in EV motion and performance simulations, these routes were derived from actual EV motion. This allows one to compare the real characteristics of the movement of the electric vehicles.

Table 1. Control cycle parameters and test results electric vehicle.

Route	Distance, km	Average Speed, km/h	Energy in the Cycle, kW	Recovery Energy, kWh	Energy Consumption, kW·h/km
First route	12.95	34.85	13.9	3.11	1.11
Second route	16.1	27.66	22.44	2.94	1.38

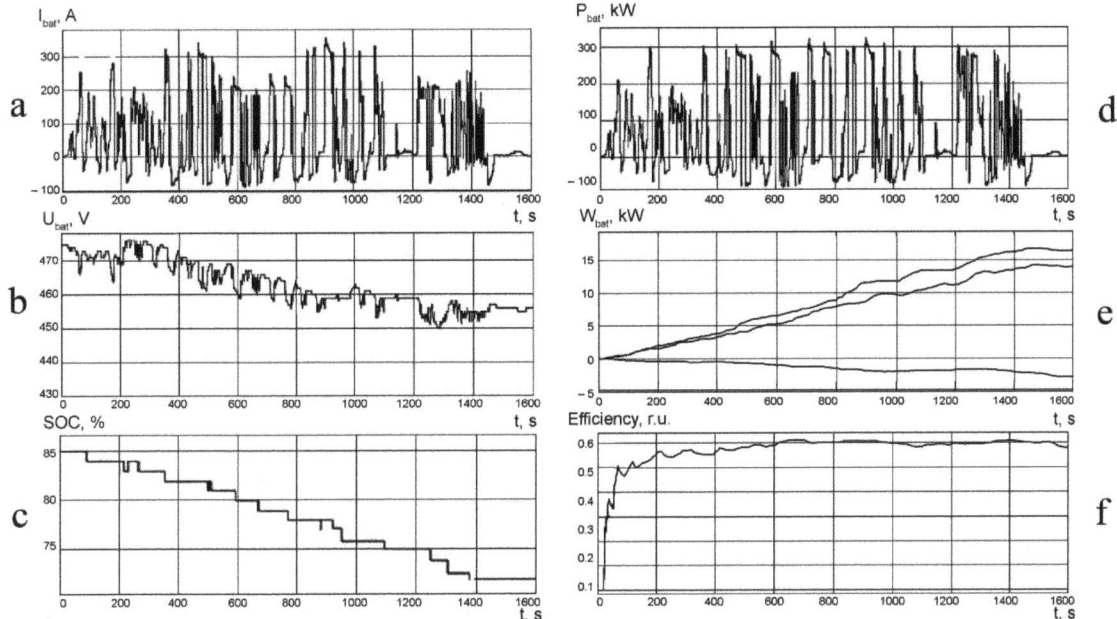

Figure 2. Energy characteristics of the electric vehicle of the first route: (**a**) I_{bat} is the battery current; (**b**) U_{bat} is the battery voltage; (**c**) *SOC* is the degree of charge; (**d**) P_{bat} is the battery power; (**e**) W_{bat} is the energy given off by the battery; (**f**) efficiency is the efficiency of the system.

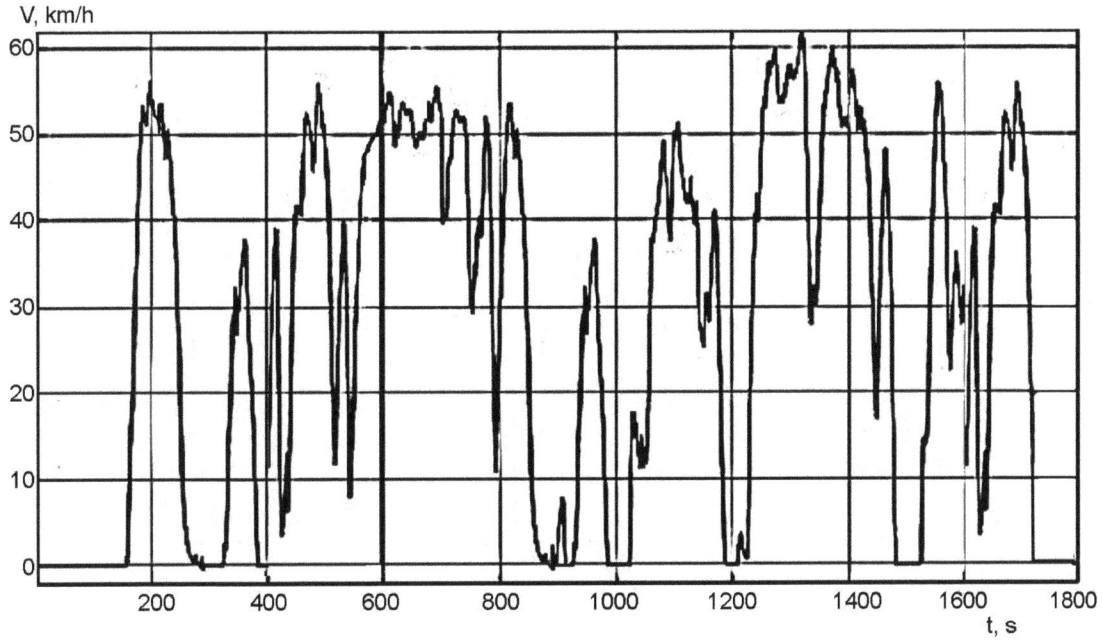

Figure 3. The graph of the change in the speed of an electric vehicle depending on the time along the second route.

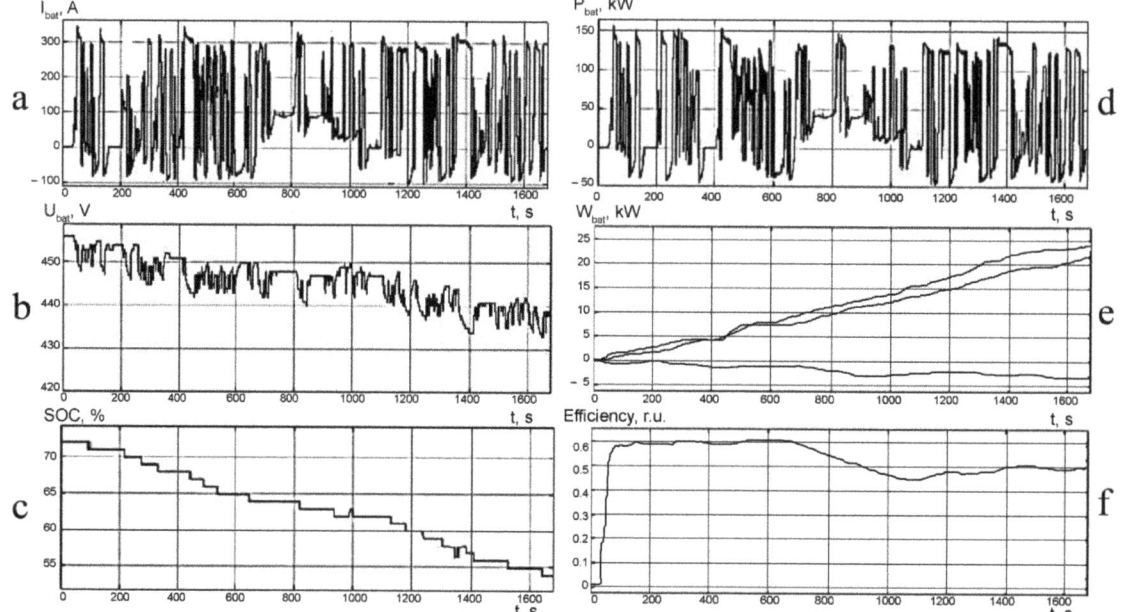

Figure 4. Energy characteristics of the electric vehicle of the second route: (**a**) I_{bat} is the battery current; (**b**) U_{bat} is the battery voltage; (**c**) SOC is the degree of charge; (**d**) P_{bat} is the battery power; (**e**) W_{bat} is the energy given off by the battery; (**f**) efficiency is the efficiency of the system.

As a result of the tests, it was determined that the electric vehicle is capable of moving along the two routes under consideration with a fully charged battery, with a partial load. The depth of discharge was 20%.

4. Simulation of the Processes Occurring in the Battery during the Charge/Discharge Modes

For mathematical modeling of the processes of charge, discharge and charge transfer in the battery, a model of a separate battery was used, which reflects the following processes of energy conversion in the battery:

1. Processes of charge and discharge;
2. Processes of charge transfer between batteries in the battery;
3. Charge, discharge and charge transfer control processes.

When constructing equations relating the parameters of the mathematical model, the authors proposed an equivalent electrical circuit (Figure 5), which connects active and reactive elements, each of which imitates the physicochemical parameters of the battery under study or a structural element of an electrochemical battery [13].

The circuit is a series connection of an EMF source E_m and four active resistances R_1, R_2, R_p, R_0. To take into account the inertia of the discharge/charge process parallel to the resistor R_1 included the electrical capacitance R_1 required to simulate transient processes when the load is turned on and off.

The characteristics of the battery for the purposes of this study are according to the manufacturer's specification. The main characteristics of a single battery are:

1. The chemical composition of the cathode material of the NMC battery (lithium-nickel-manganese-cobalt-oxide battery ($LiNi_xMn_yCo_zO_2$, NMC)—an alloy of nickel oxide, manganese, cobalt, and lithium;
2. Cell nominal voltage—3.8 V;
3. Lower voltage level 2.4 V;

4. The upper limit (depending on the charging or discharging process) of the battery open-circuit voltage is 4.2 V, at a temperature of 25 °C [14].

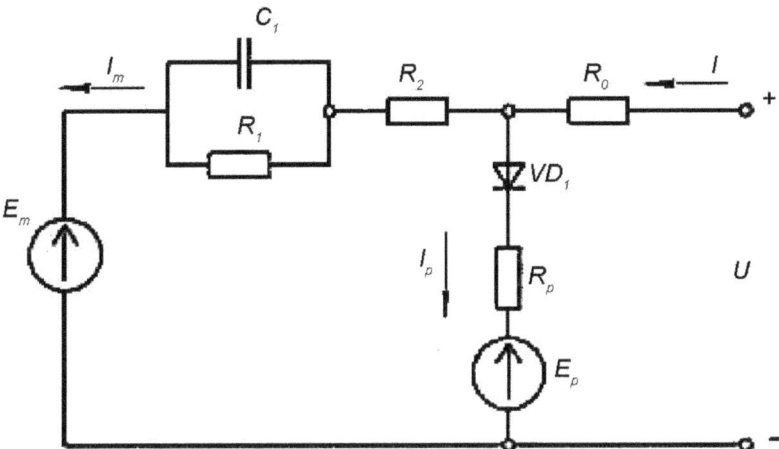

Figure 5. Equivalent electrical circuit of the model of the electrochemical cell of the battery.

Below is the system of Equation (1) for the equivalent circuit of the battery shown in Figure 5. The mathematical model in this case is used to simulate the parameters of the battery, including the main branch, parasitic branch, capacity and electrolyte temperature:

$$\left. \begin{array}{l} E_m = E_{m0} - K_E(273 + \theta) \cdot (1 - SOC) \\ R_1 = -R_{10} \ln(DOC) \\ C_1 = \frac{\tau_1}{R_1} \\ R_2 = R_{20} \frac{e^{(A_{21} \cdot (1-SOC))}}{1 + e^{(A_{22} \frac{I_m}{I^*})}} \\ I_p = V_{pn} G_{p0} e^{\left(\frac{V_{pn}}{(V_\rho s + 1)} + A_\rho \left(1 - \frac{\theta}{\theta_{\rho f}} \right) \right)} \\ Q_e(t) = Q_{e\_init} + \int\limits_0^t -I_m(\tau) d\tau \\ C(I, \theta) = \frac{K_c C_{0*} K_t}{1 + (K_c - 1) \left(\frac{I}{I^*}\right)^\delta} \\ SOC = 1 - \frac{Q_e}{C(0,\theta)} \\ DOC = 1 - \frac{Q_e}{C(I_{avg}, \theta)} \\ I_{avg} = \frac{I_m}{(\tau_1 s + 1)} \\ \theta(t) = \theta_{e\_init} + \int\limits_0^t \frac{\left(P_s - \frac{\theta - \theta_a}{R_\theta}\right)}{C_\theta} d\tau \\ P_s = \frac{V_{R_1}^2}{R_1} + I^2 R_0 + I^2 R_2 \\ R_0 = R_{00}[1 + A_0(1 - SOC)] \end{array} \right\} \quad (1)$$

The equivalent circuit depends on the battery current and nonlinear circuit elements. The above system of battery equations describes the components within the battery cell block. The system does not model the internal chemical processes of lithium-ion batteries directly—the equivalent circuit empirically approximately describes the processes occurring inside [15,16].

The equivalent circuit consists of two main parts: the main branch (elements $R_1 C_1, R_0$), which approximately describes the dynamics of the battery under most conditions, and the parasitic branch R_p, which describes the behavior of the battery at the end of the charge. The following is a detailed description of each of the circuit branches.

For the main branch, Equation (2) describes the internal electromotive force (EMF) or no-load voltage E_m of one battery pack [17]. It is assumed that the value E_m will be constant when the battery is fully charged. Voltage E_m depends on temperature and battery state of charge (SOC) and is defined as follows:

$$E_m = E_{m0} - K_E(273 + \theta) \cdot (1 - SOC) \tag{2}$$

where E_{m0} is the no-load voltage at full charge, V; K_E is the temperature coefficient, V/°C; θ is the electrolyte temperature, °C; SOC is the state of charge of the battery, p.u.

$$R_1 = -R_{10} \ln(DOC) \tag{3}$$

Equation (3) describes the change in resistance with charge depth. The resistance increases exponentially as the battery starts to exhaust itself during discharge [18].

The main resistance of the main branch is R_1, where R_{10} is constant, Ohm and DOC is the battery charge depth, r.u.

Equation (4) describes the transient process if the battery current has changed. Capacity of the main branch C_1:

$$C_1 = \frac{\tau_1}{R_1} \tag{4}$$

where τ_1 = time constant of the main branch, s.

Resistance R_2 is determined by the formula

$$R_2 = R_{20} \frac{e^{(A_{21} \cdot (1-SOC))}}{1 + e^{(A_{22} \frac{I_m}{I^*})}} \tag{5}$$

where R_{20} is resistance under normal conditions, Ohm; A_{21}, A_{22} are constants, r.u.; I_m is the current of the main branch, A; I^* is the current rated current of the battery, A.

The resistance across the terminals of one battery cell can be expressed as

$$R_0 = R_{00}[1 + A_0(1 - SOC)] \tag{6}$$

where R_{00} is resistance R_0 with a completely infected battery (SOC = 1); A_0 is a constant determined by the accumulator parameters [19].

It is assumed that the resistance is constant at different temperatures and depends on the degree of charge [20].

The parasitic branch in the battery equivalent circuit consists of three components: a diode, resistance R_p of the parasitic branch, and a constant voltage source [21]. If the voltage across the diode exceeds the forward voltage V_{pn}, then the diode behaves as a linear resistor with a low resistance R_p. If the voltage across the diode is less than the forward voltage, then the diode behaves as a linear resistor with a low turn-off G_p. A constant voltage source maintains a constant voltage V_{p0} at its output terminals regardless of the current flowing through the source. The output voltage is determined by the battery constant voltage parameter and can be any real value [22].

The value of the discharge current in the parasitic branch I_p is expressed by the dependence

$$I_p = V_{pn} G_{p0} e^{\left(\frac{\frac{V_{pn}}{(\tau_p s + 1)}}{V_{p0}} + A_p \left(1 - \frac{\theta}{\theta_{pf}}\right)\right)} \tag{7}$$

where V_{pn} is voltage on the parasitic branch, V; G_{p0} is a constant determined by the accumulator parameters, s; τ_p is the time constant of the parasitic branch, s; V_{p0} is constant, V; A_p is constant, r.u.; θ is the electrolyte temperature, °C; θ_f is the freezing temperature of the electrolyte, °C.

Equation (7) describes parasitic current losses that occur during battery charging [22]. The current depends on the temperature of the electrolyte and the voltage of the parasitic branch. The current Ip is small under most conditions, with the exception of charging at high SOC [23]. It should be noted that the constant G_{p0}, as a rule, takes values greater than 1 s; the value of G_{p0} is very small, on the order of 1012 s.

Equation (5) tracks the amount of charge drawn from the battery. It is determined by integrating the current flowing in the main branch in both directions, i.e., during battery charging and discharging [24]. Battery level Q_e

$$Q_e(t) = Q_{e\_init} - \int_0^t I_m(t)dt \tag{8}$$

where Q_{e_init} is the initial value of the charge, A s; I_m is the current in the main branch, A; t is the integration time variable, s.

Equation (9) describes the state of the battery capacity C based on the values of the discharge current and electrolyte temperature [25]:

$$C(I, \theta) = \frac{K_c C_{0*} K_t}{1 + (K_c - 1)\left(\frac{I}{I^*}\right)^\delta} \tag{9}$$

where K_c is a constant; C_{0*} is battery capacity without load at 0 °C, A s; K_t is the temperature coefficient; θ is the electrolyte temperature in °C; I is the discharge current in amperes; I^* is the current rated current of the battery, A; δ is a constant determined by the accumulator parameters, r.u.

However, the dependence of capacitance on current is determined only during the discharge. During charging, the discharge current is set to zero in Equation (5) to calculate the total capacity [26].

Equations (10) and (11) describe the state of charge of the SOC and the depth of charge of the DOC of the battery. SOC determines the amount of remaining battery charge, and DOC determines the useful fraction of the remaining charge, taking into account the average discharge current. Large discharge currents cause a premature decrease in battery charge; thus, the DOC parameter is always less than or equal to the SOC parameter [27]:

$$SOC = 1 - \frac{Q_e}{C(0, \theta)} \tag{10}$$

$$DOC = 1 - \frac{Q_e}{C(I_{avg}, \theta)} \tag{11}$$

where Q_e is the battery charge, A s; C is the battery capacity, A s; θ is the electrolyte temperature, °C; I_{avg} is the current average discharge current, A.

The current average discharge current I_{avg} is determined by the formula

$$I_{avg} = \frac{I_m}{(\tau_1 s + 1)} \tag{12}$$

where I_{avg} is the current average discharge current, A; I_m is the current of the main branch, A; τ_1 is the time constant of the main branch, s.

The change in temperature of the electrolyte in the battery θ is due to resistive losses P_s, taking into account the ambient temperature θ_a.

The thermal model consists of a first-order differential equation, with parameters for thermal resistance and capacitance:

$$\theta(t) = \theta_{e\_init} + \int_0^t \frac{\left(P_s - \frac{\theta - \theta_a}{R_\theta}\right)}{C_\theta} d\tau \qquad (13)$$

where θ_a is the ambient air temperature, °C; θ_{init} is the initial temperature of the battery (assumed to be equal to the ambient air temperature), °C; C_θ is the thermal capacity, J/°C; τ is the integration time variable, s; R_θ is the thermal resistance, °C/W; t is the simulation time, s; P_s is the power loss at active resistances R from R_0 and R_2, W.

$$P_s = \frac{V_{R_1}^2}{R_1} + I^2 R_0 + I^2 R_2 \qquad (14)$$

The power loss P_s is the heat loss power with which the battery heating can be calculated. On the basis of the described mathematical equations, a mathematical model will be further formed, which allows taking into account the processes occurring in the battery and taking into account the current profile [28].

The simulation and solution were obtained using the Matlab program (Appendix A). In the section Appendix A.1, we provide a description of the block diagram of the mathematical model in Matlab, the data that we used in the model, the basic equations of the blocks and the implementation of the thermal model in the Matlab program.

In the section Appendix A.2, we provide a logic diagram for switching motion cycles, as well as a charge cycle.

5. Discussion

One of the main criteria for optimizing battery life is temperature. As mentioned earlier, battery operation at low temperatures leads to a sharp decrease in the number of cycles. At positive temperatures, the recommended operating range is 10–35 °C. During the operating mode, the walls of the battery are heated. To determine the temperature after the operation cycle, it is necessary to carry out mathematical modeling. The current profile during a driving cycle on route 1 is reduced to one battery by dividing the total current by the number of cells connected in parallel (Figure 6).

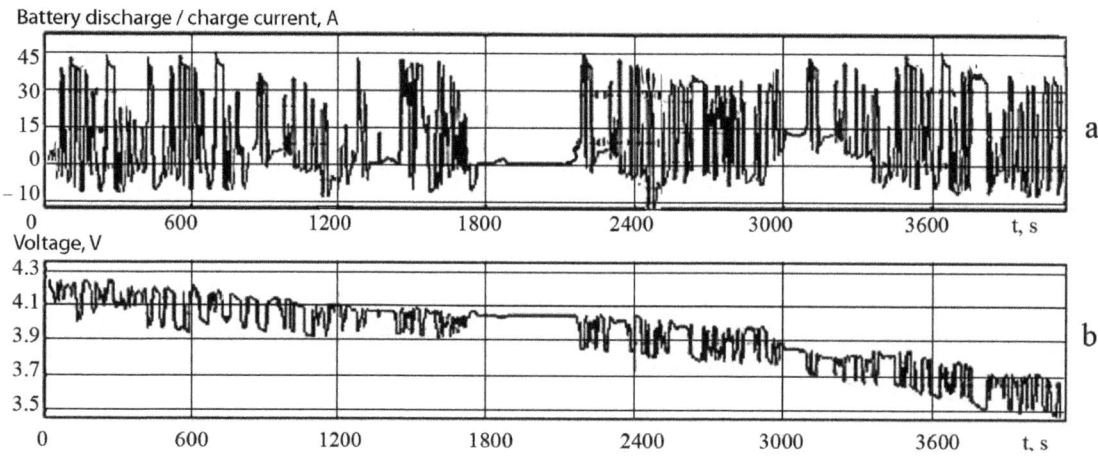

Figure 6. Cont.

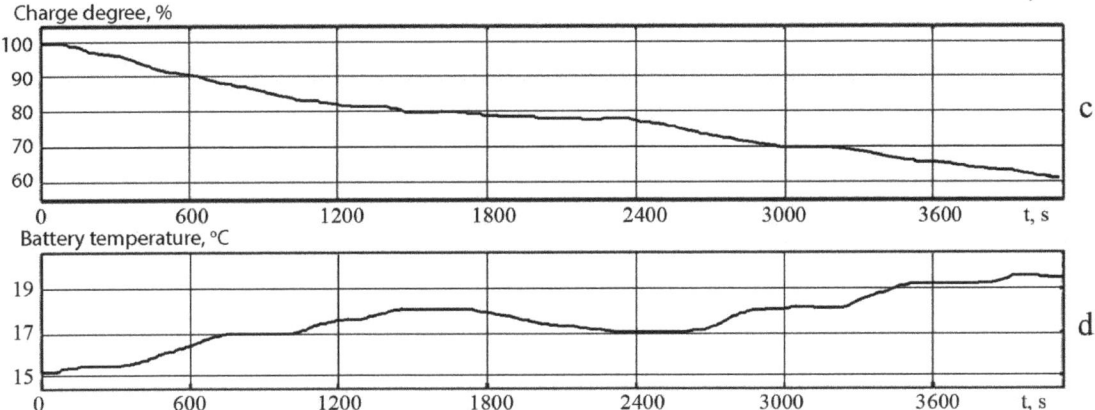

Figure 6. Time waveforms of the main characteristics of the battery in cycles at an ambient temperature of 15 °C: (a) battery discharge/charge current; (b) voltage of battery; (c) battery charge degree; (d) battery temperature.

The results of modeling the main characteristics of the battery at an ambient temperature of 15 °C are shown in Figure 6.

The maximum heating of the battery was 25 °C after a driving cycle on routes 1 and 2. The difference between the initial temperature and the temperature at the end of the cycle was 50 °C. The degree of charge decreased from 100% to 60%. The battery voltage dropped to 3.6 V.

Similar studies were carried out for 25 and 35 degrees, respectively, as the most typical operating temperatures for traction batteries. The results are shown in Figures 7 and 8.

As can be seen from the graph, the heating for the cycle was 29.2 °C, and the difference between the initial temperature and the temperature at the end of the cycle was 4.2 °C. This is due to the decrease in internal resistance as the temperature of the battery rises. Another calculation was made at an ambient temperature of 35 °C. This temperature range is critical for the battery; the resource under such conditions is significantly reduced. At an initial temperature of 40 °C, the heating per cycle was 39 °C, and the difference between the initial temperature of the cell and the temperature at the end of the cycle was 4 °C (Figure 9).

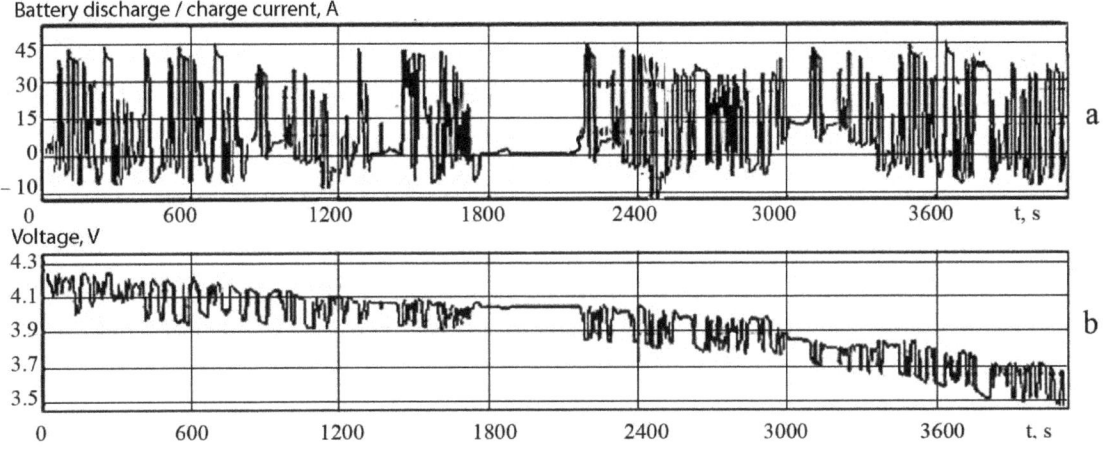

Figure 7. Cont.

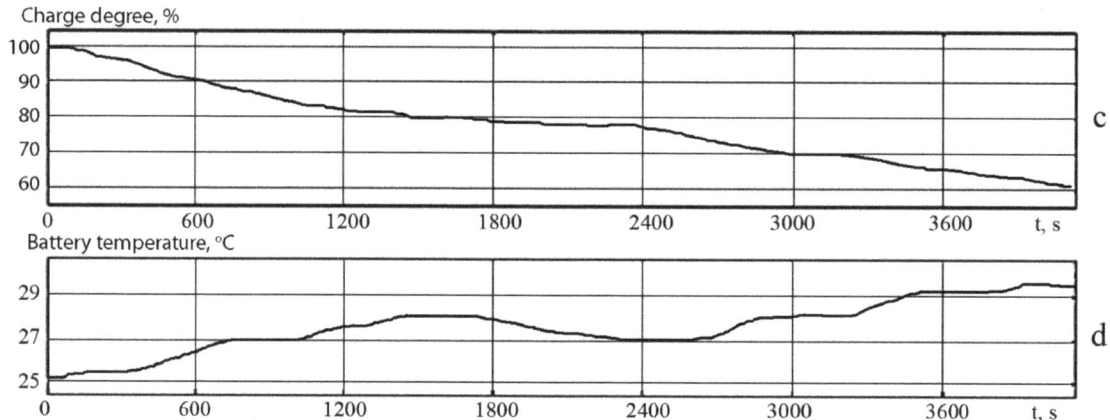

Figure 7. Graphs of the main characteristics of the battery in cycles at an ambient temperature of 25 °C: (**a**) battery discharge/charge current; (**b**) voltage of battery; (**c**) battery charge degree; (**d**) battery temperature.

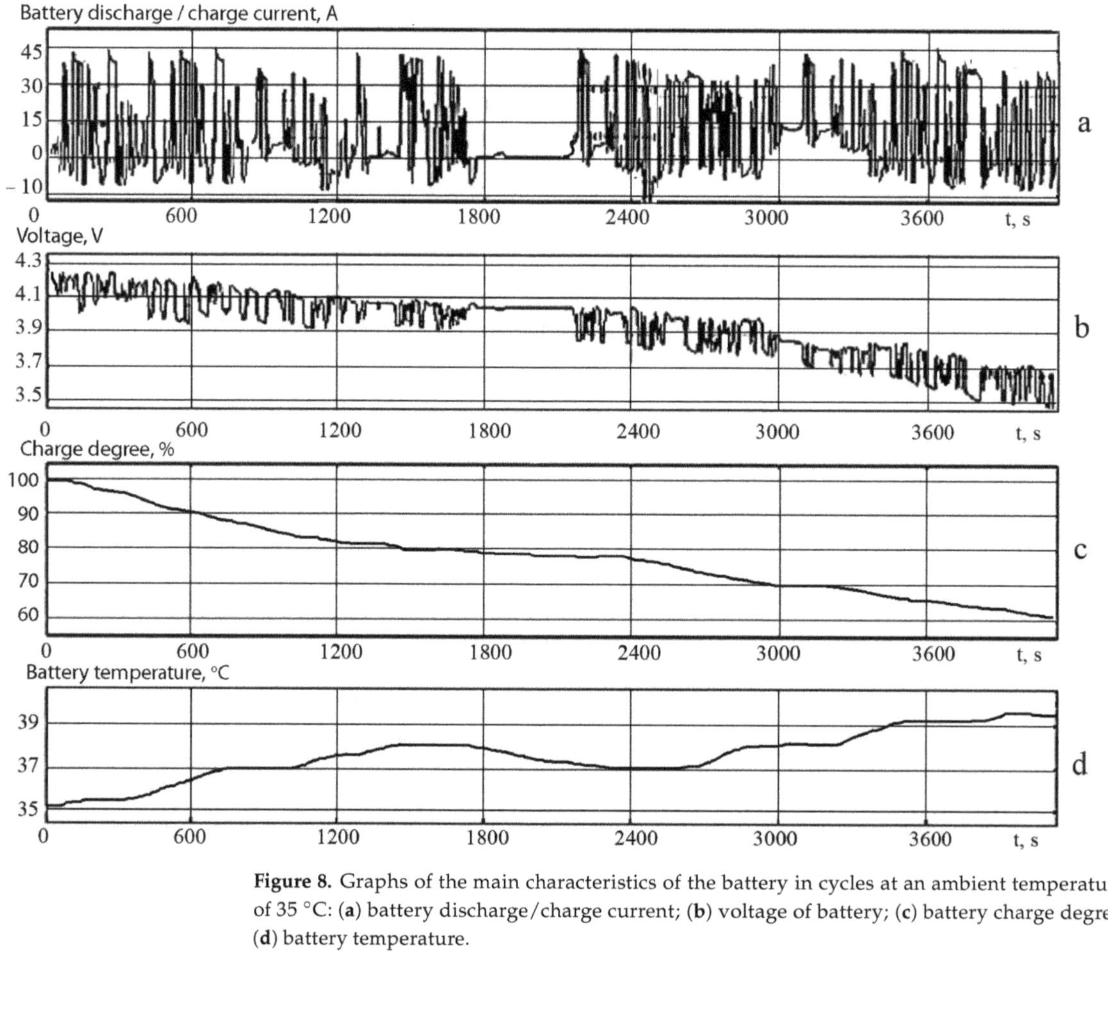

Figure 8. Graphs of the main characteristics of the battery in cycles at an ambient temperature of 35 °C: (**a**) battery discharge/charge current; (**b**) voltage of battery; (**c**) battery charge degree; (**d**) battery temperature.

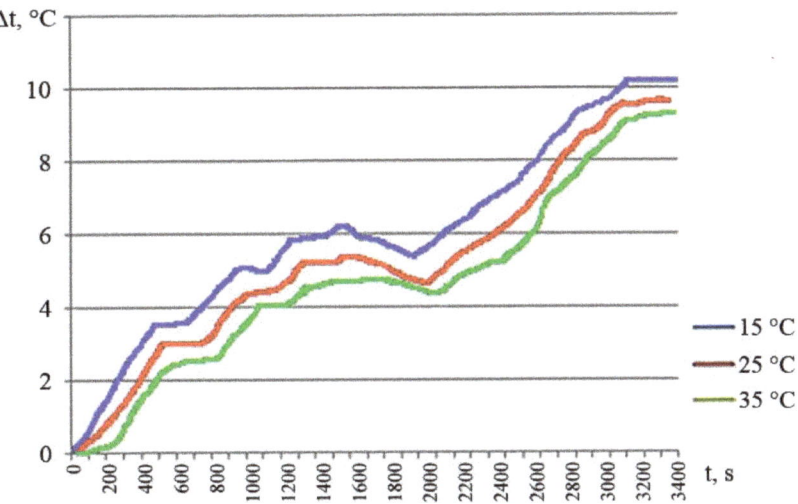

Figure 9. Battery heating at different ambient temperatures.

The results obtained show the heating of the battery in the absence of air conditioning in the cooling system. Figure 9 shows the temperature difference achieved by the battery when alternating routes 1 and 2.

6. Conclusions

Based on the battery equivalent circuit and the system of equations, a simulation model was created to calculate the electrical and thermal characteristics. The equivalent circuit includes active and reactive elements, each of which imitates the physicochemical parameter of the battery under study or the structural element of the electrochemical battery.

The input signals of the mathematical model are the current and ambient temperatures obtained during the tests of the electric vehicle, and the output signals are the voltage, electrolyte temperature and degree of charge. The resulting equations make it possible to assign values of internal resistance to a certain temperature value and a certain value of the degree of charge.

As a result of simulation modeling, the heating of the battery was determined at various values of the ambient temperature. It was found that at an ambient temperature of 20 °C, the heating at the end of the cycle was 25 °C; at 30 °C, the same parameter took on a value of 4.2 °C; and at 35 °C was 4 °C.

The absence of a forced cooling system in the model makes it possible to determine the degree of heating not only of the outer walls of the accumulator, but also inside the cell. When the charge/discharge cycle is continuous, the temperature of the element does not fall to its original state but, on the contrary, continues to increase. When determining battery life, a temperature range of more than 45 °C is not calculated, as battery life is significantly reduced. As a result of simulation modeling, a tool was obtained for calculating battery heating under various environmental conditions. During operation, the maximum heating will not exceed 50 °C, which allows us to conclude that the range recommended for battery operation will be provided.

The simulation results will be useful to researchers and developers of autonomous power sources for electric vehicles and other devices for the implementation of real-time systems that take into account the processes of changing the characteristics of traction batteries. The developed mathematical model can be used in battery management systems to monitor the state of charge and battery degradation using the assessment of the state of charge (SOC) and the state of health (SOH). This is especially important when designing and operating a smart battery management system (BMS) in virtually any lithium-ion battery application, providing information about how long the device will run before it needs to be charged (SOC value) and when the battery should be replaced due to loss of battery capacity (SOH value). Based on the equivalent circuit of the battery and the system of equations, a simulation model was created to calculate the electrical and thermal characteristics.

Author Contributions: Conceptualization, B.V.M. and N.V.M.; methodology, S.N.S. and E.A.E.; software, M.Q.; validation, S.N.S. and B.V.M.; formal analysis, N.V.M.; investigation, E.A.E.; resources, E.A.E.; data curation, S.N.S.; writing—original draft preparation, B.V.M.; writing—review and editing, N.V.M.; visualization, M.Q. All authors have read and agreed to the published version of the manuscript.

Funding: This research received no external funding.

Institutional Review Board Statement: Not applicable.

Informed Consent Statement: Not applicable.

Data Availability Statement: The data presented in this study are available from the corresponding authors upon reasonable request.

Conflicts of Interest: The authors declare no conflict of interest.

Appendix A Modeling Blocks and Data Used in the Matlab Program

Appendix A.1 Block Diagram of the Mathematical Model of the Battery

In this section, we provide a description of the block diagram of the mathematical model in Matlab, the data that we used in the model, the basic equations of the blocks, and the implementation of the thermal model in the Matlab program.

The mathematical model of processes in the battery was developed in Matlab (Figure A1). The capacitance model calculates the state of charge (SOC) and depth of charge of the battery, the voltage model calculates the resistance as a function of the state of charge and the depth of charge, and the thermal model calculates the internal temperature [29].

The input signals of the mathematical model are current and ambient temperature, and the output signals are voltage, electrolyte temperature and degree of charge [30].

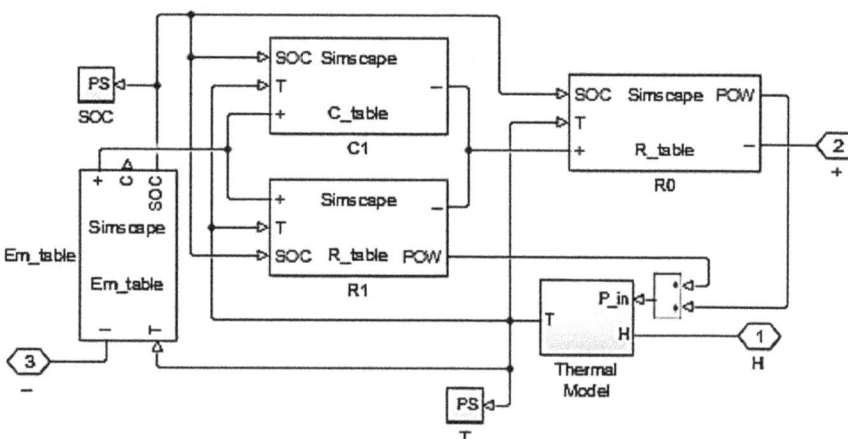

Figure A1. Block diagram of the mathematical model in Matlab.

The description of the blocks of the mathematical model of the battery is presented below.
Em_Table—EMF source, its input and output parameters:
inputs
T = {293.15,'K'} % T:right

end
outputs
C = {31,'A*hr'} %C:left
SOC = {1,'1'} %SOC:left
end

Basic Block Equations:
equations
% Charge deficit calculation, preventing SOC > 1
If Qe < 0 && i > 0 Qe.der == 0;
else Qe.der == -i;
end
% Perform the capacity table lookup C == tablelookup(Temp_Table,C_Table,T,...
interpolation = linear,extrapolation = nearest)
% SOC Equation SOC == 1−Qe/C;
% Electrical equation by table lookup v == tablelookup(SOC_Table,Temp_Table,Em_Table,SOC
interpolation = linear,extrapolation = nearest)
end

Block C1—equivalent capacitance of the RC circuit, input and output parameters:
inputs
T = {293.15,'K'}; %T:left SOC = {1,'1'}; %SOC:left
end
parameters (Size = variable) C_Table = {ones(5,3),'F'}
% Matrix of capacitance values, C(SOC,T) SOC_Table = {[0;0.1;0.5;0.9;1],'1 '} % State of charge (SOC) breakpoints Temp_Table = {[273.15 293.15 313.15],'K'} % Temperature (T) breakpoints
end

Block equations:
equations let
% Perform the table lookup C =
tablelookup(SOC_Table,Temp_Table,C_Table,SOC,T,...
interpolation = linear,extrapolation = nearest)
in
% Electrical equation i == C * v.der; end end

Block R1—equivalent resistance of the RC circuit, input and output parameters:
inputs
T = {293.15,'K'}; %T:left SOC = {1,'1'}; %SOC:left
end
outputs pow = {0,'W'}; %POW:right
end

Block Equations:
let
% Perform the table lookup
R = tablelookup(SOC_Table,Temp_Table,R_Table,SOC,T, ...
interpolation = linear,extrapolation = nearest)
in
% Electrical Equations
v == i*R; pow == v*i;
end
end

Block R0 is equivalent internal resistance; input and output parameters are similar to block R1.

The resulting equations allow for assigning internal resistance values to a certain temperature value and a certain value of the degree of charge (Table A1).

Table A1. Parameters of a typical lithium-ion battery based on NMC technology.

Cell Temperature	0 °C	30 °C	45 °C	State of Charge (SZ), %
Available battery capacity	18.008	17.625	17.639	
Battery voltage at different states of charge	2.52	2.73	2.54	0
	2.63	2.82	2.45	10
	2.87	3.12	3.21	25
	3.39	3.51	3.54	50
	3.71	3.9	3.91	75
	4	4	4	90
	4.1	4.2	4.31	100

Figure A2 shows a graph of the open-circuit voltage versus the battery state of charge.

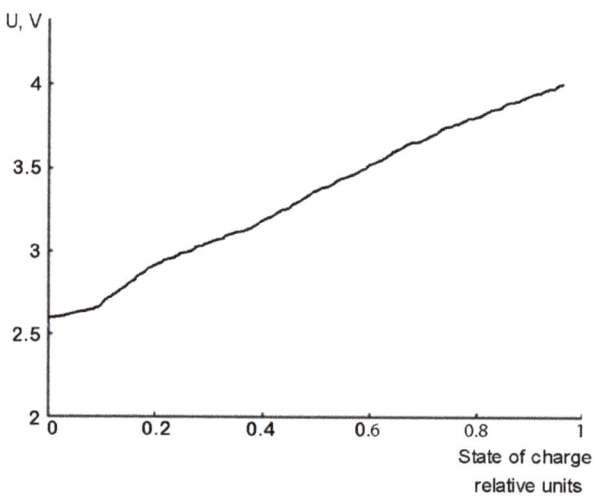

Figure A2. Battery open-circuit voltage (OCV).

In order to apply the TAB current obtained during the operation of the electric vehicle to the input, it is necessary to form data tables in which the values of the currents in the cycle will be indicated, depending on the time. In addition, a full cycle of movement along the route provides for one stop after passing the halfway point.

The thermal model block represents a thermal model that contains a heat source, a temperature sensor and a battery mass; see Figure A3.

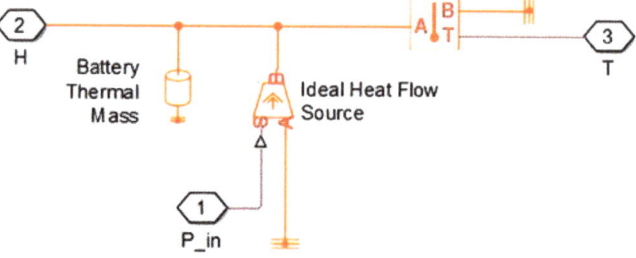

Figure A3. Thermal model in the MATLAB program.

The description of the blocks is given for the thermal model below:

inputs
S = { 0, 'K' }; %S:bottom
end nodes
A = foundation.thermal.thermal; %A:bottom B = foundation.thermal.thermal; %B:top
end
variables (Access = private) Q = { 0, 'J/s' }; T = { 0, 'K' };
end branches
Q : BQ -> AQ;
end equations
T == BT–AT;
T == S;
end
end

The thermal model allows for calculating the heating of the battery, the air intake for cooling of which is carried out directly from the environment. The ambient temperature corresponds to the temperature of the interaction of the battery walls with the air of the cooling system. When using an air conditioner in the cooling system, the temperature of the blown air will correspond to the temperature of the conditioned air. In this case, the airflow rate must be selected in accordance with the power and operating mode of the compressor. In this work, the flow rate was not taken into account; this allows one to more accurately determine the temperature inside the cell, because there is no forced heat extraction. Based on the obtained mathematical model, data on the heat release of the battery cell in the motion cycle are calculated [31].

Appendix A.2 Modeling Blocks of the Charge/Discharge Modes of an Electric Vehicle Battery

We provide a logic diagram for switching motion cycles, as well as a charge cycle. To simulate the load cycle of the battery, it is necessary to set the alternation of several charge/discharge modes in accordance with the routes of the electric vehicle. The alternation of cycles is necessary to determine the time intervals during which it is necessary to charge the battery. With the help of the alternation of motion cycles, the heating and the depth of discharge of the battery are determined [32]. When the set depth of discharge is reached, the battery is charged with the rated current. These algorithms will allow simulating the movement of an electric vehicle, taking into account a fast charge at the final stops, as well as choosing the best option for installing charging stations along the route to improve the battery resource [33]. The logic diagram for switching between cycles is shown in Figure A4.

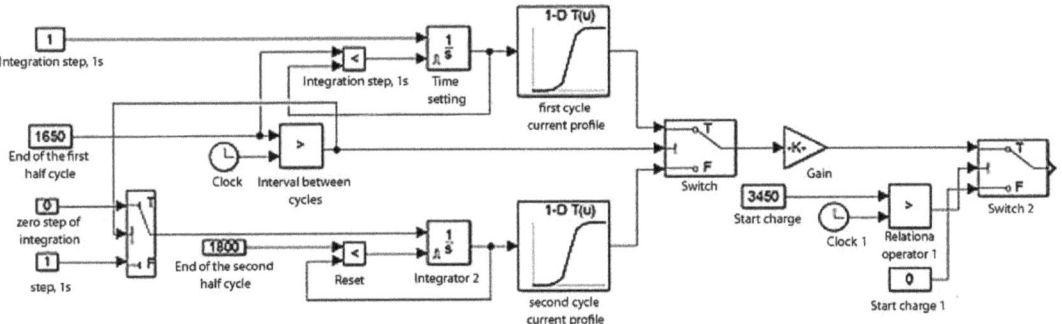

Figure A4. Logic diagram for switching motion cycles, as well as the charge cycle.

The logical scheme presents two main cycles, according to which the calculation of battery modes is performed. The current profiles correspond to the first and second routes.

The formation of time intervals can be done using the "clock" block, but if the cycles alternate one after the other, a timer must be used. The timer consists of an integrator block whose step is one second (Figure A5).

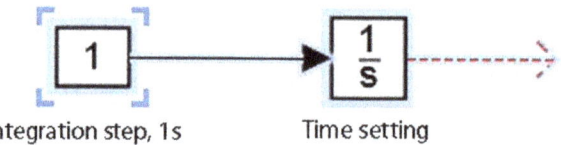

Figure A5. Forming a timer in Matlab.

The block also contains a counter reset, the input of which accepts a logical zero and a logical one, in the case of setting the reset of the counter for the time interval for which the battery is discharged. The exit from the time setting block is compared with the required end time of the first cycle (end of the first and second route).

In the logic diagram, the battery charge current and the duration of its operation can be configured. In the real tests, the charge was formed after two complete cycles of movement.

References

1. Waldmann, T.; Kasper, M.; Fleischhammer, M.; Wohlfahrt-Mehrens, M. Temperature dependent aging mechanisms in Lithium—Ion batteries—A Post—Mortem study. *J. Power Sources* **2014**, *363*, 129–135. [CrossRef]
2. Isametova, M.E.; Nussipali, R.; Martyushev, N.V.; Malozyomov, B.V.; Efremenkov, E.A.; Isametov, A. Mathematical Modeling of the Reliability of Polymer Composite Materials. *Mathematics* **2022**, *10*, 3978. [CrossRef]
3. Xia, B.; Wang, S.; Tian, Y.; Sun, W.; Xu, Z.; Zheng, W. Experimental research on the linixcoymnzo2 lithium-ion battery characteristics for model modification of SOC estimation. *Inf. Technol. J.* **2014**, *13*, 2395–2403. [CrossRef]
4. Shchurov, N.I.; Myatezh, S.V.; Malozyomov, B.V.; Shtang, A.A.; Martyushev, N.V.; Klyuev, R.V.; Dedov, S.I. Determination of Inactive Powers in a Single-Phase AC Network. *Energies* **2021**, *14*, 4814. [CrossRef]
5. Li, X.; Jiang, J.; Zhang, C.; Wang, L.Y.; Zheng, L. Robustness of SOC estimation algorithms for EV lithium-ion batteries against modeling errors and measurement noise. *Math. Probl. Eng.* **2015**, *2015*, 719490. [CrossRef]
6. Tian, Y.; Xia, B.; Wang, M.; Sun, W.; Xu, Z. Comparison study on two model-based adaptive algorithms for SOC estimation of lithium-ion batteries in electric vehicles. *Energies* **2014**, *7*, 8446–8464. [CrossRef]
7. Tseng, K.-H.; Liang, J.-W.; Chang, W.; Huang, S.-C. Regression models using fully discharged voltage and internal resistance for state of health estimation of lithium-ion batteries. *Energies* **2015**, *8*, 2889–2907. [CrossRef]
8. Shchurov, N.I.; Dedov, S.I.; Malozyomov, B.V.; Shtang, A.A.; Martyushev, N.V.; Klyuev, R.V.; Andriashin, S.N. Degradation of Lithium-Ion Batteries in an Electric Transport Complex. *Energies* **2021**, *14*, 8072. [CrossRef]
9. Hafsaoui, J.; Sellier, F. Electrochemical model and its parameters identification tool for the follow up of batteries aging. *World Electric. Veh. J.* **2010**, *4*, 386–395. [CrossRef]
10. Prada, E.; Di Domenico, D.; Creff, Y.; Sauvant-Moynot, V. Towards advanced BMS algorithms development for (p)hev and EV by use of a physics-based model of Li-Ion Battery Systems. *World Electric. Veh. J.* **2013**, *6*, 807–818. [CrossRef]
11. Varini, M.; Campana, P.E.; Lindbergh, G. A semi-empirical, electrochemistry-based model for Li-ion battery performance prediction over lifetime. *J. Energy Storage* **2019**, *25*, 100819. [CrossRef]
12. Ashwin, T.R.; McGordon, A.; Jennings, P.A. Electrochemical modeling of li-ion battery pack with constant voltage cycling. *J. Power Sources* **2017**, *341*, 327–339. [CrossRef]
13. Somakettarin, N.; Pichetjamroen, A. A study on modeling of effective series resistance for lithium-ion batteries under life cycle consideration. *IOP Conf. Ser. Earth Environ. Sci.* **2019**, *322*, 012008. [CrossRef]
14. Kuo, T.J.; Lee, K.Y.; Chiang, M.H. Development of a neural network model for SOH of LiFePO4 batteries under different aging conditions. *IOP Conf. Ser. Mater. Sci. Eng.* **2019**, *486*, 012083. [CrossRef]
15. Davydenko, L.; Davydenko, N.; Bosak, A.; Bosak, A.; Deja, A.; Dzhuguryan, T. Smart Sustainable Freight Transport for a City Multi-Floor Manufacturing Cluster: A Framework of the Energy Efficiency Monitoring of Electric Vehicle Fleet Charging. *Energies* **2022**, *15*, 3780. [CrossRef]
16. Mamun, K.A.; Islam, F.R.; Haque, R.; Chand, A.A.; Prasad, K.A.; Goundar, K.K.; Prakash, K.; Maharaj, S. Systematic Modeling and Analysis of On-Board Vehicle Integrated Novel Hybrid Renewable Energy System with Storage for Electric Vehicles. *Sustainability* **2022**, *14*, 2538. [CrossRef]
17. Chao, P.-P.; Zhang, R.-Y.; Wang, Y.-D.; Tang, H.; Dai, H.-L. Warning model of new energy vehicle under improving time-to-rollover with neural network. *Meas. Control.* **2022**, *55*, 1004–1015. [CrossRef]
18. Pusztai, Z.; K'orös, P.; Szauter, F.; Friedler, F. Vehicle Model-Based Driving Strategy Optimization for Lightweight Vehicle. *Energies* **2022**, *15*, 3631. [CrossRef]

19. Mariani, V.; Rizzo, G.; Tiano, F.; Glielmo, L. A model predictive control scheme for regenerative braking in vehicles with hybridized architectures via aftermarket kits. *Control Eng. Pract.* **2022**, *123*, 105142. [CrossRef]
20. Hensher, D.A.; Wei, E.; Liu, W. Battery electric vehicles in cities: Measurement of some impacts on traffic and government revenue recovery. *J. Transp. Geogr.* **2021**, *94*, 103121. [CrossRef]
21. Li, S.; Yu, B.; Feng, X. Research on braking energy recovery strategy of electric vehicle based on ECE regulation and I curve. *Sci. Prog.* **2020**, *103*, 0036850419877362. [CrossRef] [PubMed]
22. Laadjal, K.; Cardoso, A.J.M. Estimation of Lithium-Ion Batteries State-Condition in Electric Vehicle Applications: Issues and State of the Art. *Electronics* **2021**, *10*, 1588. [CrossRef]
23. Arango, I.; Lopez, C.; Ceren, A. Improving the Autonomy of a Mid-Drive Motor Electric Bicycle Based on System Efficiency Maps and Its Performance. *World Electric. Veh. J.* **2021**, *12*, 59. [CrossRef]
24. Mei, J.; Zuo, Y.; Lee, C.H.; Wang, X.; Kirtley, J.L. Stochastic optimization of multi-energy system operation considering hydrogen-based vehicle applications. *Adv. Appl. Energy* **2021**, *2*, 100031. [CrossRef]
25. Wu, X. Research and Implementation of Electric Vehicle Braking Energy Recovery System Based on Computer. *J. Phys. Conf. Ser.* **2021**, *1744*, 022080. [CrossRef]
26. Istomin, S. Development of a system for electric power consumption control by electric rolling stock on traction tracks of locomotive depots. *IOP Conf. Ser. Mater. Sci. Eng.* **2020**, *918*, 012157. [CrossRef]
27. Domanov, K.; Shatohin, A.; Nezevak, V.; Cheremisin, V. Improving the technology of operating electric locomotives using electric power storage device. *E3S Web Conf.* **2019**, *110*, 01033. [CrossRef]
28. Debelov, V.V.; Endachev, D.V.; Yakunov, D.M.; Deev, O.M. Charging balance management technology for low-voltage battery in the car control unit with combined power system. *IOP Conf. Ser. Mater. Sci. Eng.* **2019**, *534*, 012029. [CrossRef]
29. Widmer, F.; Ritter, A.; Duhr, P.; Onder, C.H. Battery lifetime extension through optimal design and control of traction and heating systems in hybrid drivetrains. *ETransportation* **2022**, *14*, 100196. [CrossRef]
30. Liu, X.; Zhao, M.; Wei, Z.; Lu, M. The energy management and economic optimization scheduling of microgrid based on Colored Petri net and Quantum-PSO algorithm. *Sustain. Energy Technol. Assess.* **2022**, *53*, 102670. [CrossRef]
31. Tormos, B.; Pla, B.; Bares, P.; Pinto, D. Energy Management of Hybrid Electric Urban Bus by Off-Line Dynamic Programming Optimization and One-Step Look-Ahead Rollout. *Appl. Sci.* **2022**, *12*, 4474. [CrossRef]
32. Zhou, J.; Feng, C.; Su, Q.; Jiang, S.; Fan, Z.; Ruan, J.; Sun, S.; Hu, L. The Multi-Objective Optimization of Powertrain Design and Energy Management Strategy for Fuel Cell–Battery Electric Vehicle. *Sustainability* **2022**, *14*, 6320. [CrossRef]
33. Martyushev, N.V.; Malozyomov, B.V.; Khalikov, I.H.; Kukartsev, V.A.; Kukartsev, V.V.; Tynchenko, V.S.; Tynchenko, Y.A.; Qi, M. Review of Methods for Improving the Energy Efficiency of Electrified Ground Transport by Optimizing Battery Consumption. *Energies* **2023**, *16*, 729. [CrossRef]

Disclaimer/Publisher's Note: The statements, opinions and data contained in all publications are solely those of the individual author(s) and contributor(s) and not of MDPI and/or the editor(s). MDPI and/or the editor(s) disclaim responsibility for any injury to people or property resulting from any ideas, methods, instructions or products referred to in the content.

Article

Optimal Sizing of a Photovoltaic Pumping System Integrated with Water Storage Tank Considering Cost/Reliability Assessment Using Enhanced Artificial Rabbits Optimization: A Case Study

Abdolhamid Mazloumi [1], Alireza Poolad [2], Mohammad Sadegh Mokhtari [3], Morteza Babaee Altman [4], Almoataz Y. Abdelaziz [5] and Mahmoud Elsisi [6,7,*]

1. Department of Electrical Engineering, Gorgan Branch, Islamic Azad University, Gorgan 4917954834, Iran
2. Department of Electrical Engineering, Bushehr Branch, Islamic Azad University, Bushehr 7515895496, Iran
3. Department of Computer Science, University of Antwerp, 2000 Antwerp, Belgium
4. Department of Energy Engineering, Amirkabir University of Technology (Tehran Polytechnic), Tehran 1591634311, Iran
5. Faculty of Engineering and Technology, Future University in Egypt, Cairo 11835, Egypt
6. Department of Electrical Engineering, National Kaohsiung University of Science and Technology, Kaohsiung 807618, Taiwan
7. Department of Electrical Engineering, Faculty of Engineering at Shoubra, Benha University, Cairo 11629, Egypt
* Correspondence: mahmoudelsisi@nkust.edu.tw

Citation: Mazloumi, A.; Poolad, A.; Mokhtari, M.S.; Altman, M.B.; Abdelaziz, A.Y.; Elsisi, M. Optimal Sizing of a Photovoltaic Pumping System Integrated with Water Storage Tank Considering Cost/Reliability Assessment Using Enhanced Artificial Rabbits Optimization: A Case Study. Mathematics 2023, 11, 463. https://doi.org/10.3390/math11020463

Academic Editors: Udochukwu B. Akuru, Ogbonnaya I. Okoro and Yacine Amara

Received: 12 November 2022
Revised: 9 January 2023
Accepted: 11 January 2023
Published: 15 January 2023

Copyright: © 2023 by the authors. Licensee MDPI, Basel, Switzerland. This article is an open access article distributed under the terms and conditions of the Creative Commons Attribution (CC BY) license (https://creativecommons.org/licenses/by/4.0/).

Abstract: In this paper, optimal sizing of a photovoltaic (PV) pumping system with a water storage tank (WST) is developed to meet the water demand to minimize the life cycle cost (LCC) and satisfy the probability of interrupted water (p_{IW}) constraint considering real region data. The component sizing, including the PV resources and the WST, is determined optimally based on LCC and p_{IW} using a new meta-heuristic method named enhanced artificial rabbits optimization (EARO) via a nonlinear inertia weight reduction strategy to overcome the premature convergence of its conventional algorithm. The WST is sized optimally regarding the lack of irradiation and inaccessibility of the pumping system so that it is able to improve the water supply reliability. The LCC for water extraction heights of 5 and 10 m is obtained at 0.2955 M$ and 0.2993 M$, respectively, and the p_{IW} in these two scenarios is calculated as zero, which means the complete and reliable supply of the water demand of the customers using the proposed methodology based on the EARO. Also, the results demonstrated the superior performance of EARO in comparison with artificial rabbits optimization (ARO) and particle swarm optimization (PSO); these methods have supplied customers' water demands with higher costs and lower reliability than the proposed EARO method. Also, during the sensitivity analysis, the results showed that changes in the irradiance and height of the water extraction have a considerable effect on the cost and ability to meet customer demand.

Keywords: photovoltaic pumping system; optimal sizing; cost/reliability assessment; probability of interrupted water; enhanced artificial rabbits optimization

MSC: 49K10; 68T20

1. Introduction

Today, the use of clean energy resources as distributed generation generators is of great interest [1]. One of the most common renewable resources is photovoltaic (PV) energy, which has been widely used in various energy production applications [2]. One of its applications is in water-pumping systems to extract customers' drinking water and also in the agricultural industry for irrigation [3], in such a way that the electricity required by electrical motor pumps is supplied by PV arrays. The water pumps generally depend on electricity or diesel generators to provide electricity. The pumped water is stored in the tanks for use by customers during the day, at night, or in the absence of radiation. The tank operates as a storage system, and the battery is generally not used to store PV power; however, it can be used reliably for specific needs. The use of diesel has a high fuel cost and also increases environmental pollution. PV water pump systems are not environmentally friendly and have low maintenance costs. Optimal design and optimal sizing of a water-pumping system are desirable features. To supply water to customers based on a system to minimize the cost, the size of the components must be determined optimally [4,5]. In recent years, many studies have been performed on the use of new energy sources and their optimization based on intelligent optimization methods [6–9]. The main reason for the acceptance of new energy sources is the available energy of this type of resource. On the other hand, the use of these resources is very useful for feeding electric pumps in places far from electricity networks. Studies have been conducted in the field of designing PV water-pumping systems to supply the water demand of customers. In the economic index, the cost of energy production of the system is considered, which is related to the cost of purchasing, maintaining, and replacing the cost of components. In the technical index, the water supply is also incorporated. In [10], a design method is presented for the sizing of photovoltaic panels to feed the pumping system in Spain. In [11], an optimization method is used for determining the size of PV resources for Turkish weather conditions, and the optimal size of PV energy components and the optimal electrical structure of the system components are identified. In [12], the sizing of a PV water-pumping system using a storage tank is performed with reference to the loss of energy probability and considering lifetime cost for Algeria. In [13], the load loss probability method is applied to optimize the size of PV water-pumping systems. In [14], the sizing of a PV water pump system with daily radiation profiles is presented, and the results show that the proposed method reduces the capital cost and removes battery storage dependency. In [15], a method is proposed for sizing a PV water-pumping system based on optimal control with the aim of daily pumped water maximization by optimizing motor efficiency. In [16], the sizing of a PV water-pumping system integrated with a battery storage system is presented, considering net present cost minimization and satisfying constraints regarding shortage of power supply probability. In [17], a framework for sizing a PV water-pumping system with battery storage is proposed to minimize the cost of energy considering the loss of power probability. In [18], the optimal size of a reliable PV water-pumping system integrated with a diesel generator is presented to minimize the cost of energy. In [19], the effectiveness of a PV water-pumping system integrated with battery storage is presented with peak-shaving minimization. In [20], an algorithm is proposed for sizing a photovoltaic water-pumping system with a diesel generator to minimize the LCC. In [21], the sizing of a photovoltaic pumping system with battery storage and diesel backup is developed to minimize the cost of energy. In [22], the net present cost is minimized in designing a photovoltaic water-pumping system integrated with battery storage. In [23], PV and wind arrays are applied to supply a water-pumping system with electricity. The goal of system sizing is considered to determine the size of the system components to minimize the costs of the system and satisfy the LPSP. In [24], PV resources and storage systems are applied to meet the demand of an electric water-pumping system for power. Excess photovoltaic energy that is not consumed by the electric pump is stored in a battery.

Due to the lack of grid power in remote areas, PV water-pumping systems are one of the most cost-effective methods to supply the drinking water of customers. Using a PV water-pumping system has become popular due to the importance of available electricity and rising diesel costs. In these systems, the amount of extracted water depends on the available PV radiation and sizing the system optimally. The literature review found that in the sizing studies of PV water-pumping systems, battery storage or fuel cells are applied to compensate for the shortage of power due to oscillation of the irradiance and especially the lack of radiation at night, while the use of these storage devices increases the system cost significantly. Moreover, the literature review found that using optimization methods with high convergence rates and accuracy helped to identify the components' sizing optimally and reduce the water extraction costs. Given that the optimization methods work well in some optimization problems but may not be suited for implementation in other problems' solutions, today there is still a need to use more powerful optimization methods [25,26]. A summary of the literature review is presented in Table 1.

Table 1. Summary of literature review.

Ref.	Configuration	Objective Function	Reliability	Enhanced Optimizer
[11]	PV+MP+WST+Battery	Energy efficiency	✗	✗
[12]	PV+MP+WST	LCC	✗	✗
[13]	PV+MP+Battery	Net present cost	✓	✗
[14]	PV+MP+Battery	Environmental impacts	✗	✗
[15]	PV+MP	System efficiency	✗	✓
[16]	PV+MP+Battery	Net present cost	✓	✗
[17]	PV+MP+Battery	Levelized Cost of Energy	✓	✗
[18]	PV+MP+Diesel	Cost of Energy	✗	✗
[19]	PV+MP+Battery	Peak shaving	✗	✗
[20]	PV+MP+Diesel	LCC	✓	✗
[21]	PV+MP+Battery+Diesel	Cost of Energy	✗	✗
[22]	PV+MP+Battery	Net present cost	✓	✗
[23]	PV+WT+Battery	Cost of Energy	✓	✗

* MP: Motor pump, WT: Wind turbine, ✗ refers to not included and ✓ refers to included.

In this paper, the sizing of a PV water-pumping system with a water storage tank (WST) is performed to minimize the life cycle cost (LCC) and satisfying a reliability constraint regarding the probability of interrupted water (pIW) for remote area application considering real regional data. Decision variables include the number of PV arrays and WSTs, and these variables are determined using an enhanced artificial rabbits optimization (EARO) method. The PV arrays are used to supply the electrical energy needed by the motor pump. The required electrical power of the motor pump is optimized. The conventional ARO [27] is enhanced with a nonlinear inertia weight reduction strategy [28] to remove premature convergence and enhance the ARO by preventing local optimum trapping. The capability of EARO to solve the problem is compared with those of traditional ARO and PSO. The impact of several important factors on the system design has also been investigated. Highlights of the research are as follows:

- Sizing of a photovoltaic water-pumping system for the Gorgan region in Iran;
- Sizing framework considering reliability/cost assessment;
- Using a new enhanced artificial rabbits optimization (EARO) method with a nonlinear inertia weight reduction strategy;
- Considerable effect of changes in irradiance and water extraction height of the system sizing;
- Superior performance of EARO compared with conventional ARO and PSO.

The PV water pump system is mathematically modeled in Section 2. Section 3 describes the method of determining the system size considering economic and technical indices. In Section 4, the EARO method and its processes for solving the problem are demonstrated. Section 5 presents the simulation results of sizing the PV pumping system and the sensitivity analysis results, and finally, Section 6 concludes the research findings.

2. Modeling of PV Pumping System

Electrical water pumps for drinking applications represent an important field of reliable PV systems. As shown in Figure 1, these systems typically include a PV generator, a water tank, and a DC pump.

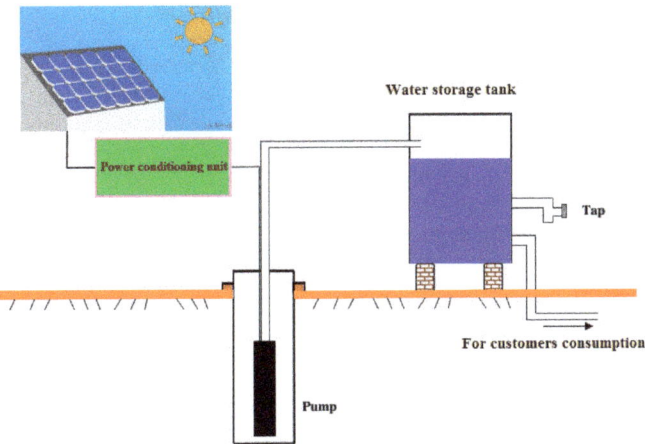

Figure 1. Schematic of the PV water-pumping system.

In the studied system, a water storage tank (WST) is used instead of a battery bank. In systems involving batteries, a lack of power for the electric pump is counteracted by discharging the battery, while in a PV pump system equipped with a WST, the tank plays the role of a battery through the management of water consumption. The operation of the proposed system is as follows:

- If the amount of pumped water is more than the required water at that hour, the excess water is saved in the WST, and the new state is determined when the tank is full. In this case, the amount of remaining water is not stored.
- If the amount of water demanded per hour is less than the amount of pumped water, the WST is applied to fully satisfy the water demand. The new charge status of the WST is determined. If the WST is depleted, the shortage is expressed as the probability of interrupted water.

2.1. PV Model

The PV module and array power based on solar radiation and temperature is computed by [7,11]:

$$p_{PV}(t) = P_{PV,Rated} \times \eta_{MPPT} \times \frac{\alpha_{PV}(t)}{\alpha_{PV,Ref}} \times (1 + \mu \times (\xi_c(t) - \xi_{ref})) \quad (1)$$

$$\xi_c(t) = \xi_a(t) + (\frac{\chi - 20}{800}) \times \alpha_{PV}(t) \quad (2)$$

$$P_{PV}(t) = N_{PV} \times p_{PV}(t) \quad (3)$$

where, $p_{PV}(t)$ is the PV module-generated power at time t, $P_{PV,Rated}$ is the nominal power of the PV module, η_{MPPT} is the efficiency of the PV maximum power point tracking (98.83%), $\alpha_{PV}(t)$ and $\alpha_{PV,Ref}$ are the instantaneous irradiance at time t and solar radiation in standard condition (1000 W/m^2), respectively, μ is the temperature coefficient of the PV system (-3.7×10^{-3} (1/°C)), ξ_{ref} indicates the temperature of the module at time t (°C), ξ_{ref} refers to the reference temperature of the PV system, χ is the nominal operating cell temperature (°C), N_{PV} refers to the PV module number, and $P_{PV}(t)$ is the power of the PV array.

MPPT of PV System

In this study, the maximum power point tracking (MPPT) of the PV system is based on a meta-heuristic algorithm to maximize the output power of the PV system and maximize PPV (d) by optimizing the duty cycle (d) of the DC/DC converter. The duty cycle range of the converter is $d_{min} < d < d_{max}$, where d_{min} and d_{max} respectively represent the minimum and maximum values of the duty cycle, i.e., 0 and 1. The EARO method is used as a direct control method to optimally adjust the duty cycle of the DC/DC converter of the PV system and reduce the steady-state fluctuations of the system. First, the EARO algorithm information is entered, including the number of the population (here 10) and the maximum iterations (here 100), and the minimum and maximum duty cycle intervals are also applied. For each population, the duty cycle algorithm is randomly selected in its allowed range by the EARO method, and voltage, current, and, as a result, PV power are calculated for it. The member of the population corresponding to the previously obtained best photovoltaic power is selected as the best result of the algorithm. Then, the population set of the algorithm is updated. For the updated population (selection of new cycles), the objective function, i.e., photovoltaic power, is calculated. The best member of the population with the maximum previously obtained power is selected as the representative of the population. If the solution is better compared to the previous value, it will be replaced. If the convergence condition is estimated, which is to achieve the maximum power and execute the maximum iterations of the EARO algorithm, by determining the optimal duty cycle, the algorithm is stopped; otherwise, the above steps are repeated until the optimal work cycle is determined. In this study, the tracking efficiency of the photovoltaic system is 98.83%.

2.2. Pump Model

The below model is considered for the water flow Q against the input power P of pumping system and the height h [10–12]:

$$P(Q,h) = \alpha(h) \times Q^3 + \beta(h) \times Q^2 + \Omega(h) \times Q + \phi(h) \quad (4)$$

where the coefficients depend on the water height and are defined as follows [12]:

$$\alpha(h) = \alpha_0 + \alpha_1 \times h + \alpha_2 \times h^2 + \alpha_3 \times h^3 \quad (5)$$

$$\beta(h) = \beta_0 + \beta_1 \times h + \beta_2 \times h^2 + \beta_3 \times h^3 \quad (6)$$

$$\Omega(h) = \Omega_0 + \Omega_1 \times h + \Omega_2 \times h^2 + \Omega_3 \times h^3 \quad (7)$$

$$\phi(h) = \phi_0 + \phi_1 \times h + \phi_2 \times h^2 + \phi_3 \times h^3 \quad (8)$$

The Q corresponding to the P is determined by Equation (9), with $\gamma > P(Q)$. In the iteration k, Q can be presented as [12]:

$$Q_k = Q_{k-1} - \frac{\Im(Q_{k-1})}{\Im'(Q_{k-1})} \quad (9)$$

where

$$\Im(Q_{k-1}) = \alpha \times Q^3_{k-1} + \beta \times Q^2_{k-1} + \delta \times Q_{k-1} + \gamma - P(Q_{k-1}) \quad (10)$$

where $\Im'(Q_{k-1})$ is derived from $\Im(Q_{k-1})$.

2.3. WST Model

The size of the WST is found to meet the water requirement for a period when there is no energy source; this period is called system adequacy. Depending on the production of the PV array and the total load demand, the water charge state (WCHS) can be computed as follows [7,11]:

- Charge of WST

$$WCHS(t) = WCHS(t-1) + (P_{PV}(t) - P_{WD}(t)/\eta_{Inv}) \times \Delta t \times \eta_{WTS} \quad (11)$$

- Discharge of WST

$$WCHS(t) = WCHS(t-1) - (P_{WD}(t)/\eta_{Inv} - P_{PV}(t)) \times \Delta t \tag{12}$$

where $WCHS(t)$ and $WCHS(t-1)$ refer to the charge condition of the WST (Wh) at time t and (t − 1), respectively; $P_{PV}(t)$ refers to the power generated using the PV array (W); $P_{WD}(t)$ refers to the hydraulic demand at t (W); and η_{WTS} is the tank efficiency in the charge state (equal to 1). Also, $WCHS(t) = N_{WTS} \times VOL_{WTS}$ is considered for water storage demanded, in which N_{WTS} is the WST number, and VOL_{WTS} shows the WST volume (m^3). Each WST can transmit 1 m^3 to customers in one hour. The charge state of the WST is limited by

$$0 \leq WCHS(t) \leq WCHS_{max} \tag{13}$$

3. Sizing Methodology

For determination of the optimal component size, the system is optimized to meet the water demand and evaluated with economic and technological indices. In this study, the proposed approach to system evaluation is based on the two concepts of project lifetime cost for economic evaluation (LCC) and a reliability index called the probability of interrupted water (p_{IW}). Combining the system with the lowest LCC and the best p_{IW} is the optimal combination and provides the desired reliability.

Currently, excessive consumption of non-renewable energies such as coal, gas, and oil for the traditional production of electrical energy has caused serious environmental threats, including climate change and increased air pollution. For this reason, adopting an approach towards using energy sources that are compatible with the environment's health is necessary. After that, it is necessary to pay attention to the environmental costs of electricity production, which can have destructive effects on natural resources. In the energy sector, external costs that are imposed on society and the environment cause water and air pollution, reduction of freshwater resources, etc. By employing new and renewable energies, we can help reduce these costs. Among the mentioned renewable energies, solar energy can be mentioned as an endless source of energy that solves many problems in the field of energy and environment. A circular economy is a closed-loop system in which a product is not thrown away after use. Thus, there will be no waste at the end of the production cycle, making it more efficient in the long run. By implementing circular economy models, it is possible to significantly reduce the amount of production waste and increase economic growth by creating new industries around better management of production waste. In this research, instead of using energy sources based on fossil fuels with waste, photovoltaic renewable energy sources have been used to supply the required power of the water pump system to minimize the environmental effects and waste resulting from it, reduce the cost of the energy project over its lifetime, and reduce CO_2 emissions by using a circular economy.

3.1. Objective Function

LCC includes the costs of components such as PV arrays, motor pump sets, tanks, and inverters. According to the system under study, the cost of LCC is expressed as the investment cost (C_{cap}), the maintenance cost ($C_{O\&M}$), and the cost of interrupted water (C_{WRW}). It should be noted that the LCC only considers the costs incurred during the project's lifetime, and the impact of end-of-life system components' value (considering circular economy) has not been considered. Therefore, LCC is defined as follows [1,5,10–12]:

$$\text{Minimize } LCC = C_{cap} + C_{O\&M} + C_{IW} + C_{WRW} \tag{14}$$

- Initial Investment Cost

The component capital cost includes the cost of components, the cost of construction, and the cost of installation of the components. The construction and installation costs are considered equal to 40% of the PV array cost and equal to 20% of the cost of the motor pump set. The capital cost (C_{cap}) is expressed as follows:

$$C_{cap} = N_{PV} \cdot C_{Unit,PV} + N_{WST} \cdot C_{Unit,WST} \\ + N_{pump} \cdot C_{Unit,pump} + N_{Inv} \cdot C_{Inv} \quad (15)$$

where N_{PV} and $C_{Unit,PV}$ refer to the PV number and the PV unit cost, respectively. N_{WST} and $C_{Unit,WST}$ are the tank number and the tank unit cost, respectively. N_{pump} and $C_{Unit,pump}$ are the pump motor number and the unit cost, respectively, and N_{Inv} and C_{Inv} are the inverter number and unit cost, respectively.

- *Operation and Maintenance Cost*

The cost of operation and maintenance ($C_{O\&M}$) is defined by [11–13]

$$C_{O\&M} = \begin{cases} C_{(O\&M)_1} \times \left[\frac{1+\tau_1}{\psi-\tau_1}\right] \times \left[1 - \frac{1+\tau_1}{1-\psi}\right]^\kappa, & \text{for } \psi \neq \tau_1 \\ C_{(O\&M)_1} \times \kappa, & \text{for } \psi = \tau_1 \end{cases} \quad (16)$$

where τ_1 is the inflation rate, ψ refers to the annual interest rate, and κ indicates the system lifetime. $C_{(O\&M)0}$ indicates $C_{O\&M}$ in the first year, which can be defined in terms of ∂ as part of the capital cost (C_{cap}). $C_{(O\&M)1}$ is computed as follows:

$$C_{(O\&M)_1} = \partial \times C_{cap} \quad (17)$$

- *Cost of Water Reliability Weakness*

The water reliability weakness of customers is equal to the amount of water not supplied by the system multiplied by the cost per liter of water, which is defined as follows:

$$C_{WRW} = \sum_{t=1}^{T}(P_{WD}(t) \times p_{IW} \times C_{IW}) \quad (18)$$

where p_{IW} is the probability of interrupted water, and C_{IW} is the cost of not supplying each liter of water demanded by customers in terms of U.S. dollars.

3.2. Reliability Constraint

The p_{IW} index is a technical index for finding the size of PV water-pumping system components for a system equipped with a water tank. When the pumped water amount is more than the amount of water consumed by the customers, the excess water is saved in the WST, and the amount of water stored in hour t is obtained from Equation (11). In case that the amount of water required by the customers is more than the amount of pumped water, the water shortage of the customers will be compensated by the discharge of the WST, in which case the amount of water in the tank at hour t is defined by Equation (12). However, if the amount of water in the reservoir is not able to fully meet the water demand of the customers, then the amount of interrupted water per hour ($IW(t)$) is defined as follows [5,7,9]:

$$IW(t) = P_{WD}(t) \times \Delta t - (P_{PV}(t) \times \Delta t + WCHS(t-1)) \quad (19)$$

$$p_{IW} = \frac{\sum_{t=1}^{T} IW(t)}{\sum_{t=1}^{T} P_{WD}(t)} \quad (20)$$

where the value of p_{IW} is between 0 and 1. A value of 0 means that all the water customers demand has been supplied, and a value of 1 means that the total water demand has not been supplied.

3.3. Proposed Optimizer (EARO)

The sizing of the PV pump system with the aim of minimizing LCC and p_{IW} satisfaction using enhanced artificial rabbits optimization (EARO) is presented. The optimization variables include the PV array number and the WST number determined using the EARO method.

3.3.1. Inspiration

The ARO algorithm is modeled based on the survival strategies of rabbits in the wild [27]. The ARO algorithm uses strategies of foraging and hiding and reduces their energy to exchange among these strategies to solve an optimization problem.

3.3.2. Searching for Shortcut Food (Exploration)

Rabbits tend to look for food far away, so they are not interested in looking for food in nearby places; in other words, they are not satisfied with the grass in their area and search far away, which is called detour foraging. In the ARO algorithm, each rabbit has a number d of hiding places in its own area. Rabbits randomly consider the position of other rabbits to search for food. In this way, rabbits may gather around a food source to obtain enough food while searching for food. So, detour foraging means that each searcher is interested in updating its position towards each other searcher by adding a disturbance. The detour foraging model is presented as follows [27]:

$$\vec{v}_i(t+1) = \vec{x}_j(t) + R.(\vec{x}_i(t) - \vec{x}_j(t)) + round(0.5.(0.05 + r_1)).n_1, \tag{21}$$

$i, j = 1, ..., n$ and $j \neq i$

$$R = L.c \tag{22}$$

$$L = (e - e^{(\frac{t-1}{T})^2}).\sin(2\pi r_2) \tag{23}$$

$$c(k) = \begin{cases} 1 & if\ k == g(l),\ k = 1,...,d\ and\ l = 1,...,\lceil r_3.d \rceil \\ 0 & else \end{cases} \tag{24}$$

$$g = rand\ perm(d) \tag{25}$$

$$n_1 \sim N(0,1) \tag{26}$$

where $\vec{v}_i(t+1)$ represents the ith candidate rabbit position at time $t+1$, $\vec{x}_i(t)$ refers to the ith rabbit position at time t, n represents a rabbit population size, d is the number of dimensions of the problem, and T represents the maximum iterations number, $\lceil . \rceil$ indicates the ceiling function, *rand perm* represents the random order of integers 1 to d, r_1 to r_3 represents three random numbers in the range (1, 0), L is the length of the run, and n_1 is bound to the normal distribution.

3.3.3. Random Hiding (Exploitation)

In each iteration, a rabbit generates a number of hiding places (d) around each dimension of the search space and considers one of those hiding places to hide. In this way, it reduces the possibility of being hunted. The jth hiding place of the ith rabbit is defined as follows [27]:

$$\vec{b}_{i,j}(t) = \vec{x}_i(t) + H.g.\vec{x}_i(t),\ i = 1,...,n\ and\ j = 1,...,d \tag{27}$$

$$H = \frac{T-t+1}{T}.r_4 \tag{28}$$

$$n_2 \sim N(0,1) \tag{29}$$

$$g(k) = \begin{cases} 1 & if\ k == j,\ k = 1,...,d \\ 0 & else \end{cases} \tag{30}$$

During each dimension, d number of hiding places is produced in the neighborhood of a rabbit. H represents the hidden parameter, which goes from 1 to 1/T based on a random disturbance during the repetitions in a linear way.

In order to hide from hunters and not be hunted, rabbits are not interested in choosing one of the hiding places randomly. Random hiding behavior is defined as follows [27]:

$$\vec{v}_i(t+1) = \vec{x}_i(t) + R.(r_4.\vec{b}_{i,r}(t) - \vec{x}_i(t)), \ i = 1,...,n \tag{31}$$

$$g_r(k) = \begin{cases} 1 & if \ k == \lceil r_5.d \rceil, \ k = 1,...,d \\ 0 & else \end{cases} \tag{32}$$

$$\vec{b}_{i,r}(t) = \vec{x}_i(t) + H.g_r.\vec{x}_i(t) \tag{33}$$

where $\vec{b}_{i,r}$ refers to a hideout considered randomly to hide in from d number of hideouts, and r_4 and r_5 represent numbers between 0 and 1, randomly selected. According to the above equations, the i-th searching person tries to update his position with respect to a random hideout considered from the d available hideouts.

The position of the i-th rabbit is updated as follows [27]:

$$\vec{x}_i(t+1) = \begin{cases} \vec{x}_i(t) & f(\vec{x}_i(t)) \leq f(\vec{v}_i(t+1)) \\ \vec{v}_i(t+1) & f(\vec{x}_i(t)) > f(\vec{v}_i(t+1)) \end{cases} \tag{34}$$

3.3.4. Energy Reduction (Transition from Exploration to Exploitation)

In the ARO algorithm, rabbits tend to engage in detour foraging behavior repeatedly, while they engage in random hiding behavior in the later stage of iterations. Therefore, over time, a rabbit's energy decreases. Therefore, the energy factor is presented as follows [27]:

$$A(t) = 4(1 - \frac{t}{T}) \ln \frac{1}{r} \tag{35}$$

where r represents a number between 0 and 1. When A(t) > 1, the rabbit is subject to random exploration, and detour foraging occurs. When A(t) $\leq$ 1, the rabbit is not interested in randomly using its hiding places, and in this condition random hiding occurs. The search structure according to factor A is shown in Figure 2.

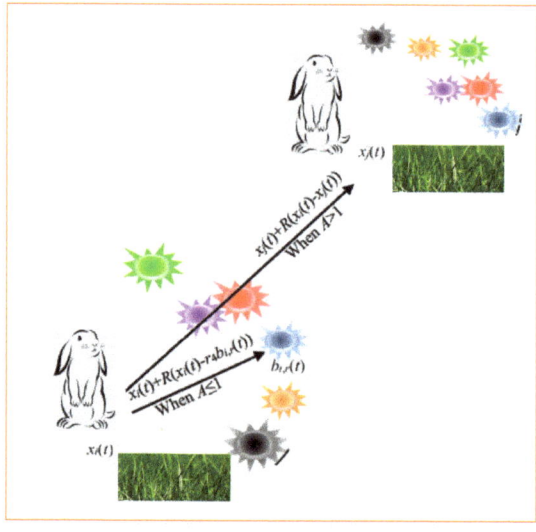

Figure 2. Search structure according to factor A.

Therefore, the ARO algorithm randomly generates a population of rabbits as candidate answers in the search space. In each iteration, a rabbit updates its position relative to a random rabbit from the population or a random rabbit taken from the hiding places. As the repetitions increase, factor A goes through a decreasing process, so that each rabbit in the population is forced to perform a transfer operation. The algorithm is updated until it reaches the convergence criterion of the algorithm to obtain the best response. The pseudo-code of the ARO algorithm is presented in Algorithm 1.

Algorithm 1 Pseudo-Code of ARO

Initiate a rabbits set $_i$ and calculate the fitness (Fit$_i$) and X$_{best}$
While the convergence criteria is not met do
  For each individual X$_i$ do
  Compute the A operator via Equation (35)
  If A > 1
    Select a rabbit from individuals randomly
  Compute R via Equations (23)–(26)
  Implement detour foraging using Equation (21)
  Compute the Fit$_i$
  Update the present individual position via Equation (34)
  Else
    Produce d burrows and pick hiding randomly via Equation (33)
  Implement the hiding randomly via Equation (31)
  Compute the Fit$_i$
  Update the individual position via Equation (34)
  End If
  Update the best solution determined X$_{best}$
  End For
End While
Return X$_{best}$

3.3.5. Overview of EARO

In the optimization process, the optimal selection of the inertia weight is very effective in solving the problem. A big value of the inertia weight makes the algorithm perform better in global search. However, a small value of the inertia weight makes the algorithm perform better in local search. In the ARO algorithm, the value of the inertia weight is chosen as equal to one. Therefore, to strengthen the performance of the algorithm in preventing premature convergence, it is better to consider the inertia weight dynamics to accelerate the achievement of the global optimum. In solving the optimization problem based on the ARO algorithm, to improve the convergence and prevent premature convergence, the nonlinear inertia weight reduction method [28] is applied as follows:

$$IW(t) = IW_L + (0.5(1 + \cos(\frac{\pi t}{T})))^{\psi} \times (IW_U - IW_L) \qquad (36)$$

where IW_L and IW_U refer to lower and upper amounts of IW, respectively (ψ = 10 [28]).
According to Equation (36), Equations (34) and (31) are updated:

$$\vec{x}_i(t+1) = \begin{cases} IW(t).\vec{x}_i(t) & f(\vec{x}_i(t)) \leq f(\vec{v}_i(t+1)) \\ \vec{v}_i(t+1) & f(\vec{x}_i(t)) > f(\vec{v}_i(t+1)) \end{cases} \qquad (37)$$

$$\vec{v}_i(t+1) = IW(t).\vec{x}_i(t) + R.(r_4.\vec{b}_{i,r}(t) - IW(t).\vec{x}_i(t)), \ i = 1, ..., n \qquad (38)$$

3.3.6. The EARO Implementation

The PV water pump system sizing is developed using the EARO method. The optimization variables are optimally determined by EARO. The algorithm iterations number is considered to be 100, and the population number is selected as 50 according to the trial-and-error method and the authors' experience. The flowchart of the EARO to implement the problem is depicted in Figure 3. The steps of EARO performing sizing solving are presented as follows:

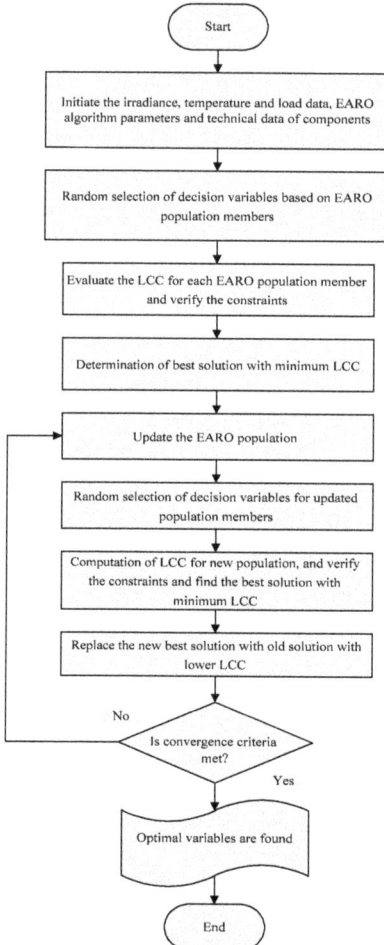

Figure 3. Flowchart of the EARO implementation for sizing problem solving.

Step 1: Insert the data related to the irradiance and temperature considering real regional data and produce the initial population for $x = [N_{PV}, N_{Tank}]$ as a decision variables vector.

Step 2: The variables set are identified randomly after considering the search space for members of the population.

Step 3: The LCC is calculated for each variable set via ARO, and the corresponding lowest LCC is determined as the best member of the population.

Step 4: The population position is updated, and then the LCC is computed for the new population. If the member corresponding to the best value of the cost is better than the LCC gained in Step 4, it replaces the older value.

Step 5: In the enhanced ARO phase, the algorithm position is updated with a nonlinear inertia weight reduction strategy, and then the OF is computed for the updated population.

Step 6: The optimal variable set is replaced by the best set obtained in Step 5 if it has better results than the cost achieved in Step 4.

Step 7: The convergence criteria are evaluated. If these criteria are met, go to Step 8; otherwise, return to Step 4.

Step 8: Stop the EARO and print the best variables.

4. Simulation Results

In this paper, the sizing of the PV water pump system is performed by considering the possibility of water supply to the customers using EARO considering real regional data. The simulation of the studied system is performed in different scenarios of water height. The simulation results include the optimal components capacity and system cost for the full supply of customer demand. It should be noted that in order to confirm the efficiency of the EARO method, the problems with traditional ARO and PSO, which have shown their ability in recent years to optimize power engineering problems, have been evaluated, and the results are compared. Finally, the effect of some system parameters on optimization is evaluated.

4.1. Sizing Parameters

The proposed framework is applied for sizing the PV water pump system separate from the network with the aim of providing drinking water to customers. The cost data of the system are presented in Table 2. The efficiency of the inverter is considered to be 95% [11–13]. The technical data of the PV array and pump motor are given in Tables 3 and 4, respectively.

Table 2. The cost data of the system [11–13].

Component	Capital Cost (U.S. Dollars)	Maintenance Cost (U.S. Dollars)	τ_1 (%)	ψ	κ
PV Array	294.91	2.95	4	8	25
Pump Motor	210	2.1	4	8	10
Water Tank	42,000	420	4	8	25
Inverter	50.057	0.5	–	–	10

Table 3. Technical data of PV array [12,13].

P_{max} (W)	q	k	n	Rs	In	Vnoc	Isc	Vmax	Imax
55	1.6×10^{-19}	1.38×10^{-23}	1.5	0.012	6.5	21.7	3.4	17.4	3.16

Table 4. Technical characteristics of motor pump [12,13].

Motor Type	Nominal Power (W)	Voltage Range (V)	Maximum Current (A)
DC	400	0–48	13

The amount of water consumed in each hour during a 24 h period is equal to 24.10 cubic meters. Each water tank has a maximum capacity of 1 m^3, and the capacity of each at the beginning of the program is considered 0.25 m^3. The values of the coefficients of Equations (4)–(8) are given in Table 5. The changes in radiation as well as temperature for a full day and night are shown in Figures 4 and 5. The solar radiation and temperature are of the Gorgan area (latitude 37°24′ and longitude 55°15′) in Iran.

Table 5. Coefficients value of motor pump [12,13].

Coefficient	Value	Coefficient	Value
$\alpha(h)$	$\alpha_0 = -214.42$ $\alpha_1 = 108.43$ $\alpha_2 = -9.9276$ $\alpha_3 = 0.2201$	$\Omega(h)$	$\Omega_0 = -152.82$ $\Omega_1 = 72.369$ $\Omega_2 = -6.5469$ $\Omega_3 = 0.1499$
$\beta(h)$	$\beta_0 = 470.5$ $\beta_1 = -157.19$ $\beta_2 = 15.038$ $\beta_3 = -0.339$	$\phi(h)$	$\Phi_0 = 16.79$ $\Phi_1 = -2.8140$ $\Phi_2 = 0.7072$ $\Phi_3 = -0.0158$

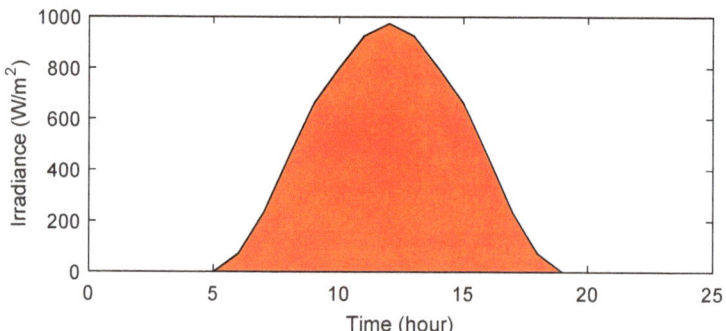

Figure 4. Irradiance during a day for Gorgan region.

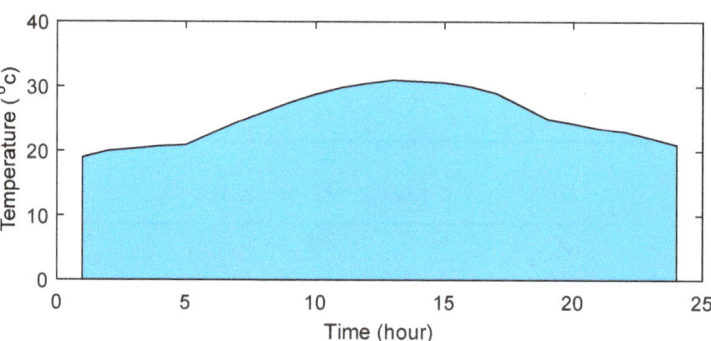

Figure 5. Ambient temperature during a day for Gorgan region.

The cost of the photovoltaic array from the investment point of view is USD 249.91 per kilowatt, and the cost from the maintenance point of view is 1% of that, equal to USD 2.49 per hour. The purchase cost of each water tank is USD 42,000, and the maintenance cost is USD 420 [12,13]. The sizing of the PV water pump system is considered using EARO, taking into account the possibility of non-supply of water to p_{IW} customers. System costs are presented as an economic index, and p_{IW} as a technical index for the system's ability to supply water to customers. The simulations are implemented in two scenarios as follows:

Scenario #1: Sizing of a PV water pump system for a water extraction height of 5 m with LCC minimization and satisfying the p_{IW} constraint;

Scenario #2: Sizing of a PV water pump system for a water extraction height of 10 m with LCC minimization and satisfying the p_{IW} constraint.

4.2. Simulation Results of the First Scenario

Based on Scenario 1, the solar pumping system is designed considering a water height of 5 m. The convergence process of the different algorithms is shown in Figure 6. This figure shows that the system costs using the EARO method are lower than for other methods. In Table 6, the results are given. Based on Scenario 1, the PV array and water tank numbers are set to 5 and 7, respectively. The system cost is 0.2955 M$. Using EARO, the LCC is lower than for the ARO and PSO methods, and the system was able to fully supply the water consumed by the customers. However, when using the ARO and PSO methods, system costs were equal to 3119 and 0.3078 M$, respectively, and on the other hand, these methods did not provide 4.18 and 4.16% of the water required by customers, respectively. Therefore, the obtained results confirm the better capability of EARO in terms of achieving the lowest cost and the highest reliability of water supply to customers.

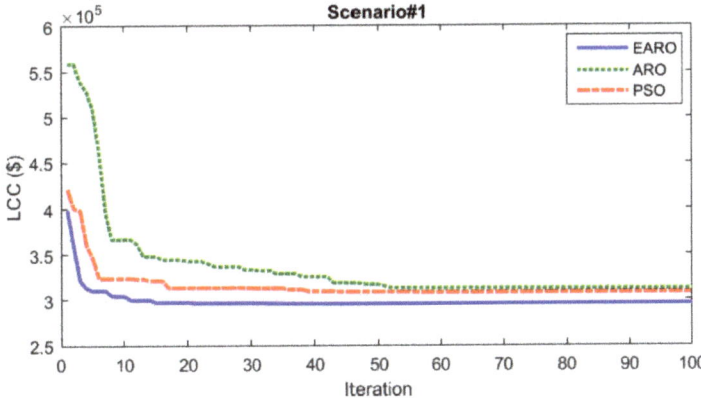

Figure 6. The EARO, ARO, and PSO convergence curves in the first scenario.

Table 6. System optimization results in the first scenario.

Method/Parameter	N_{PV}	N_{WST}	LCC (M$)	p_{IW} (%)
EARO	5	7	0.2955	0
ARO	5	6	0.3119	4.18
PSO	4	6	0.3078	4.16

Table 7 shows the sensitivity of p_{IW} compared to the number of PV arrays based on the EARO method; in other words, it can be seen that with the increase in the PV number, unsupplied water for customers decreases, and they are supplied with higher reliability.

Table 7. p_{IW} sensitivity to the number of PV arrays in the first scenario.

N_{PV}	1	2	3	4	5
p_{IW} (%)	54.16	29.16	16.66	4.16	0

Figure 7 illustrates the variations in p_{IW} compared to the PV number; with the increase in the PV number, the water demand of the customers is met with a higher level of reliability.

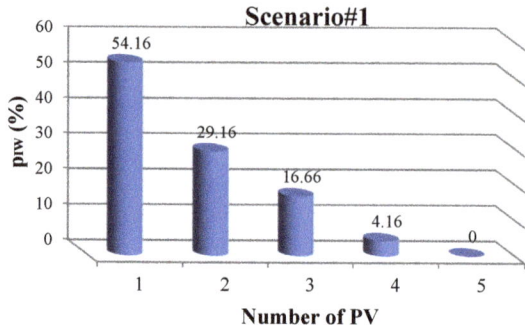

Figure 7. Change compared to PV number in Scenario 1.

4.3. Results of Scenario 2

The results Scenario 2, with height of 10 m, are given in Table 8. Figure 8 shows the convergence process of the different algorithms and demonstrates that EAROs obtain the optimal component sizing with lowest LCC.

Table 8. System optimization results in the second scenario.

Method/Parameter	N_{PV}	N_{WST}	LCC (M$)	p_{IW} (%)
EARO	6	7	0.2993	0
ARO	6	6	0.3157	4.47
PSO	6	6	0.3134	4.23

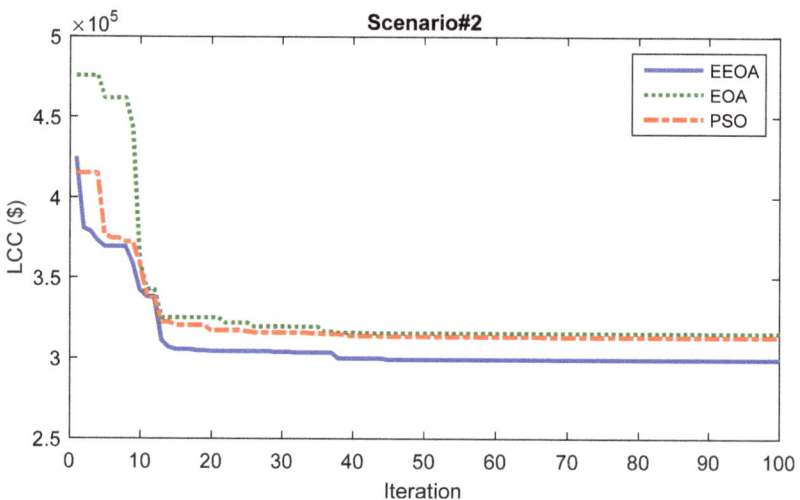

Figure 8. The EARO, ARO, and PSO convergence curves in the second scenario.

In the second scenario, the PV array number and tank number of the EARO method are 6 and 7, respectively. The LCC obtained using EARO is 0.2993 M$. For ARO and PSO, N_{PV} and N_{WST} are equal to 6. Using ARO and PSO, the LCC is 0.3157 M$ and 0.3134 M$, respectively. Therefore, according to the results, it can be said that the EARO method has optimized the system at a lower cost and also has provided the total water demanded by customers with p_{IW} equal to zero. On the other hand, the ARO and PSO methods did not provide 4.47 and 4.23% of the total water required by the customers, respectively, which increased the cost by imposing the cost of WNS by these methods. Therefore, the proposed method of EARO is a cost-effective and reliable method compared to other methods.

In Table 9 and Figure 9, the variations in p_{IW} compared to the PV array number are plotted; it is clear that the p_{IW} value decreases as the PV array number increases.

Table 9. Sensitivity to the number of PV arrays in the second scenario.

N_{PV}	1	2	3	4	5	6
p_{IW} (%)	83.33	45.83	29.16	20.83	8.33	0

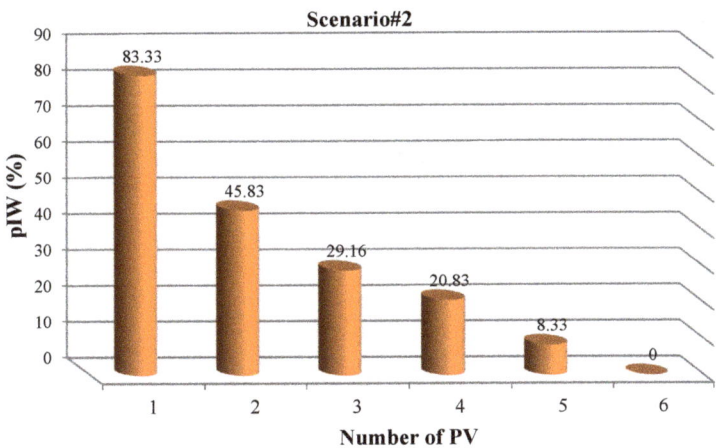

Figure 9. Changes relative to PV number in the Scenario 2.

4.4. Comparison of Scenario Results

The results of Scenarios 1 and 2 obtained via EARO are compared with each other. The required power of the motor pump is supplied by photovoltaic sources, and the required power of the motor pump is optimized. When the height of water extraction increases, in this case, the motor pump needs more power to extract the same volume of water in the base state (base height of 5 m). Based on Table 10, it is clear that with the increase of water height from 5 to 10 m and thus the increase in the height of water extraction, the optimization program has considered more photovoltaic panels or more photovoltaic power to supply the required power of the motor pump. On the other hand, the optimization program has guaranteed the complete and reliable supply of 100% (with p_{IW} = 0) of the water demand of the customers under the conditions of changing the height of water extraction, satisfying the water reliability index. As a result, it can be seen that the LCC of the system has increased with the increase in the height of water extraction due to the increase in photovoltaic power required by the motor pump.

Table 10. Results of Scenarios 1 and 2.

Method/Parameter	N_{PV}	N_{WST}	LCC (M$)	p_{IW} (%)
Scenario #1	5	7	0.2955	0
Scenario #2	6	7	0.2993	0

In the following, the effects of some effective technical parameters such as changes in the intensity of PV radiation and changes in temperature and the water demand of customers on system optimization—in other words, on the optimal capacity of components and system costs—have been evaluated. It should be noted that the simulations in this section have been performed at a height of 10 m.

4.5. Sensitivity Analysis

4.5.1. Effect of Height Changes

In Table 11 and Figure 10, the effect of increasing the height of water extraction on the number of photovoltaic panels is evaluated. With the increase in the height of water extraction, the required power of the motor pump has increased the number of photovoltaic panels used.

Table 11. Changes in the number of PV arrays relative to water extraction height.

Height (m)	5	10	15
N_{PV}	5	6	7

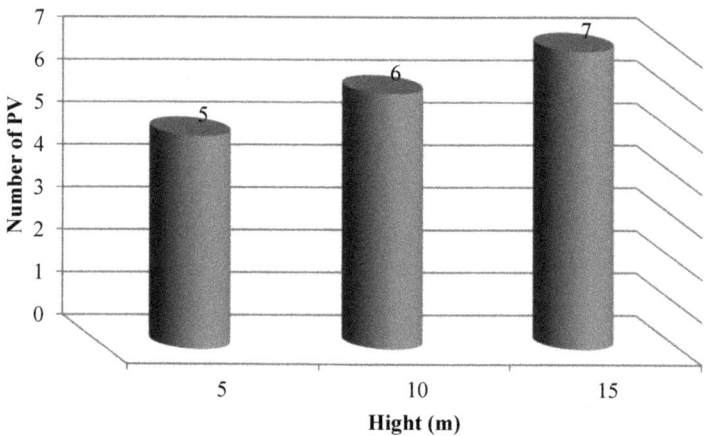

Figure 10. Curve of changes in the number of PV arrays relative to water extraction height.

4.5.2. Effect of Changes in the Number of Tanks

The results of changes in the number of water storage tanks in relation to the height are presented in Table 12. In a 24 h period, the water demand of the customers is completely satisfied. Therefore, the optimization program under the conditions of changes in the height of water extraction has tried to fully supply the same volume of water required by the customers as in the base height scenario by increasing the required power of the motor pump. So, with the increase in the height of water extraction, the number of water tanks required by the customers has remained unchanged.

Table 12. Results of scenarios 1 and 2.

Height (m)	5	10	15
N_{PV}	7	7	7

4.5.3. Effect of Changes in Irradiance

The irradiance is decreased by 25% and increased by 25%, and the results are compared with a height of 10 m. The results of irradiance changes are presented in Table 13. As we know, with the increase of irradiance (compared to the basic state), the output power of a photovoltaic panel increases. Therefore, in conditions of increased irradiance, the amount of extracted power is higher. As can be seen, with the 25% increase in irradiance, the number of photovoltaic panels has decreased from seven to five. In other words, the optimization program considers the power produced by five photovoltaic panels in the condition of a 25% increase in irradiance equivalent to the power produced by six photovoltaic panels in the basic irradiance condition. Also, the optimization program has selected the power produced by nine photovoltaic panels in the conditions of 25% reduction of irradiance, equivalent to the power generated by six photovoltaic panels in the basic irradiance. On the other hand, it is clear that with the increase (decrease) of the number of photovoltaic panels in the conditions of 25% reduction (increase) of irradiance, the cost of the system has increased (decreased). Also, the results show that the decrease in irradiance has weakened the reliability of the customers' water supply.

Table 13. Effect of changes in PV radiation intensity on system optimization.

Irradiance/Parameter	N_{PV}	N_{WST}	LCC (M$)	p_{IW} (%)
25% decrease	9	7	0.3326	2.67
Nominal	6	7	0.2993	0
25% increase	5	7	0.2922	0

4.5.4. Effect of Temperature Changes

The effect of temperature variations (10% decrease and increase compared to the baseline) on the optimal capacity of system components and the cost to meet the total water demand is presented in Table 14. An increase in temperature causes very small changes in the output power of photovoltaic panels. This effect is mostly considered in low-power applications of photovoltaic modules, and it is ignored in high-power applications. Based on the obtained results, it is clear that 10% changes in temperature did not affect the production of photovoltaic power and ultimately the cost and water supply of the customers.

Table 14. Effect of temperature changes on system optimization.

Temperature/Parameter	N_{PV}	N_{WST}	LCC (M$)	p_{IW} (%)
10% decrease	6	7	0.2993	0
Nominal	6	7	0.2993	0
10% increase	6	7	0.2993	0

4.5.5. Effect of Water Demand Changes

The effect of changes in the percentage of water consumed by customers on system optimization has been evaluated in this section. The results are given in Table 15. As the water demand of the customers increases, the pump motor needs more power to extract more water and vice versa. Therefore, more photovoltaic panels are needed to supply the required power for the motor pump. Therefore, the increase in the water demand of the customers increases the cost. Similarly, reducing the water demand of the customers will reduce the number of photovoltaic panels and reduce the cost. On the other hand, in the conditions of increased water needs, the optimization program may not be able to fully meet the needs of the customers.

Table 15. Optimization results from water demand changes.

Water Demand/Parameter	N_{PV}	N_{WST}	LCC (M$)	p_{IW} (%)
20% decrease	4	7	0.2937	0
Nominal	6	7	0.2993	0
20% increase	8	7	0.3289	3.4

4.5.6. Effect of Considering Replacement Cost

In this section, the effect of considering the replacement cost of system components according to the project lifetime of the system (25 years) and operation period of the equipment is presented in the solutions of Scenarios 1 and 2 according to Table 16. In this condition the replacement cost considered for the motor pump and also inverter is added to Equation (14). So, the replacement cost is defined as $C_R = C_{rep} \times SFF(i, y_{rep})$, and $SFF = i/(1+i)^y - 1$ (C_{rep} is the replacement cost of a motor pump and inverter (U.S. dollars), i is the annual real interest rate, and y_{rep} is the lifetime of the motor pump and inverter). In this case, based on Table 2, the lifespan of the project is considered to be 25 years according to the lifespan of the photovoltaic sources and the water tank, and the useful lifespan of the pump motor and inverter is also considered to be 10 years. In this situation, the cost of replacing the equipment is considered similar to the investment cost of each piece of equipment, which is naturally included in the cost function (LCC) of Equation (14) based on the annual operating period of this study, and the simulation results using the EARO optimization method are presented in Table 16. The LCC for Scenario 1 and Scenario 2 is obtained 0.3073 M$ and 0.3111 M$, respectively.

Table 16. Results of effect of considering replacement cost for Scenarios 1 and 2.

Method/Parameter	N_{PV}	N_{WST}	LCC (M$)	p_{IW} (%)
Scenario #1 With replacement cost	5	7	0.3073	0
Scenario #2 With replacement cost	6	7	0.3111	0

5. Comparison of the Results

In battery-based water-pumping systems, the water demand is met online. By receiving radiation, PV resources supply the electrical energy of the motor pump. In other words, the water-related electricity needed by customers to extract water is delivered to the pump motor, and the extra PV power is saved in the batteries for hours without PV radiation and thus without PV electricity. The battery is discharged, and the pump motor is provided with power for the extraction of water for delivery to customers. In other words, the warehouse is electric. In a tank-based photovoltaic water pump, the battery is deleted, and the water tank replaces it. For the method of EARO, according to the annual cost of the system for 10 m of irrigation equal to 0.3073 M$ and the consumption of 10 m^3 of water by the customers every day, the cost of the water supplied to subscribers per liter (CWS) is equal to USD 0.0841. The numerical comparison of the CWS is given in Table 17. It can be seen that the EARO method with the PV pump system based on the water storage tank has obtained a lower cost per liter compared to other methods based on the battery storage system [29,30]. The configuration provided along with the proposed methodology is simpler compared to other methods, the life of the storage system based on the water storage tank is longer, and the cost of the water supplied to the subscribers is lower.

Table 17. Comparison of the EARO performance with previous studies.

Algorithm	System	CWS ($)
EARO	PV/Pump/WST	0.0841
[29]	PV/Pump/Battery	0.375
[30]	PV/Pump/Diesel	0.261

6. Conclusions

In this paper, the sizes of PV pumping system components are determined to meet the water demand of customers to minimize LCC and satisfy a reliability constraint, optimally for the Gorgan region. The sizes of components such as PV arrays and the number of water storage tanks are calculated optimally via EARO. Also, the capability of EARO is compared with those of the traditional ARO and PSO. The LCC values for water extraction heights of 5 m and 10 m are 0.3073 M$ and 0.3111 M$, respectively, using EARO. The results demonstrated that the water-pumping system can supply the customers' demand fully based on continuous provision. The results showed that EARO, in designing the PV water pump system compared to the ARO and PSO methods, has a lower cost with higher reliability. The effect of some important factors on system design is also evaluated. As PV energy size increases, system reliability and cost increase. As the water extraction depth increases, the PV energy required to supply the pump motor as well as the system cost increase. As the irradiance increases, the PV number and consequently the system cost decrease, and vice versa. The results also demonstrated that increasing the temperature does not have a significant effect on system optimization, optimal component sizes, or the reliability index. Also, the proposed method performance was confirmed, compared to the previous methods, to have a lower cost of water extraction per liter. Assessing the uncertainty of PV power generation and water consumption of customers in designing a PV water pump system is suggested for future work.

Author Contributions: A.M.: conceptualization, methodology, software, and writing; A.P.: conceptualization, methodology, software, and writing; M.S.M.: methodology and software; M.B.A.: software, optimization, and writing—original draft; A.Y.A. and M.E.: investigation, supervision, validation, and writing—review & editing. All authors have read and agreed to the published version of the manuscript.

Funding: This research received no external funding.

Data Availability Statement: Not applicable.

Conflicts of Interest: The authors declare no conflict of interest.

References

1. Naderipour, A.; Abdul-Malek, Z.; Nowdeh, S.A.; Kamyab, H.; Ramtin, A.R.; Shahrokhi, S.; Klemeš, J. Comparative evaluation of hybrid photovoltaic, wind, tidal and fuel cell clean system design for different regions with remote application considering cost. *J. Clean. Prod.* **2021**, *283*, 124207. [CrossRef]
2. Arabi, N.S.A.; Saftjani, P.B.; Abdul-Malek, Z.; Mustafa, M.W.B.; Kamyab, H.; Davoudkhani, I.F. Deterministic and probabilistic multi-objective placement and sizing of wind renewable energy sources using improved spotted hyena optimizer. *J. Clean. Prod.* **2021**, *286*, 124941.
3. Jafar-Nowdeh, A.; Babanezhad, M.; Arabi-Nowdeh, S.; Naderipour, A.; Kamyab, H.; Abdul-Malek, Z.; Ramachandaramurthy, V.K. Meta-heuristic matrix moth–flame algorithm for optimal reconfiguration of distribution networks and placement of PV and wind renewable sources considering reliability. *Environ. Technol. Innov.* **2020**, *20*, 101118. [CrossRef]
4. Errouha, M.; Derouich, A.; Motahhir, S.; Zamzoum, O.; El Ouanjli, N.; El Ghzizal, A. Optimization and control of water pumping PV systems using fuzzy logic controller. *Energy Rep.* **2019**, *5*, 853–865. [CrossRef]
5. Nowdeh, S.A.; Ghahnavieh, A.A.; Khanabdal, S.A.H.E.B. *PV/FC/Wind Hybrid System Optimal Sizing Using PSO Modified Algorithm*; Tomul LVIII (LXII), Fasc, 4; Universitatea Tehnică „Gheorghe Asachi" din Iasi: Cluj-Napoca, Romania, 2012.
6. Davoodkhani, F.; Nowdeh, S.A.; Abdelaziz, A.Y.; Mansoori, S.; Nasri, S.; Alijani, M. A new hybrid method based on gray wolf optimizer-crow search algorithm for maximum power point tracking of photovoltaic energy system. In *Modern Maximum Power Point Tracking Techniques for Photovoltaic Energy Systems*; Springer: Cham, Switzerland, 2020; pp. 421–438.
7. Naderipour, A.; Abdul-Malek, Z.; Vahid, M.Z.; Seifabad, Z.M.; Hajivand, M.; Arabi-Nowdeh, S. Optimal, Reliable and Cost-Effective Framework of Photovoltaic-Wind-Battery Energy System Design Considering Outage Concept Using Grey Wolf Optimizer Algorithm—Case Study for Iran. *IEEE Access* **2019**, *7*, 182611–182623. [CrossRef]
8. Abdul-Malek, Z.; Noorden, Z.A.; Davoudkhani, I.F.; Nowdeh, S.A.; Kamyab, H.; Ghiasi, S.M.S. Carrier wave optimization for multi-level photovoltaic system to improvement of power quality in industrial environments based on Salp swarm algorithm. *Environ. Technol. Innov.* **2021**, *21*, 101197.
9. Arabi-Nowdeh, S.; Nasri, S.; Saftjani, P.B.; Naderipour, A.; Abdul-Malek, Z.; Kamyab, H.; Jafar-Nowdeh, A. Multi-criteria optimal design of hybrid clean energy system with battery storage considering off-and on-grid application. *J. Clean. Prod.* **2021**, *290*, 125808. [CrossRef]
10. Paredes-Sánchez, J.P.; Villicaña-Ortíz, E.; Xiberta-Bernat, J.P.V. water pumping system for water mining environmental control in a slate mine of Spain. *J. Clean. Prod.* **2015**, *87*, 501–504. [CrossRef]
11. Dursun, M.; Ozden, S. Application of PV powered automatic water pumping in Turkey. *Int. J. Comput. Electr. Eng.* **2012**, *4*, 161. [CrossRef]
12. Bakelli, Y.; Arab, A.H.; Azoui, B. Optimal sizing of photovoltaic pumping system with water tank storage using LPSP concept. *PV Energy* **2011**, *85*, 288–294. [CrossRef]
13. Bakelli, Y.; Kaabeche, A. Optimal size of photovoltaic pumping system using nature-inspired algorithms. *Int. Trans. Electr. Energy Syst.* **2019**, *29*, e12045. [CrossRef]
14. Maddalena, E.T.; da Silva Moraes, C.G.; Bragança, G.; Junior, L.G.; Godoy, R.B.; Pinto, J.O.P. A battery-less photovoltaic water-pumping system with low decoupling capacitance. *IEEE Trans. Ind. Appl.* **2019**, *55*, 2263–2271. [CrossRef]

15. Errouha, M.; Derouich, A.; Motahhir, S.; Zamzoum, O. Optimal control of induction motor for photovoltaic water pumping system. *Technol. Econ. Smart Grids Sustain. Energy* **2020**, *5*, 6. [CrossRef]
16. Khiareddine, A.; Salah, C.B.; Mimouni, M.F. Power management of a photovoltaic/battery pumping system in agricultural experiment station. *Sol. Energy* **2015**, *112*, 319–338. [CrossRef]
17. Yang, J.; Olsson, A.; Yan, J.; Chen, B. A hybrid life-cycle assessment of CO2 emissions of a PV water pumping system in China. *Energy Procedia* **2014**, *61*, 2871–2875. [CrossRef]
18. Liu, B.; Wang, Z.; Feng, L.; Jermsittiparsert, K. Optimal operation of photovoltaic/diesel generator/pumped water reservoir power system using modified manta ray optimization. *J. Clean. Prod.* **2021**, *289*, 125733. [CrossRef]
19. Sarmas, E.; Spiliotis, E.; Marinakis, V.; Tzanes, G.; Kaldellis, J.K.; Doukas, H. ML-based energy management of water pumping systems for the application of peak shaving in small-scale islands. *Sustain. Cities Soc.* **2022**, *82*, 103873. [CrossRef]
20. Nikzad, A.; Chahartaghi, M.; Ahmadi, M.H. Technical, economic, and environmental modeling of solar water pump for irrigation of rice in Mazandaran province in Iran: A case study. *J. Clean. Prod.* **2019**, *239*, 118007. [CrossRef]
21. Muhsen, D.H.; Khatib, T.; Haider, H.T. A feasibility and load sensitivity analysis of photovoltaic water pumping system with battery and diesel generator. *Energy Convers. Manag.* **2017**, *148*, 287–304. [CrossRef]
22. Mukherjee, S.; Chattaraj, S.; Prasad, D.; Singh, R.P.; Khan, M.I. MPPT-Based PV Powered Water Pumping With RMS: Augmentation of IoE Technology. In *Role of IoT in Green Energy Systems*; IGI Global: Hershey, PA, USA, 2021; pp. 194–224.
23. Ma, T.; Yang, H.; Lu, L.; Peng, J. Optimal design of an autonomous PV–wind-pumped storage power supply system. *Appl. Energy* **2014**, *160*, 728–736. [CrossRef]
24. Kaldellis, J.K.; Kapsali, M.; Kondili, E.; Zafirakis, D. Design of an integrated PV-based pumped hydro and battery storage system including desalination aspects for the Island of Tilos-Greece. In Proceedings of the International Conference on Clean Electrical Power (ICCEP), Sardinia, Italy, 11–13 June 2013.
25. Ma, T.; Yang, H.; Lu, L.; Peng, J. Technical feasibility study on a standalone hybrid PV-wind system with pumped hydro storage for a remote island in Hong Kong. *Renew. Energy* **2014**, *69*, 7–15. [CrossRef]
26. Ma, T.; Yang, H.; Lu, L. Feasibility study and economic analysis of pumped hydro storage and battery storage for a renewable energy powered island. *Energy Convers. Manag.* **2014**, *79*, 387–397. [CrossRef]
27. Wang, L.; Cao, Q.; Zhang, Z.; Mirjalili, S.; Zhao, W. Artificial rabbits optimization: A new bio-inspired meta-heuristic algorithm for solving engineering optimization problems. *Eng. Appl. Artif. Intell.* **2022**, *114*, 105082. [CrossRef]
28. Jahannoush, M.; Nowdeh, S.A. Optimal designing and management of a stand-alone hybrid energy system using meta-heuristic improved sine–cosine algorithm for Recreational Center, case study for Iran country. *Appl. Soft Comput.* **2020**, *96*, 106611. [CrossRef]
29. Bhayo, B.A.; Al-Kayiem, H.H.; Gilani, S.I. Assessment of stand-alone PV-Battery system for electricity generation and utilization of excess power for water pumping. *Sol. Energy* **2019**, *194*, 766–776. [CrossRef]
30. Lorenzo, C.; Almeida, R.H.; Martínez-Núñez, M.; Narvarte, L.; Carrasco, L.M. Economic assessment of large power photovoltaic irrigation systems in the ECOWAS region. *Energy* **2018**, *155*, 992–1003. [CrossRef]

Disclaimer/Publisher's Note: The statements, opinions and data contained in all publications are solely those of the individual author(s) and contributor(s) and not of MDPI and/or the editor(s). MDPI and/or the editor(s) disclaim responsibility for any injury to people or property resulting from any ideas, methods, instructions or products referred to in the content.

Article

Study of the Effect of Throttling on the Success of Starting a Line-Start Permanent Magnet Motor Driving a Centrifugal Fan

Aleksey Paramonov, Safarbek Oshurbekov, Vadim Kazakbaev, Vladimir Prakht and Vladimir Dmitrievskii *

Department of Electrical Engineering, Ural Federal University, 620002 Yekaterinburg, Russia
* Correspondence: vladimir.dmitrievsky@urfu.ru; Tel.: +7-909-028-49-25

Abstract: Direct-on-line synchronous motors are a good alternative to induction motors in fluid machinery drives due to their greater energy efficiency but have the significant disadvantage of limiting the maximum moment of inertia of the loading mechanism to ensure their successful and reliable start-up. This disadvantage is critical in centrifugal fans with a massive steel impeller. In this article, using a mathematical model, the dynamics of starting and synchronizing a permanent magnet synchronous motor fed directly from the mains as part of a fan drive are studied. The simulation results show the possibility of increasing the maximum moment of inertia of the load at the successful start-up of a direct-on-line synchronous motor by adjusting the hydraulic part of the fan pipeline by means of throttling. The conclusions of this paper can be used when selecting an electric motor to drive industrial fans and can contribute to wider use of energy-efficient synchronous motors with direct start-up from the mains.

Keywords: centrifugal fans; electric motors; energy efficiency class; energy saving; line-start permanent magnet synchronous motor; motor starting

MSC: 00A06

1. Introduction

The main function of industrial fans and blowers is to supply a large flow of air or gas at low pressure to various parts of a building or process system. They are usually driven by squirrel cage induction motors (Figure 1a). Such motors have a simple and reliable design; however, due to their operating principle, they also have a relatively high power loss, which limits their energy efficiency class [1]. For low-power induction motors in the 0.55–7.5 kW range, there are difficulties in achieving the IE4 energy efficiency class according to IEC standard 60034-30-1 "Rotating Electrical Machines—Part 30-1: Efficiency classes of line operated AC motors (IE code)" [2].

An alternative to induction motors is synchronous motors with or without permanent magnets and with a direct mains supply. This article discusses line-start permanent magnet synchronous motors (LSPMSMs, Figure 1b) [3,4]. Such motors cannot replace induction motors in the entire range of applications but have already firmly occupied some niches such as fan and pump drives [5–8]. LSPMSMs can meet the requirements of the IE4 class while retaining the dimensions of IE3 efficiency class induction motors [9].

In recent decades, the requirements for the eco-design of industrial equipment of European regulations have been tightening, namely, the requirements for the energy efficiency of electric motors [10] and the reduction of CO_2 emissions [11] are growing. The use of energy-efficient motors in industrial applications is a necessary step to achieve these goals. The goal of tightening the requirements is to achieve a climate-neutral economy [12]. Many countries have already introduced requirements for a mandatory minimum IE3 motor class: The European Union (0.75–375 kW, from 2009); Switzerland and Turkey (0.75–375 kW, from 2017); US (0.75–200 kW, since 2017); Canada (0.75–150 kW, since 2017); Mexico (0.75–375 kW, since 2010); South Korea (0.75–200 kW, since 2017); Singapore (0.75–375 kW, since

Citation: Paramonov, A.; Oshurbekov, S.; Kazakbaev, V.; Prakht, V.; Dmitrievskii, V. Study of the Effect of Throttling on the Success of Starting a Line-Start Permanent Magnet Motor Driving a Centrifugal Fan. *Mathematics* **2022**, *10*, 4324. https://doi.org/10.3390/math10224324

Academic Editors: Udochukwu B. Akuru, Ogbonnaya I. Okoro and Yacine Amara

Received: 7 October 2022
Accepted: 16 November 2022
Published: 18 November 2022

Copyright: © 2022 by the authors. Licensee MDPI, Basel, Switzerland. This article is an open access article distributed under the terms and conditions of the Creative Commons Attribution (CC BY) license (https://creativecommons.org/licenses/by/4.0/).

2013); Japan (0.75–375 kW, since 2014); Saudi Arabia (0.75–375 kW, since 2018); and Brazil (0.75–185 kW, since 2017) [13].

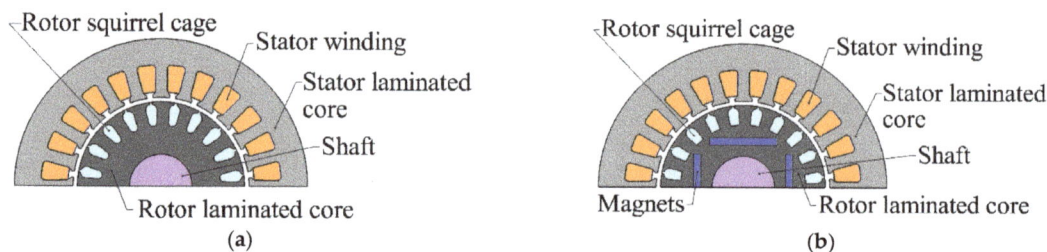

Figure 1. Motor sketches: (**a**) Induction motor (IM); (**b**) line-start permanent magnet synchronous motor (LSPMSM) [1].

LSPMSMs are commercially available from WEG [14], SEW [15], and Bharat Bijlee [16]. There are also serially produced compressors driven by LSPMSMs [17].

Induction motors usually do not have the problem of a failed start. Research on starting induction motors is usually devoted to reducing the starting time in order to reduce the motor overheating [18–20] and reducing the starting current in order to reduce the load on the network [21–23].

The identification of motor model parameters to correctly determine the starting time is often carried out by fitting the parameters using an automatic optimization method when comparing the model response with experimental starting waveforms [24]. Based on the analysis of starting waveforms and starting time, diagnostics of motor faults can also be carried out [25].

The disadvantage of LSPMSMs in comparison with induction motors is a significant limitation of the moment of inertia of the loading mechanism, which is usually indicated in the motor catalog [14,26]. This value indicates the maximum moment of inertia of the load at which the LSPMPM successfully starts and synchronizes at the rated load torque. If starting conditions are worse (exceeding load inertia and/or load torque), successful motor synchronization is not possible and, in the steady state, the motor speed is oscillatory, while the stator current exceeds the rated one by 5–7 times. Induction motors do not have such a limitation in most typical applications (fans, pumps, and compressors). The load inertia limitation requires an evaluation of the success of the synchronization during the LSPMSM start-up. There are several methods for assessing the success of the LSPMSM synchronization using: (1) Finite element model; (2) lumped-circuit model; and (3) synchronization trajectories [27,28].

The start-up simulation using the finite element model requires knowledge of the full internal structure of the LSPMSM [29], which is not always possible. The parameterization of the lumped model requires standard motor tests, such as the no-load test, short-circuit test, and load test [1,28]. The method of synchronization trajectories requires the same motor parameters for analysis as the lumped model [28]. The method of synchronization trajectories and the finite element model method have the serious drawback of being poorly suited for simulating the motor start-up in the drive of a specific mechanism. Depending on the type of driven mechanism, both the load torque and the moment of inertia reduced to the motor shaft can change during start-up. For this reason, when studying the motor start-up in the drive of a certain mechanism, it is better to use the lumped parameter model.

In most of the literature, experimental studies of LSPMSMs are carried out using laboratory benches with a load electric machine [30,31]. However, this approach does not consider the dynamic processes occurring in real applications, such as features of the mechanical characteristic of the loading mechanism (for example, the braking torque of fans and pumps is quadratically dependent on their rotation speed) and the non-linearity of the inertia of the mechanism and fluid. Due to the lack of information on the dynamic processes in real mechanisms, when designing a drive, the motor is often oversized, which leads to a decrease in its efficiency and an increase in its cost. There are not as many articles devoted to the simulation of the operation of an LSPMSM as part of a specific mechanism, for example, a piston pump [29]. Moreover, a literature overview shows that there are no studies covering the LSPMPM start-up as part of a fan drive.

This article discusses the mathematical model of the LSPMSM as part of a centrifugal fan with a belt drive and control of the inlet vanes, as well as transients, when starting the LSPMSM powered by the mains applying fan throttling. It is known that the use of throttling in the process of starting the motor can reduce its mechanical load [27]. This article shows that this effect may be of key importance for LSPMSMs, since the impellers of the fans have a high moment of inertia, and the success of motor synchronization depends on both the moment of inertia and the braking torque of the loading mechanism. By reducing the braking torque of the fan by adjusting the inlet vanes, it is possible to ensure the start-up of the LSPMSM with a higher moment of inertia.

2. Problem Statement

Figure 2 shows the block diagram of the simulated system.

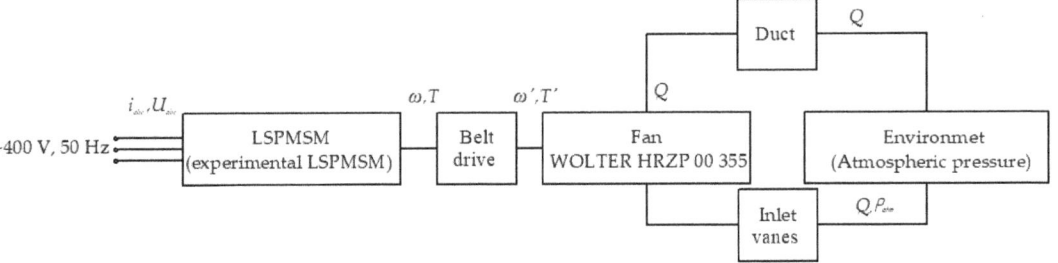

Figure 2. The block diagram of the centrifugal fan with the LSPMSM drive.

The system consists of an LSPMSM powered directly by a three-phase AC network with 400 V at 50 Hz and a serially produced centrifugal fan WOLTER HRZP 00 355 [32], the shaft of which is connected to the motor shaft through a belt drive, an inlet valve, and an air duct, through which air enters the fan from the environment and returns. This scheme is typical for separately operating fluid machines that move gas along a horizontal plane.

When modeling the system, the following assumptions were made:
- Only one fan is operating in the hydraulic system.
- All hydraulic resistances including inlet vanes (such as rooms, air ducts, air duct connections, filters, etc.) are reduced to one hydraulic diameter and reduced to one lumped element.
- There are no leaks.
- The gas is ideal and its properties are the same at every point of the duct.
- Belt drive losses are not considered.

The belt drive makes it possible to change the fan's rotational speed according to the speed of the motor [27,31]. This allows us to move the motor load closer to the rate one since the hydraulic power of the fan is proportional to the rotation speed of the third power. However, the negative side of the use of a belt drive is an additional loss [27]. Changing the fan speed causes a change in its flow–pressure curve and its dependence on the braking power on the flow, according to the laws of similarity [33]:

$$Q_2 = Q_1(n_2/n_1); \tag{1}$$

$$P_2 = P_1(n_2/n_1)^2; \tag{2}$$

$$N_{shaft\ 1} = N_{shaft\ 2}(n_2/n_1)^3, \tag{3}$$

where Q is the flow of the air; P is the static pressure rise produced by the fan; N_{shaft} is the mechanical (braking) power on the fan shaft; n is the rotational speed; index 1 denotes values without the belt drive at the fan speed indicated in the catalogue; and index 2 denotes the values after the application of the belt drive.

Throttling with inlet vanes allows for the adjustment of the operating point on the flow–pressure characteristic during the fan operation. Furthermore, this method can make it easier to start the LSPMSM, as will be shown below. The negative side is that, similar to a belt drive, throttling with inlet vanes creates an additional loss [27].

3. Mathematical Models of System Elements

3.1. LSPMSM's Mathemetical Model

For the LSPMSM mathematical model, the following assumptions are made:
- Magnetic fluxes generated by the stator and rotor windings have a sinusoidal distribution along the air gap.
- The magnetic permeability of steel is constant.
- Stator and rotor windings are symmetrical.
- Each winding is powered by a separate source.
- The supply voltage value does not depend on the load of the electrical machine.
- Losses in the magnetic core are not considered.

At each pole pitch, the LSPMSM rotor houses a permanent magnet. The axis coinciding with the direction of the magnetic flux of the magnet is denoted by q (quadrature axis). The axis perpendicular to this flux is denoted by d (direct axis). The magnet has a low value of magnetic permeability (close to 1). Therefore, the magnetic conductivity of the q-axis is less compared to the magnetic conductivity of the d-axis (Figure 3).

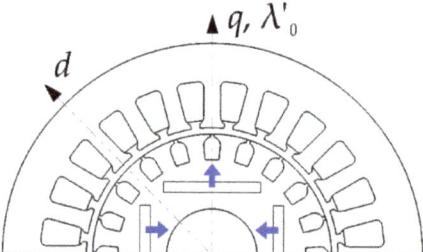

Figure 3. Directions of the d- and q-axes relative to the direction of magnetization of the permanent magnets in the rotor (shown by blue arrows).

The equations of the electrical circuits of the rotor and stator in the coordinate system associated with the rotor have the following form [34,35]:

$$\begin{aligned} d\lambda_{sd}/dt - \lambda_{sq} \cdot \omega_r + R_s \cdot I_{sd} &= U_{sd}; \\ d\lambda_{sq}/dt + \lambda_{sd} \cdot \omega_r + R_s \cdot I_{sq} &= U_{sq}; \\ d\lambda'_{rd}/dt + r'_d \cdot I'_{rd} &= 0; \\ d\lambda'_{rq}/dt + r'_q \cdot I'_{rq} &= 0, \end{aligned} \quad (4)$$

where U_{sd} and U_{sq} are the stator voltages along d- and q-axes; I_{sd}, I_{sq}, I'_{rd}, and I'_{rq} are the stator and rotor currents; λ_{sd}, λ_{sq}, λ'_{rd}, and λ'_{rq} are the stator and rotor flux linkages; ω_r is the rotor electrical angular frequency; and R_s is the stator resistance.

The rotor resistance is anisotropic, and its values along the rotor axes are designated as r'_{rd} and r'_{rq}. All rotor variables refer to the stator and therefore have a stroke in the designation. The electrical equations are supplemented by mechanical ones:

$$\begin{aligned} J \cdot d\omega_m/dt &= T - T_{load}; \\ d\varphi/dt &= \omega_m, \end{aligned} \quad (5)$$

where $T = 3/2 \cdot Z_p \cdot (\lambda_{sd} \cdot I_{sq} - \lambda_{sq} \cdot I_{sd})$ represents the motor torque; T_{load} is the load torque; J is the total moment of inertia; $\omega_m = \omega_r/Z_p$ represents the mechanical angular frequency; φ is the mechanical rotational angle; and Z_p is the number of motor pole pairs.

The magnetic circuit equations describe the self-induction, as well as the mutual induction of the rotor and stator. Note that, due to the presence of permanent magnets, the flux linkages of the stator and rotor along the q-axis are non-zero even at zero currents [34]:

$$\begin{aligned} \lambda'_{sd} &= L_{sd} \cdot I_{sd} + M_d \cdot I'_{rd}; \\ \lambda'_{rd} &= M_d \cdot I_{sd} + L_{rd} \cdot I'_{rd}; \\ \lambda_{sq} &= L_{sq} \cdot I_{sq} + M_q \cdot I'_{rq} + \lambda'_0; \\ \lambda_{rq} &= M_q \cdot I_{sq} + L_{rq} \cdot I'_{rq} + \lambda'_0, \end{aligned} \qquad (6)$$

where L_{sd}, L_{sq}, L_{rd}, and L_{rq} are the stator and rotor total inductances; M_d and M_q are the stator and rotor mutual inductances; and λ'_0 is the permanent magnet flux linkage referred to as the stator.

Term λ'_0 turns out to be a free term in the expression for λ_{rq}. The free term in the expression for λ'_{rq} does not play any role, since (4) contains only the derivative of λ'_{rq}. For convenience, we will assume that the free term in the expression for λ'_{rq} is also λ'_0.

The motor is described by the system of Equations (4) and (5) with respect to ω_r, φ, λ_{sd}, λ_{sq}, λ'_{rd}, and λ'_{rq}, while the currents in (4) are considered dependencies on λ_{sd}, λ_{sq}, λ'_{rd}, and λ'_{rq}, which can be expressed in (6). Note that the currents and flux linkages of the rotor are not available for either direct or indirect measurement by simple means, such as current and voltage sensors. Rotor resistances r'_d and r'_q are not available for direct measurement with an ohmmeter. Upon substituting $\lambda_{rd} = \lambda'_{rd}/k$ and $I_{rd} = k \cdot I'_{rd}$, we obtain $r'_d = k^2 \cdot r_d$, $L'_{rd} = k^2 L'_{rd}$, $M'_d = k \cdot M_d$. Similarly, upon making the substitution $\lambda_{rq} = \lambda'_{rq}/k$ and $I_{rq} = k \cdot I'_{rq}$, we obtain $r'_q = k^2 \cdot r_q$, $L'_{rq} = k^2 \cdot L'_{rq}$, $M'_d = k \cdot M_q$, where k is a constant.

Thus, the set of parameters of the motor model is redundant, and changing the parameters by scaling the rotor currents and fluxes does not lead to a change in the dynamics of the observed values. To eliminate redundancy, we take $M_d = L_{sd}$, $M_q = L_{sq}$. Therefore, we also introduce the rotor leakage inductances $L_{\sigma d} = L_{rd} - L_{sd}$, $L_{\sigma q} = L_{rq} - L_{sq}$ only. Then (6) takes the form:

$$\begin{aligned} \lambda_{sd} &= L_{sd} \cdot I_{sd} + L_{sd} \cdot I'_{rd}; \\ \lambda'_{rd} &= L_{sd} \cdot I_{sd} + (L_{sd} + L_{\sigma d}) \cdot I'_{rd}; \\ \lambda_{sq} &= L_{sq} \cdot I_{sq} + L_{sq} \cdot I'_{rq} + \lambda'_0; \\ \lambda'_{rq} &= L_{sq} \cdot I_{sq} + (L_{sq} + L_{\sigma q}) \cdot I'_{rq} + \lambda'_0. \end{aligned} \qquad (7)$$

In this case, the dependencies of currents on flux links are:

$$\begin{aligned} I'_{rd} &= (\lambda'_{rd} - \lambda_{sd})/L_{\sigma d}; \\ I'_{rq} &= (\lambda'_{rq} - \lambda_{sq})/L_{\sigma q}; \\ I_{sd} &= \lambda_{sd}/L_{sd} - I'_{rd}; \\ I_{sq} &= (\lambda_{sq} - \lambda'_0)/L_{sq} - I'_{rq}. \end{aligned} \qquad (8)$$

The initial conditions for the electric circuit of Equation (4) are $I'_{rd} = I'_{rq} = I_{sd} = I_{sq} = 0$, which is expressed in terms of independent variables as follows: $\lambda'_{rd} = \lambda_{sd} = 0$; $\lambda'_{rd} = \lambda_{sd} = \lambda'_0$. Thus, upon eliminating the ambiguity, we can write the system of ordinary differential equations for the LSPMSM complemented with the algebraic expressions to be solved in the form:

$$\begin{aligned}
&d\lambda_{sd}/dt - Z_p \cdot \lambda_{sq} \cdot d\varphi/dt + R_s \cdot I_{sd} = U_{sd}; \\
&d\lambda_{sq}/dt + Z_p \cdot \lambda_{sd} \cdot d\varphi/dt + R_s \cdot I_{sq} = U_{sq}; \\
&d\lambda'_{rd}/dt + r'_d \cdot I'_{rd} = 0; \\
&d\lambda'_{rq}/dt + r'_q \cdot I'_{rq} = 0; \\
&I'_{rd} = (\lambda'_{rd} - \lambda_{sd})/L_{\sigma d}; \\
&I'_{rq} = (\lambda'_{rq} - [\lambda_{sq} - \lambda'_0])/L_{\sigma q}; \\
&I_{sd} = \lambda_{sd}/L_{sd} - I'_{rd}; \\
&I_{sq} = (\lambda_{sq} - \lambda'_0)/L_{sq} - I'_{rq}; \\
&T = 3/2 \cdot Z_p \cdot (\lambda_{sd} \cdot I_{sq} - \lambda_{sq} \cdot I_{sd}); \\
&J \cdot d^2\varphi/dt^2 = T - T_{load}.
\end{aligned} \qquad (9)$$

Figure 4 shows how the solving of this system is implemented in Simulink.

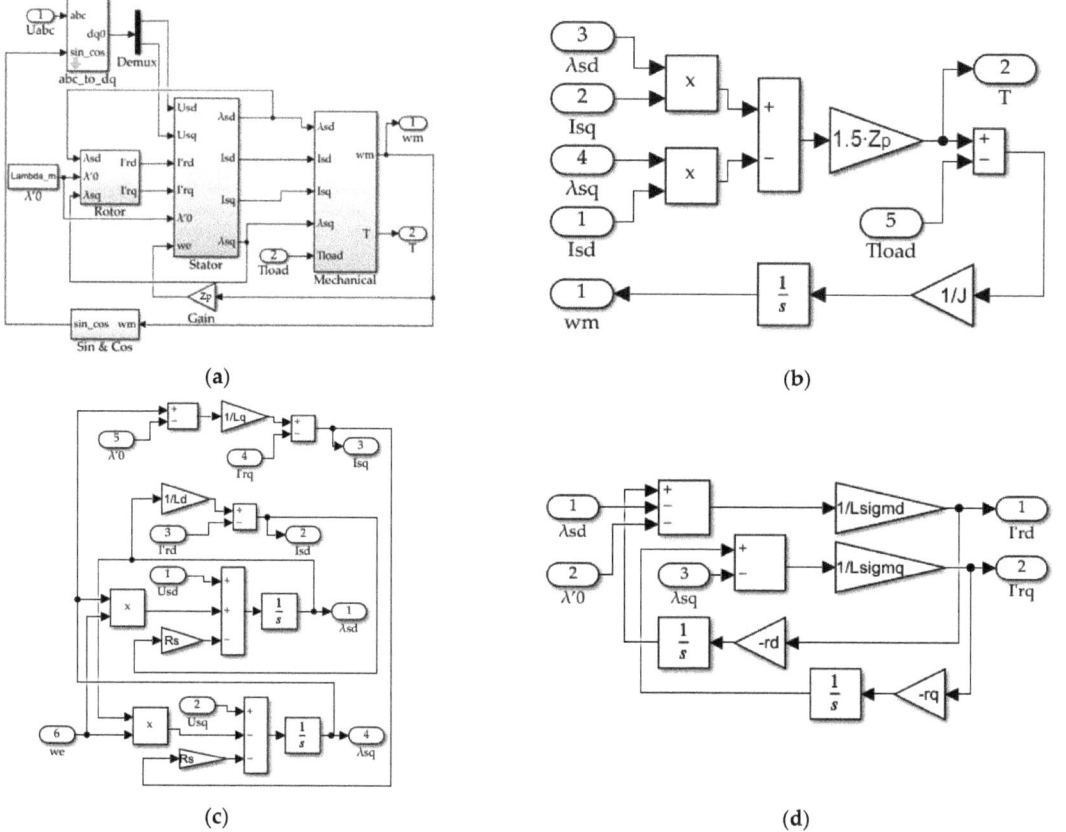

Figure 4. Simulink model of the LSPMSM motor on the d–q-axes: (**a**) General view of the model; (**b**) calculation of the motor torque and angular frequency; (**b**) calculation of stator currents; (**d**) calculation of rotor currents.

3.2. Modeling the Fan, Throttle, and Duct

The main purpose of the simulation is to determine the conditions for successful start-up and synchronization of the LSPMSM in the fan drive. Therefore, when modeling, it is important to take into account all the factors that affect the dynamics of the load at rotational speeds close to the synchronous one. At these speeds, the phase of the current with respect to the phase of the supply voltage changes slowly, and the motor mode is replaced by a generator mode and vice versa until synchronism is achieved. Therefore, in this case, it is sufficient to use a quasi-steady model, that is, a model in which, although the hydrodynamic variables in the elements of the fluid network are assumed to be time dependent, the relationships connecting them do not depend on the previous state and do not contain time derivatives.

To simulate the hydraulic system, models from the Simscape/Fluids/Gas library of the Matlab Simulink environment were used, designed to simulate the flow of an ideal gas. The presented model uses the following gas dynamics simulation blocks: Fan (G), Pipe (G), and Fluids Local Restriction (G) [36–38].

The Fan (G) block models an adiabatic isentropic flow produced by the fan. In other words, the air can be heated due to reversible adiabatic compression by the fan. Air heating due to dissipative, irreversible processes such as viscous friction and mechanical losses of the fan are considered in the fan efficiency as decelerating torque applied to the fan shaft. The mass of air inside the fan is assumed to be negligible, and therefore inlet and outlet mass flow rates are assumed to be equal. The parameters of this block are the static pressure rise and the fan efficiency as functions of the reference parameters (the reference flow and the reference rotational speed). The static pressure rises and the fan efficiency at other flow rates and rotational speeds are calculated using the similarity laws.

The Pipe (G) block [37] is used to model the duct. This block simulates the dynamics of gas flow in a long pipeline. This considers viscous friction losses and convective heat transfer with the pipeline wall. The flow is assumed to be fully developed. Friction losses and heat transfer do not include input effects. The pipeline contains a constant volume of gas, i.e., the pipeline walls are perfectly rigid. Although the block allows for considering the change in the mass of air in the pipe due to pressure changes, the inertia of this air is not considered, which is acceptable for a quasi-stationary model of air flow.

The valve is modeled using the Fluids Local Restriction (G) block [38]. This block models the pressure drops due to a local reduction in the flow area. The processes occurring in the block are adiabatic but not isentropic (irreversible). Similar to the fan model, the air mass in this block is neglected, i.e., inlet and outlet mass flow rates are assumed to be equal. The value of the hydraulic resistance of this block can be adjusted dynamically using the "Inlet vanes" control signal. Similar to the pipeline model, this block maintains the balance of mass and energy, except for heat exchange with the environment.

Thus, although the blocks under consideration are somewhat redundant, for example, the change in the air mass in the duct due to the change in air density is insignificant, these blocks provide the necessary functionality to simulate load conditions when starting the LSPMSM as part of the fan.

4. Simulation Model of the Centrifugal Fan with the LSPMSM Drive

Figure 5 shows the block diagram of the mathematical model of the fan with the LSPMSM drive, the individual blocks of which are described in the previous section. The LSPMSM block simulating the motor is created in accordance with the system of differential Equation (9) and Figure 4. The motor is powered by a three-phase symmetrical voltage system of 400 V at 50 Hz, modeled by the Discrete 3-phase Source block.

The Fan (G) block connects the hydraulic and mechanical models. Ports A and B are connected to the hydraulic circuit (purple lines), and ports C and R are connected to the mechanical circuit (green lines). The hydraulic circuit variables are the flow and pressure. The mechanical circuit variables are the rotational speed and torque.

The load torque signal measured with the Ideal Torque Sensor at the point between the Ideal Angular Velocity Source and Fan (G) blocks is input into the LSPMSM block. The motor rotational speed calculated by the LSPMSM block is sent to the input of the Ideal Angular Velocity block through the Gain block, which simulates a belt drive and multiplies the motor speed by the gear ratio i. The Angular Velocity Source block sets the value of speed in the mechanical circuit to which the Fan (G) block is connected. The torque in this circuit is calculated based on the given flow–pressure curve and the dependence of the fan efficiency on the flow, taking into account similarity laws (1)–(3).

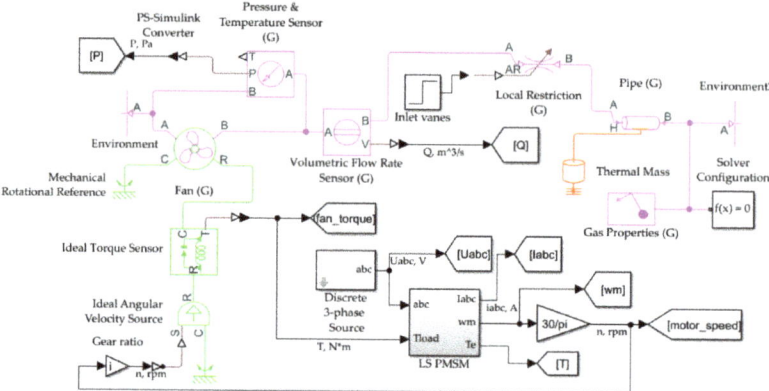

Figure 5. General view of the model of the fan with the LSPMSM drive in the Matlab Simulink.

The hydraulic part of the model consists of the fan (Fan (G) block), inlet vanes (Local Restriction block), duct (Pipe (G) block), and the environment (Environment and Environment2 blocks). The air taken from the environment under the action of the pressure rise created by the fan is regulated by the inlet vanes and, passing through the air duct, returns to the environment. Gas parameters are set in the Gas properties block. The pressure rises and flow produced by the fan are measured by the Pressure and Temperature Sensor (G) and Volumetric Flow Rate Sensor (G) blocks.

The Thermal Mass block models a thermal mass capable of accumulating internal energy [39]. The thermal mass is described by the following formula:

$$P_{thermal} = c \cdot m \cdot d\theta/dt, \tag{10}$$

where c is the specific heat capacity of a substance; m is the mass of the substance; θ is the temperature; and t is the time variable.

The blocks of this model are parameterized in such a way that the operating point of the fan in the steady state is within the flow–pressure curve reported in the manufacturer's catalog (red line in Figure 6) or within the transformed curve calculated using similarity laws (1)–(3) for a different rotational speed (for example, the blue line in Figure 6) [32].

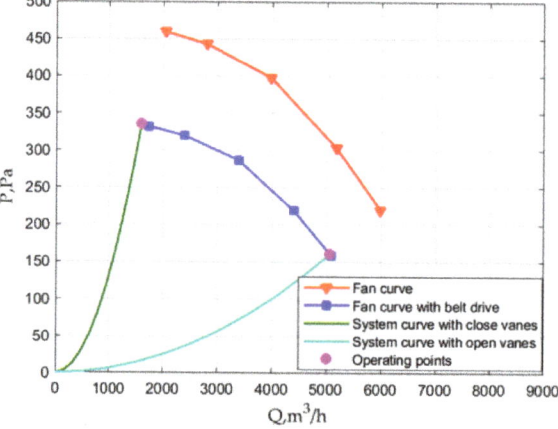

Figure 6. Flow–pressure curves of the fan with and without belt drive, as well as the duct curves with open and closed inlet vanes.

Table 1 shows the parameters of the LSPMSM with a rated power of 0.55 kW and a rated speed of 1500 rpm used for the simulation.

Table 1. LSPMSM parameters.

Parameter	Value
Rated power P_{rate}, kW	0.55
Rated line-to-line voltage U_{rate}, V	380
Rated frequency f, Hz	50
Pole pair number Z_p	2
Stator phase resistance R_s, Ohm	15.3
Total direct inductance L_{sd}, H	0.26
Total quadrature inductance L_{sq}, H	0.15
Leakage direct inductance $L_{\sigma d}$, H	0.038
Leakage quadrature inductance $L_{\sigma q}$, H	0.051
Rotor direct resistance r'_d, Ohm	9.24
Rotor quadrature resistance r'_q, Ohm	10.1
Permanent magnet flux linkage λ'_0, Wb	0.76
Motor inertia moment J_m, kg·m²	0.003
Fan impeller inertia moment J_i, kg·m²	0.06

To simulate the fan, the parameters of a commercially produced centrifugal fan WOLTER HRZP 00 355 with a moment of inertia J_i = 0.1 kg·m² and characteristics shown in Table 2 according to [32] were used. This fan involves the use of a belt drive. For this study, the gear ratio i = 0.85 of the belt drive was selected. Figure 6 shows the flow–pressure characteristics of the fan without and with the use of a belt drive, as well as the curves of the duct at various positions of the inlet vanes: $S_{valve} = S$ (fully open) and $S_{valve} = 0.25 \cdot S$ (fully closed). The cross-sectional area of the duct is equal to S_{duct} = 0.099 m², and the length of the duct is equal to l_{duct} = 30 m.

Table 2. Fan characteristics.

Flow Q, m³/h	Pressure P, Pa	Fan Hydraulic Efficiency η, %
2035	459	64
2803	442	75
3977	397	79
5170	303	78
5981	219	58

5. Simulation Results

The LSPMSM start-up simulation is carried out for two different cases:
(1) When the inlet vanes are fully open (the control signal of the inlet valve is maximum $S_{valve} = S$).
(2) When dynamically changing the valve position, corresponding to the following algorithm: $S_{valve} = 0.25 \cdot S$ at t < 4 s (this value of the control signal corresponds to the minimum fan flow from the flow–pressure curve given for the considered fan type in the catalog [32], which models the fan operation with a fully closed valve; this assumption is made due to the lack of information about the fan braking torque at lower flow rates) and $S_{valve} = S$ at t ≥ 4. Figures 7–12 show the simulation results for both cases.

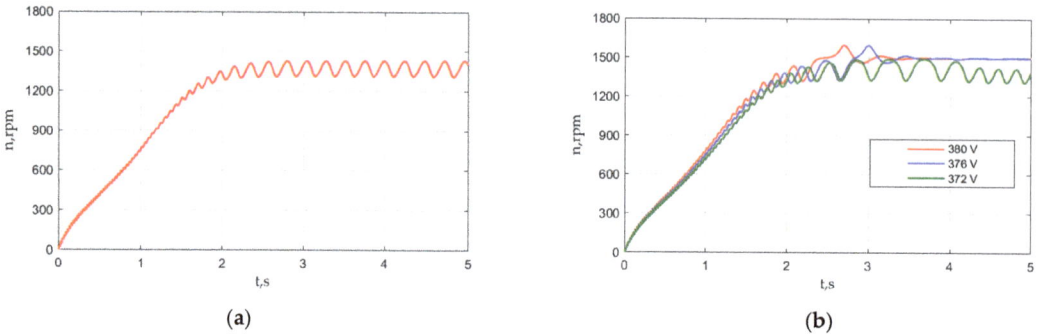

Figure 7. LSMPM rotational speed versus time: (**a**) Fully open valve, 380 V; (**b**) fully closed valve, various voltages.

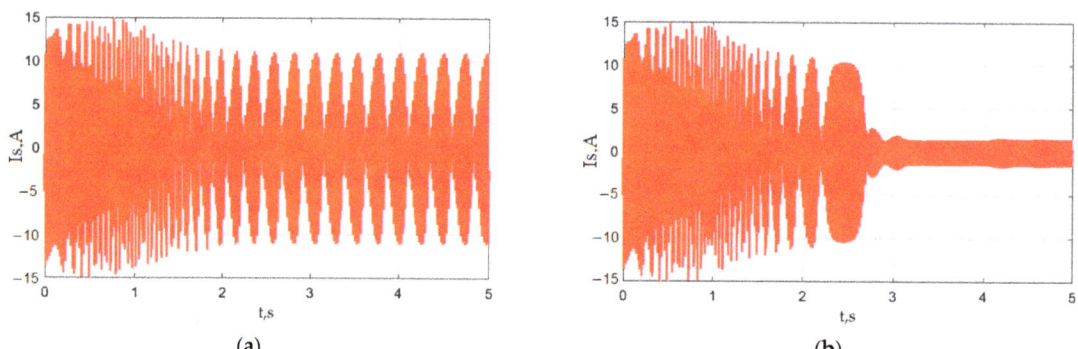

Figure 8. LSPMSM stator current versus time, 380 V: (**a**) Fully open valve; (**b**) fully closed valve.

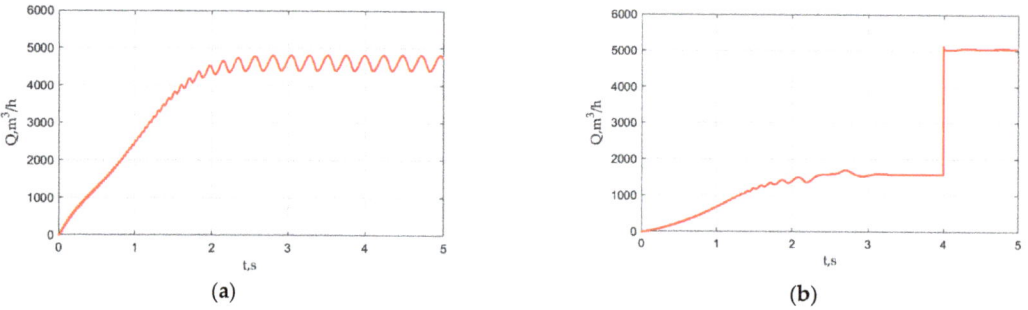

Figure 9. Flow versus time, 380 V: (**a**) Fully open valve; (**b**) fully closed valve.

Figure 7a shows that with $S_{valve} = S$, the motor reaches a speed of approximately ≈ 1350 rpm in approximately 2 s but cannot accelerate to a synchronous speed of 1500 rpm. Next, quasi-steady oscillations in speed begin with a maximum value of 1430 rpm, a minimum value of 1315 rpm, and an average value of 1408 rpm. Thus, starting the motor in this case is unsuccessful.

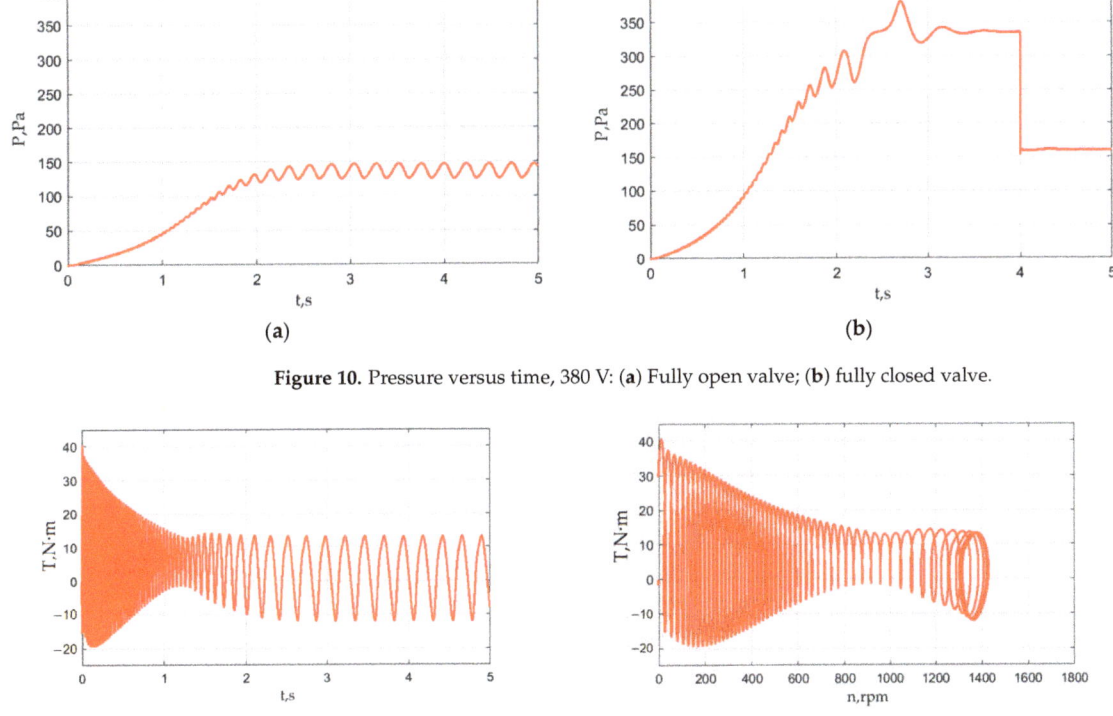

Figure 10. Pressure versus time, 380 V: (**a**) Fully open valve; (**b**) fully closed valve.

Figure 11. LSPMSM torque with fully open valve, 380 V: (**a**) Torque versus time; (**b**) torque versus speed.

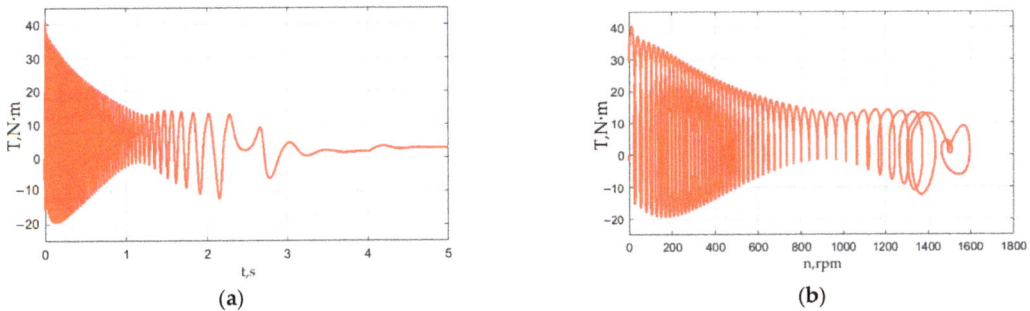

Figure 12. LSPMSM torque with fully closed valve, 380 V: (**a**) Torque versus time; (**b**) torque versus speed.

The red line in Figure 7b shows the motor speed waveform at the rated voltage when starting with the valve closed ($S_{valve} = 0.25 \cdot S$). In this case, the motor successfully synchronizes, reaching a steady speed of 1500 rpm in approximately 3.5 s. After the opening of the inlet vanes (load surge) at the time t = 4 s, the speed first decreases to 1493 rpm. Then, after 0.5 s, the synchronous speed of 1500 rpm is restored.

The blue line in Figure 7b shows the speed waveform with a voltage drop of 1%. In this case, the motor still starts successfully, although the start time is slightly longer. The green line shows the speed waveform with a voltage drop of 2%. In this case, although the motor reaches 1500 rpm, it then falls out of synchronism, and the process of being pulled in and out of synchronism is periodically repeated. After a load surge at time t = 4 s, the motor starts to operate in an asynchronous mode without reaching the synchronous speed.

Figures 8–10 show the current, flow, and pressure transients at various valve positions at the rated supply voltage. Figure 8a shows a waveform of the phase current of the motor during the failed synchronization. In this case, the steady current amplitude is variable and depends on the instantaneous speed. The maximum instantaneous current of the LSPMSM in the quasi-steady state is 10.5 A, which exceeds the current amplitude during normal operation by 6.7 times (10.6/1.56 = 6.7, see below). With the prolonged operation, such a current will certainly lead to the failure of the motor. Figure 8b shows the current waveform with the valve closed during the start-up. In this case, the start and synchronization take 3.5 s. The amplitude of the steady current after the synchronization is 1.44 A. After opening the valve, the amplitude of the current increases to 1.56 A.

Figure 9a shows the waveform of the fan volumetric air flow with the inlet vanes fully open. Since the flow is proportional to the fan speed (1), this transient follows the motor speed waveform, which also takes place due to the low inertia of the gas. The flow has a quasi-steady oscillatory character with a maximum of 4809 m^3/h, a minimum of 4408 m^3/h, and an average of 4609 m^3/h. Figure 9b shows the waveform of the flow with the valve closed during start-up. The flow rate gradually increases from 0 to 1600 m^3/h as the motor accelerates to its synchronous speed. As noted above, the presence of the non-zero flow rate at a fully closed valve is an assumption of this model, made due to the lack of information about the fan braking torque at lower flow rates, which does not affect the main conclusions of the article. After opening the valve, the flow through the fan increases several times up to 5000 m^3/h. Opening the valve after the motor has reached a synchronous speed causes a step change in the flow. This step response is due to the low inertia of the gas.

Figure 10a shows the fan pressure rise waveform with the valve open. Due to the quadratic dependence of pressure on the fan speed (2), this transient follows the motor speed waveform but has a large amplitude of the oscillatory component. The maximum value in the quasi-steady operation is 146.67 Pa, the minimum is 125.67 Pa, and the average is 136.17 Pa. Figure 10b shows the waveform of the pressure with the valve closed during start-up. Opening the inlet vanes causes a stepwise decrease in the pressure. The steady pressure before opening the valve is 335 Pa, and after opening it is 160.61 Pa.

Figures 11–14 show plots of the motor torque versus time and speeds for the various cases under consideration.

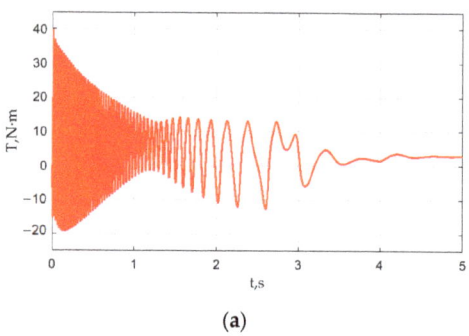
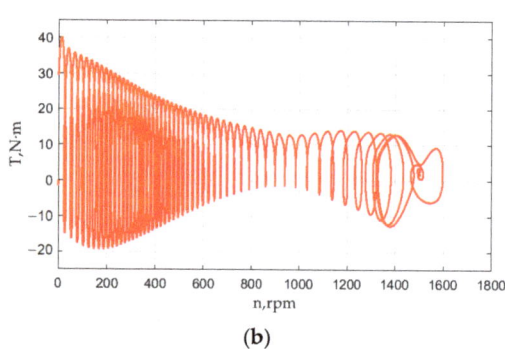

(a) (b)

Figure 13. LSPMSM torque with fully closed valve, 376 V: (**a**) Torque versus time; (**b**) torque versus speed.

Figure 11 shows the torque waveforms with a fully open valve and a 380 V supply voltage. Since the LSPMSM operates in an asynchronous mode due to unsuccessful synchronization, the torque has a large oscillatory component with a maximum of 13.1 N·m, a minimum of −11.87 N·m, and an average of 0.62 N·m. In this case, the presence of torque and speed oscillations (see the speed waveform in Figure 7a and the torque waveform in Figure 11a) is due to the fact that the LSPMSM operates in an asynchronous mode. During the initial startup phase, the LSPMSM runs in an asynchronous mode. In this mode, the positive torque, due to which the motor accelerates, is the asynchronous torque of the short-circuited cage, and the magnets create an oscillatory alternating torque. If the moment of inertia is too large, the speed oscillations are smoothed out, and the LSPMSM operates with some slip, similarly to an asynchronous motor, i.e., does not reach the synchronous speed.

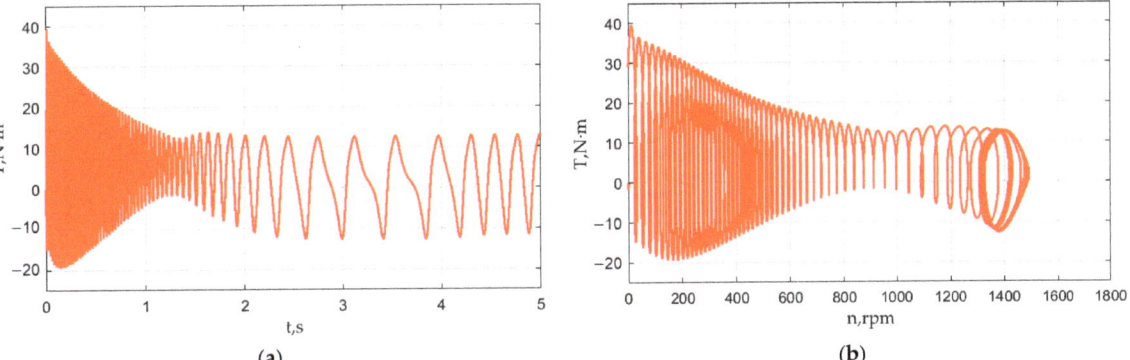

Figure 14. LSPMSM torque with fully closed valve, 372 V: (**a**) Torque versus time; (**b**) torque versus speed.

If the inertia is not too large or the load torque is low, the speed oscillations caused by torque oscillations increase. If the speed oscillations are sufficient to achieve a synchronous speed (as in the case shown in Figures 7b and 12a), then the synchronization process begins. In this process, the motor first overshoots the synchronous speed (see Figure 7b, red and blue lines at t ≈ 3 s) and then settles down, i.e., the speed stabilizes and becomes equal to the synchronous speed. The critical moment of inertia of the loading mechanism (the moment of inertia of the loading mechanism at which an LSPMSM is able to start at the rated load torque) is indicated in the catalogs of mass-produced LSPMSMs [14,15].

Figure 12 shows the torque waveform with the valve closed at start-up. After successful synchronization, the motor torque has a constant value. The steady torque before opening the valve is 3.91 N·m and is 4.9 N·m after opening the valve. Figures 13 and 14 show torque plots for starting the motor with the fully closed valve at 1% and 2% voltage drops, respectively.

6. Conclusions

In this study, a mathematical model was developed to investigate the starting performance of a line-start permanent magnet synchronous motor (LSPMSM) as part of an industrial centrifugal fan drive. This mathematical model evaluated the critical moment of inertia and the static torque of the loading mechanism, exceeding which makes successful synchronization during the start-up process impossible.

The simulation was carried out using the parameters of a 0.55 kW 1500 rpm LSPMSM and a serially produced fan. The comparison of dynamic processes and simulation results of the motor start-up in a fan drive with and without air flow control by inlet vanes was carried out. It was shown that closing the inlet valve at start-up facilitates synchronization of the LSPMSM and increases the maximum moment of inertia at which the fan starts successfully.

When the fan moment of inertia is high (exceeding the critical moment of inertia of the motor), which cannot be reduced by a belt drive, and the static load torque is close to the rated one, the LSPMSM cannot successfully synchronize. Long-term operation of LSPMSMs in asynchronous mode leads to large current overloads, overheating, and, ultimately, motor failure. Therefore, the asynchronous mode is unacceptable for the long-term operation of LSPMSMs.

However, as shown by the simulation results presented in the article, in this case, successful synchronization can be achieved by throttling the fan duct. Closing the inlet valve at start-up contributes to the successful synchronization of the LSPMSM. After the motor start-up, the valve can be opened again.

The results of this study can be used to select a more energy-efficient fan drive motor, even if the starting characteristics of the motor specified in the manufacturer's catalog are not satisfactory for a given application. In future works, the start-up of fluid-processing mechanisms employing direct-on-line synchronous motors at reduced voltages and with long cable lengths will be investigated.

Author Contributions: Conceptual approach, A.P., V.K. and V.P.; data duration, A.P. and S.O.; software, A.P., S.O. and V.K.; calculations and modeling, A.P., S.O., V.D., V.K. and V.P.; writing—original draft, A.P., S.O., V.D., V.K. and V.P.; visualization, A.P. and V.K.; review and editing, A.P., S.O., V.D., V.K. and V.P. All authors have read and agreed to the published version of the manuscript.

Funding: The work was partially supported by the Ministry of Science and Higher Education of the Russian Federation (through the basic part of the government mandate, Project No. FEUZ 2020-0060).

Data Availability Statement: Data are contained within the article.

Acknowledgments: The authors thank the editors and reviewers for careful reading and constructive comments.

Conflicts of Interest: The authors declare no conflict of interest.

References

1. Kazakbaev, V.; Paramonov, A.; Dmitrievskii, V.; Prakht, V.; Goman, V. Indirect Efficiency Measurement Method for Line-Start Permanent Magnet Synchronous Motors. *Mathematics* **2022**, *10*, 1056. [CrossRef]
2. Rotating Electrical Machines—Part 30-2: Efficiency Classes of Variable Speed AC Motors (IE-Code). IEC 60034-30-2/ IEC: 2016-12. Available online: https://webstore.iec.ch/publication/30830 (accessed on 6 October 2022).
3. Liu, C. Rotor Conductor Arrangement Designs of High-Efficiency Direct-on-Line Synchronous Reluctance Motors for Metal Industry Applications. *IEEE Trans. Ind. Appl.* **2020**, *56*, 4337–4344. [CrossRef]
4. Ding, T.; Takorabet, N.; Sargos, F.; Wang, X. Design and Analysis of Different Line-Start PM Synchronous Motors for Oil-Pump Applications. *IEEE Trans. Magn.* **2009**, *45*, 1816–1819. [CrossRef]
5. Goman, V.; Oshurbekov, S.; Kazakbaev, V.; Prakht, V.; Dmitrievskii, V. Energy Efficiency Analysis of Fixed-Speed Pump Drives with Various Types of Motors. *Appl. Sci.* **2019**, *9*, 5295. [CrossRef]
6. Baka, S.; Sashidhar, S.; Fernandes, B.G. Design of an Energy Efficient Line-Start Two-Pole Ferrite Assisted Synchronous Reluctance Motor for Water Pumps. *IEEE Trans. Energy Conv.* **2021**, *36*, 961–970. [CrossRef]
7. Ismagilov, F.; Vavilov, V.; Gusakov, D. Line-Start Permanent Magnet Synchronous Motor for Aerospace Application. In Proceedings of the 2018 IEEE International Conference on Electrical Systems for Aircraft, Railway, Ship Propulsion and Road Vehicles and International Transportation Electrification Conference (ESARS-ITEC 2018), Nottingham, UK, 7–9 November 2018; pp. 1–5. [CrossRef]
8. Kurihara, K.; Rahman, M. High-Efficiency Line-Start Interior Permanent-Magnet Synchronous Motors. *IEEE Trans. Ind. Appl.* **2004**, *40*, 789–796. [CrossRef]
9. Kazakbaev, V.; Prakht, V.; Dmitrievskii, V.; Golovanov, D. Feasibility Study of Pump Units with Various Direct-On-Line Electric Motors Considering Cable and Transformer Losses. *Appl. Sci.* **2020**, *10*, 8120. [CrossRef]
10. European Commission Regulation (EC), No. 640/2009 Implementing Directive 2005/32/ EC of the European Parliament and of the Council with Regard to Ecodesign Requirements for Electric Motors, (2009), Amended by Commission Regulation (EU) No 4/2014 of January 6, 2014. Document 32014R0004. Available online: https://eur-lex.europa.eu/legal-content/EN/TXT/?uri=CELEX%3A32014R0004 (accessed on 6 October 2022).
11. European Council Meeting (10 and 11 December 2020)—Conclusions. EUCO 22/20 CO EUR 17 CONCL 8. Available online: https://www.consilium.europa.eu/media/47296/1011-12-20-euco-conclusions-en.pdf (accessed on 6 October 2022).
12. The European Green Deal. Available online: https://ec.europa.eu/info/strategy/priorities-2019-2024/european-green-deal_en (accessed on 6 October 2022).
13. Goman, V.; Prakht, V.; Kazakbaev, V.; Dmitrievskii, V. Comparative Study of Induction Motors of IE2, IE3 and IE4 Efficiency Classes in Pump Applications Taking into Account CO_2 Emission Intensity. *Appl. Sci.* **2020**, *10*, 8536. [CrossRef]
14. WQuattro, Super Premium Efficiency Motor, Product Catalogue, WEG Group—Motors Business Unit, Cod: 50025713, Rev: 03, Date (m/y): 07/2017. Available online: https://static.weg.net/medias/downloadcenter/h01/hfc/WEG-w22-quattro-european-market-50025713-brochure-english-web.pdf (accessed on 6 October 2022).
15. Addendum to the Operating Instructions: AC Motors DR.71.J-DR.100.J with LSPM Technology, 21281793/EN, 09/2014, SEW Eurodrive. Available online: https://download.sew-eurodrive.com/download/pdf/21343799.pdf (accessed on 6 October 2022).

16. Catalogue of Super Premium Efficiency SynchroVERT LSPM Motors. Available online: https://www.bharatbijlee.com/media/14228/synchrovert_catalogue.pdf (accessed on 6 October 2022).
17. KT-420-5, Operation of Bitzer reciprocating compressors with external frequency inverters, Bitzer, 01.2022. Available online: https://www.bitzer.de/shared_media/html/kt-420/Resources/pdf/279303819.pdf (accessed on 6 October 2022).
18. Grag, A.; Tomar, A. Starting Time Calculation for Induction Motor. *J. Electr. Electron. Syst.* **2015**, *4*, 149. [CrossRef]
19. Xia, Y.; Han, Y.; Xu, Y.; Ai, M. Analyzing Temperature Rise and Fluid Flow of High-Power-Density and High-Voltage Induction Motor in the Starting Process. *IEEE Access* **2019**, *7*, 35588–35595. [CrossRef]
20. Dems, M.; Komeza, K.; Szulakowski, J.; Kubiak, W. Dynamic Simulation of High-Speed Induction Motor. *Energies* **2021**, *14*, 2713. [CrossRef]
21. Aree, P. Starting time calculation of large induction motors using their manufacturer technical data. In Proceedings of the 2016 19th International Conference on Electrical Machines and Systems (ICEMS), Chiba, Japan, 13–16 November 2016; pp. 1–5. Available online: https://ieeexplore.ieee.org/document/7837339 (accessed on 6 October 2022).
22. Digalovski, M.; Rafajlovski, G. Calculation of Starting and Breaking Times of Induction Motor Electric Drives, for Different Mechanical Loads. In Proceedings of the 2020 International Conference on Information Technologies (InfoTech), Varna, Bulgaria, 17–18 September 2020; pp. 1–4. [CrossRef]
23. Maleki, M.G.; Chabanloo, R.M.; Farrokhifar, M. Accurate coordination method based on the dynamic model of overcurrent relay for industrial power networks taking contribution of induction motors into account. *IET Gener. Transm. Distrib.* **2020**, *14*, 645–655. [CrossRef]
24. Guedes, J.J.; Castoldi, M.F.; Goedtel, A.; Agulhari, C.M.; Sanches, D.S. Parameters estimation of three-phase induction motors using differential evolution. *Electr. Power Syst. Res.* **2018**, *154*, 204–212. [CrossRef]
25. Faiz, J.; Ebrahimi, B.M.; Akin, B.; Toliyat, H.A. Finite-Element Transient Analysis of Induction Motors Under Mixed Eccentricity Fault. *IEEE Trans. Magn.* **2008**, *44*, 66–74. [CrossRef]
26. Abdoulkadri, C.; Albert, J.; Wang, S.; Wang, R. Analytical Synchronization Analysis of Line-Start Permanent Magnet Synchronous Motors. *Prog. Electromagn. Res. M* **2016**, *48*, 183–193. [CrossRef]
27. Barkeley, L. Improving Fan System Performance. A Sourcebook for Industry. U.S. Department of Energy. 2003; DOE/GO-102003-1294. Available online: https://www.nrel.gov/docs/fy03osti/29166.pdf (accessed on 6 October 2022).
28. Soulard, J.; Nee, H. Study of the synchronization of line-start permanent magnet synchronous motors. In Proceedings of the 2000 IEEE Industry Applications Conference. Thirty-Fifth IAS Annual Meeting and World Conference on Industrial Applications of Electrical Energy (Cat. No.00CH37129), Rome, Italy, 8–12 October 2000; pp. 424–431. [CrossRef]
29. Poudel, B. Line Start Permanent Magnet Synchronous Motor for Multi Speed Application. Master Thesis, University of New Orleans, New Orleans, LA, USA, 2017. Available online: https://scholarworks.uno.edu/td/2430 (accessed on 6 October 2022).
30. Kahrisangi, M.; Hassanpour, I.; Zadeh, A.; Zadeh, V.; Mohammad, S.; Mohammad, R. Line-start permanent magnet synchronous motors versus induction motors: A comparative study. *Front. Electr. Electron. Eng.* **2012**, *7*, 459–466. [CrossRef]
31. Cory, W. *Fans & Ventilation: A Practical Guide*; Elsevier Science: Amsterdam, The Netherlands, 2005.
32. Centrifugal Fans. Wolter GmbH+Co KG. M08.6. V.2006/1. Available online: https://www.wolter.eu/fileadmin/BILDER/produkte/5-radialventilatoren/pdf/Radialventilator_TRZ_HRZ_HRZP.pdf (accessed on 6 October 2022).
33. Rory, A.R.; Scott, M. Modeling Techniques for a Computational Efficient Dynamic Turbofan Engine Model. *Int. J. Aerosp. Eng.* **2014**, *2014*, 283479. [CrossRef]
34. Wang, D.; Wang, X.; Chen, H.; Zhang, R. Matlab/Simulink-Based Simulation of Line-start PMSM Used in Pump Jacks. In Proceedings of the IEEE Conference on Industrial Electronics and Applications, Harbin, China, 23–25 May 2007; pp. 1179–1181. [CrossRef]
35. Modeer, T. Modeling and Testing of Line Start Permanent Magnet Motors. Ph.D. Thesis, School of Electrical Engineering, KTH, Stockholm, Sweden, 2007. Available online: https://citeseerx.ist.psu.edu/viewdoc/download?doi=10.1.1.826.9539&rep=rep1&type=pdf (accessed on 6 October 2022).
36. Mathworks. Fan (G), Fan in Gas Network. Model Description. Available online: https://www.mathworks.com/help/hydro/ref/fang.html;jsessionid=dd48cc776badaa81163a0d6aa2ff (accessed on 6 October 2022).
37. Mathworks. Pipe (G), Rigid Conduit for Gas Flow. Model Description. Available online: https://www.mathworks.com/help/simscape/ref/pipeg.html (accessed on 6 October 2022).
38. Mathworks. Local Restriction (G), Restriction in Flow Area in Gas Network. Model Description. Available online: https://www.mathworks.com/help/simscape/ref/localrestrictiong.html (accessed on 6 October 2022).
39. Mathworks. Thermal Mass, Mass in Thermal Systems. Model Description. Available online: https://www.mathworks.com/help/simscape/ref/thermalmass.html (accessed on 6 October 2022).

Article

High Impedance Fault Detection Protection Scheme for Power Distribution Systems

Katleho Moloi * and Innocent Davidson *

Department of Electrical Power Engineering, Durban University of Technology, Durban 4000, South Africa
* Correspondence: katlehom@dut.ac.za (K.M.); innocentd@dut.ac.za (I.D.)

Abstract: Protection schemes are used in safe-guarding and ensuring the reliability of an electrical power network. Developing an effective protection scheme for high impedance fault (HIF) detection remains a challenge in research for protection engineers. The development of an HIF detection scheme has been a subject of interest for many decades and several methods have been proposed to find an optimal solution. The conventional current-based methods have technical limitations to effectively detect and minimize the impact of HIF. This paper presents a protection scheme based on signal processing and machine learning techniques to detect HIF. The scheme employs the discrete wavelet transform (DWT) for signal decomposition and feature extraction and uses the support vector machine (SVM) classifier to effectively detect the HIF. In addition, the decision tree (DT) classifier is implemented to validate the proposed scheme. A practical experiment was conducted to verify the efficiency of the method. The classification results obtained from the scheme indicated an accuracy level of 97.6% and 87% for the simulation and experimental setups. Furthermore, we tested the neural network (NN) and decision tree (DT) classifiers to further validate the proposed method.

Keywords: classification; high impedance fault; power system; support vector machine; wavelet packet transform

MSC: 49M41

1. Introduction

The power system distribution network forms an integral part of the electricity network value chain. The distribution network serves as an interlink between the power grid and the customer load segment connected into the network. Power distribution systems are prone faults. The faults occurring on the system have both technical and economic impacts. Thus, is it important to design a protection scheme that will respond efficiently to mitigate the impact of faults [1]. Over the years, overcurrent protection schemes have been successfully used to detect low impedance faults (LIF) [2]. LIFs occur when there is an insulation breakdown between the conductor phases or the conductor phases and the ground. When an LIF occurs, the fault current increases drastically, thus enabling the protection relay to detect the abnormality on the system and subsequently, tripping the circuit breaker. However, this is not the case when an HIF occurs. In case of an HIF, the current magnitude drops below a nominal current threshold value which is unlikely to be detected by conventional protection schemes. Table 1 shows the typical HIF current magnitudes on different surfaces [3]; it can be observed that the current magnitude on different surfaces may affect the protection scheme to detect the HIF accurately.

HIFs usually occur from two common cases. The first case is when a high impedance object contacts an energized power line, and the second case is when an energized power line breaks and fall on the ground. In both cases, the fault current developed is usually minimal to trigger the relay for any protective action. Unlike LIFs, HIF may cause serious damages to the environment, humans, and animals as they can cause a fire [4]. It is

Citation: Moloi, K.; Davidson, I. High Impedance Fault Detection Protection Scheme for Power Distribution Systems. *Mathematics* 2022, 10, 4298. https://doi.org/10.3390/math10224298

Academic Editors: Udochukwu B. Akuru, Ogbonnaya I. Okoro and Yacine Amara

Received: 19 September 2022
Accepted: 15 November 2022
Published: 16 November 2022

Copyright: © 2022 by the authors. Licensee MDPI, Basel, Switzerland. This article is an open access article distributed under the terms and conditions of the Creative Commons Attribution (CC BY) license (https://creativecommons.org/licenses/by/4.0/).

imperative to design a scheme that will effectively detect HIF to minimize the adverse impact of faults.

Table 1. Fault magnitude on different surfaces [3].

Surface	Current Magnitude (A)
Dry asphalt or sand	0
Dry grass	25
Wet soil	40
Wet grass	50
Reinforced concrete	75

A significant number of methods have been proposed to find an effective solution for HIF detection. These methods range from classical to heuristic approaches. Initially, current-based methods were proposed for HIF detection. In [5,6], the authors proposed an algorithm based on current magnitude detection. However, these algorithms have often failed to detect HIFs due to the minimal or no-fault current magnitude to trigger any protection operation at the point of fault. The techniques based on harmonic content for HIF detection were proposed in [7–9], where the frequency spectrum of the HIF was used to detect the variations in the third harmonic content of the current and voltage magnitude. Based on spectrum and frequency analysis, Cui [10] proposed an algorithm based on Kalman filter (KF) analysis to detect HIF in medium voltage power systems. The application of the KF was used to estimate the effect of harmonic changes on the fault current magnitude. Other methods based on KF technique were proposed to detect the high-resolution of the current magnitude during arching [11]. The application of wavelet transform (WT) for signal interpretation has been widely used to detect HIF. In [12], the WT technique was used to distinguish the signal component of HIF from other power system operations to minimize nuisance trips. A technique based on discrete wavelet transform (DWT) and frequency range was applied to analyze and detect the signature pattern emitted by HIF [13]. A technique based on DWT and evolutionary neural network (ENN) was proposed by Silva [14]. The scheme employed the DWT technique for feature extraction and ENN for HIF classification; the scheme produced high classification results. New techniques based on wavelet transform were presented in [15–17], these techniques used DWT for feature extraction and pattern recognition to detect HIF. In [18], a technique based on DWT used the residual current magnitude on medium voltage power lines to detect HIFs. In [19], a feature extraction scheme based on discreet Fourier transform (DFT) was used to select the HIF signature from a pool of signatures. The DFT output signature was fed into the extended Kalman filter scheme to detect HIF.

Nowadays, researchers are placing more emphasis on computer intelligence-based techniques to detect HIFs. In [20], a hybrid scheme based on the energy and entropy analysis of the random behavior of the fault signal to detect HIFs was proposed. Other studies conducted in [21–24] used the practical neural network (NN)-based algorithm to detect HIF. A technique based on a combination of packet wavelet transform (PWT) and support vector machine (SVM) was proposed to detect the HIF [25]. In [26,27], the decision tree (DT) algorithms were proposed to detect HIF in low voltage power networks. The proposed algorithm produced good results. However, the technique was only used in a single-line radial network. In [28], a protection scheme based on unsupervised learning and convolutional autoencoder was proposed to detect HIF. The scheme was validated using the IEEE 13-bus test system and produced good results compared to the supervised learning systems. In another study [29], a hybrid method based on DWT and probabilistic neural network (PNN) was proposed to detect HIF. The technique used the DWT for feature selection and PNN for classification of HIF from other non-fault conditions. A technique based on (WT) for signal processing and feature selection combined with convolutional neural network (CNN) classifier was proposed to detect HIF in power distribution network was proposed [30]. In [31], a protection scheme based on empirical mode decomposition

(EMD) and artificial neural network (ANN) to detect HIF was proposed. The scheme utilized the harmonic content of the HIF current signal for classification. A protection scheme on the WT and back propagation neural network (BPNN) was proposed to detect HIF [32]. The method was tested using the data from the substation practical data and an 80% detection accuracy was achieved. Although the technical challenges of HIF detection and electricity safety in the power system industry has not yet been fully achieved, there has been significant contributions presented in the literature. Studies also show a promising trend of using artificial intelligence (AI) techniques to improve the accuracy level of HIF detection schemes. The application of using signal processing abilities of the wavelet transforms (WT) for feature extraction and pattern recognition has been widely used to enhance efficiency of protection technology [33,34].

Mathematical models form an integral part of designing rigorous fault diagnostic techniques in power systems engineering. Most of the engineering solutions have been developed using mathematical approaches. In the present work, we propose the application of mathematical models using machine learning techniques to solve an engineering problem. The proposed hybrid model demonstrates the applicability of mathematical models in the engineering fraternity. Our model integrates various segments of mathematics which includes signal processing, feature extraction, optimization, and pattern recognition. For instance, at the initial stage of our model, a fault signal is decomposed using DWT to analyze the segments of interest from the unabridged fault signal spectrum; thereafter, a feature extracting technique is employed using the mathematical statistical features to select specific features from the entire data spectrum, subsequently these features are used as inputs to train, test, and validate the pattern recognition and classification mathematical algorithm using SVM. Lastly, the GA technique (Table A1) uses the biological concept to optimize the parameters of the SVM classifier (Table A2) and thus improving the classification accuracy. The main contribution of the study is to exhibit the interrelation between mathematics and engineering.

2. Feature Extraction Based on Discrete Wavelet Transform

Feature extraction can be defined as a mathematical technique used to decode high dimensional data sets into a smaller dimension without losing the content of the actual data set. Thus, feature extraction is an essential segment of the protection scheme which is used to improve the fault classification [35]. In the present work, statistical features are extracted at each level of signal decomposition. The feature includes the energy and entropy of the fault signal.

2.1. Wavelet Transforms

WT has appeared as a powerful signal processing technique for signal decomposition and feature extraction over the traditional Fourier transforms [36,37]. There are two mostly used WTs, namely the continuous wavelet transforms (CWT) and DWT techniques. The CWT of a given $x(t)$ signal can be calculated as:

$$CWT(a,b) = \frac{1}{\sqrt{a}} \int_{-\infty}^{+\infty} x(t)\psi\left(\frac{t-b}{a}\right) dt \quad (1)$$

where, a and b represents the scaling and translation factors. Similarly, the DWT can be defined mathematically as:

$$DWT(m,n) = \frac{1}{\sqrt[m]{a}} \sum_k x(k)\psi\left(\frac{k - nb_0 a_0^m}{a_0^m}\right) dt \quad (2)$$

where, a and b in Equation (1), are transformed to be the functions of integers m, n, k. The DWT technique can be used to effectively recognize non-stationary signals. The decomposition process is performed using the multiresolution analysis technique (MRA). The process is shown in Figure 1. The process begins with a signal passing through both the high and low pass filters. The low pass filter is replicated by the approximation coefficient (c_j) and the high pass filter is replicated by the detail coefficient (d_j). At each level of decomposition, the detail coefficient information from the high pass filter is discarded before the process is re-established in the next level of decomposition. The most important aspect of using WT technique is the proper selection of a mother wavelet. The approximation and detail coefficients are mathematically represented as:

$$c_j = \sum_k x(n).h(2n-k) \tag{3}$$

$$d_j = \sum_k x(n).g(2n-k) \tag{4}$$

where, c_j is the output from the low pass filter which replicates the approximation coefficients of the original signal, and d_j is the output from the high pass filter which replicates the detail coefficient of the original signal.

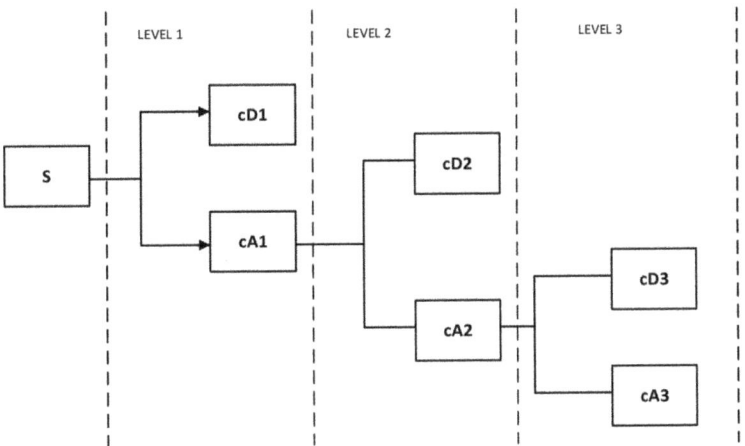

Figure 1. MRA decomposition tree.

2.2. Feature Extraction

Feature extraction techniques are valuable tools used as a foundation to most power system classification and regression problems. Feature extraction techniques are used to reduce high dimension data spectrum to a minimized sizable data spectrum without losing the essence of the original data set. In the current work, the two statistical features extracted from the reconstructed signal are the energy and the entropy of the signal. The energy of the fault current signal $y(t)$ is given by:

$$E(t_1, t_2) = \int_{t_1}^{t_2} [y(t)]^2 \, dt \tag{5}$$

where, t_1, t_2 represents the time range for the energy measurement. The entropy EN of the signal $y(t)$ such that $E(0) = 0$ is given by:

$$EN(y) = \sum_i EN(y_I) \tag{6}$$

where, y_I is the decomposed coefficient of the original signal $y(t)$, and EN is the approximated entropy. A feature matrix is formulated based on the energy and entropy measurements and subsequently used as input to the classifiers.

3. High Impedance Fault Classification Based on Support Vector Machines

Classification is a mathematical process used to identify specific features of interest from a wide range of features. This phenomenon has been widely used in power systems for condition monitoring. It is important to design protection schemes with rigorous pattern recognition abilities to enable prompt protection responses during fault conditions.

3.1. Support Vector Machines

SVMs form part of the statistical learning techniques based on structural risk minimization methods. SVMs have been successfully used in power systems for pattern recognition and classification problems [38]. The objective of using SVMs is to find a separating margin between two different classes of data called the hyperplane. The hyperplane is determined by mapping the input vectors into a high dimensional space. Generally, a hyperplane is set to be optimal under two conditions, (a) if the data classes are separated by a greater margin, and (b) if the distance between the closest data class and the hyperplane is maximal [39]. Suppose we are given the input training data $(x_1, y_1), (x_2, y_2), \cdots (x_l, y_l)$ $x_i \in R^n$, $y_i \in [+1, -1]$. x_i indicates the input patterns, and y_i is the desired target output, for instance, y_i belongs to the -1 class when the data value is -0.1, and y_i belongs to the $+1$ class when the data value is $+0.1$. The input data is mapped into a feature space by means of a non-transformation function Φ.

$$\Phi : R^n \rightarrow F^m \quad x_i \rightarrow \Phi.(x_i) \quad (7)$$

The data are subsequently separated by using the function f in the high dimensional space. The dimensional space (Y^2) where $(Y \in \{+1, -1\})$ is mapped by applying the function f and is given by:

$$f : F^m \rightarrow Y^2 \quad \Phi.(x_i) \rightarrow f(\Phi.(x_i)) \quad (8)$$

Suppose the data are linearly separable, the vector dimension is given $w \in R^N$ and the scaler dimension is given by $b \in R$, such that the desired output $y_i(w.x_i + b) \geq 1 \; \forall$ data parameters within the training set $(i = 1, 2, \cdots l)$. Consequently, the hyperplane can be determined by computing $w.x + b = 1$ for the data points nearer to the plane on one side and, $w.x + b = -1$ for the data points nearer on the other side of the plane. Thus, the optimal hyperplane can be computed by solving the quadratic programming problem given by:

$$\min \frac{1}{2} ||w^2|| + C \left(\sum_{I=1}^{L} \varepsilon_I \right) \quad (9)$$

Subject to $y_i(w.x_i + b) \geq 1 - \varepsilon_I, \varepsilon_I \geq 0 \forall i$.

The optimized problem formulation is solved by using the Langrangian multipliers defined as:

$$\lambda(w, b, a) = \frac{1}{2} ||w^2|| - \sum_{l=1}^{l} \alpha(y_i(w.x_i + b) - 1 \quad (10)$$

where, w, b are the primal variables, and α is the dual variable, which are used to minimize the Langrangian function. The solution vector in terms of the training design is computed by using the Karush–Kuhn–Tucker (KKT) conditions. Consequently, upon solving the optimization problem, the training points with $\alpha_i > 0$ are referred to as support vectors (SVs). The primal variables w and b can be calculated as:

$$w = \sum_{i=1}^{l} \alpha_i y_i x_i \quad (11)$$

$$b = y_{sv} - \sum_{i=1}^{l} \alpha_i y_i . k(x, x_i) \quad (12)$$

where, x represents the test vector. Suppose that $\alpha_i \neq 0$ and sgn is the signal function, the optimal decision function can be expressed mathematically as:

$$f(x) = sgn \left(\sum_{i=1}^{l} \alpha_i y_i . k(x, x_i) + b \right) \tag{13}$$

where, k is the kernel function and it is computed by determining the inner product $\langle \Phi(x_i), \Phi x_j \rangle$ in the feature space as a function of the input data set. The classification of HIF and non-fault conditions is achieved by using different kernel functions, such as the linear, quadratic, and radial bias function 38.

3.2. Decision Tree

The decision (DT) algorithm has been widely used in pattern recognition and classification purposes. The main quality of the DT algorithm is its ability to maximize and fix the data division 39. When using the DT algorithm, the data set is spilt into numerous branches recursively. This process is repeated until the classification efficiency is achieved. Subsequently, the DT algorithm is mathematically defined as:

$$\overline{X} = \{X_1, X_2, \cdots X_m\}^T \tag{14}$$

$$X_i = \{x_1, x_2, \cdots, x_{ij}, \cdots, X_{in}\} \tag{15}$$

$$S = \{S_1, S_2, \cdots, S_i, \cdots, S_m\} \tag{16}$$

where, m, n, and S, represents the number of observations, the independent variable number and the dimensional vector of the variables forecasted from $\overline{X}$. The ith component of the is represented by X_i. The $x_1, x_2, \cdots, x_{ij}, \cdots, X_{in}$ represents the autonomous variable of the component vector, and T is the transpose natation vector. The fundamental desired output of using the DT algorithm is to predict the S, based on the $\overline{X}$ variables. The challenge of using the DT algorithm is to obtain a best possible tree for efficient classification due to high space dimension. It is for such reason that an optimal DT algorithm tree T_{k0} is designed by solving the optimization problem defined mathematically by:

$$\hat{R}(T_{k0}) = \min_k \{\hat{R}T_k\}, k = 1, 2, 3 \ldots, K \tag{17}$$

$$\hat{R}(T) = \sum_{t \in \hat{T}} \{r(t)p(t)\} \tag{18}$$

where, $\hat{R}(T)$ represents the misclassification error of T_k, T_{k0} is the optimal DT algorithm to curtail the classification error, T is the binary tree $\in \{T_1, T_2, T_3 \ldots, T_k, t_1\}$, k,t and t_1 represent the tree index, the tree node and the root node, respectively, $r(t)$ represent the estimate of classification error in node t and $p(t)$ represents the probability of any case that may drop into node t. Generally, the implementation of the DT algorithm is simple and produces good results for classification purposes.

4. Proposed Protection Scheme for High Impedance Fault Detection

The detection of HIF with high accuracy has been a technical problem for protection engineers over the years [39]. The difficulties with HIF detection are well documented and several techniques have been proposed. In this section, the proposed method for HIF detection using DWT-GA-SVM scheme is discussed. The method uses the first cycle of the fault current measured at the source terminal after the occurrence of the fault. Subsequently, the measured current is passed through both the high and low pass filters by means of the DWT technique. The signal is analyzed, thus obtaining the detail and approximation coefficients using the MRA technique. The decomposition process is computed until the fourth level. Subsequently, the statistical features (energy and entropy) are extracted from the detail coefficient of the reconstructed signal at level 4. The GA technique is used to optimize the parameters of the extracted features used to train and test the SVM and DT for classification of HIF and other power system operations. The logic architecture of the proposed method is given in Figure 2.

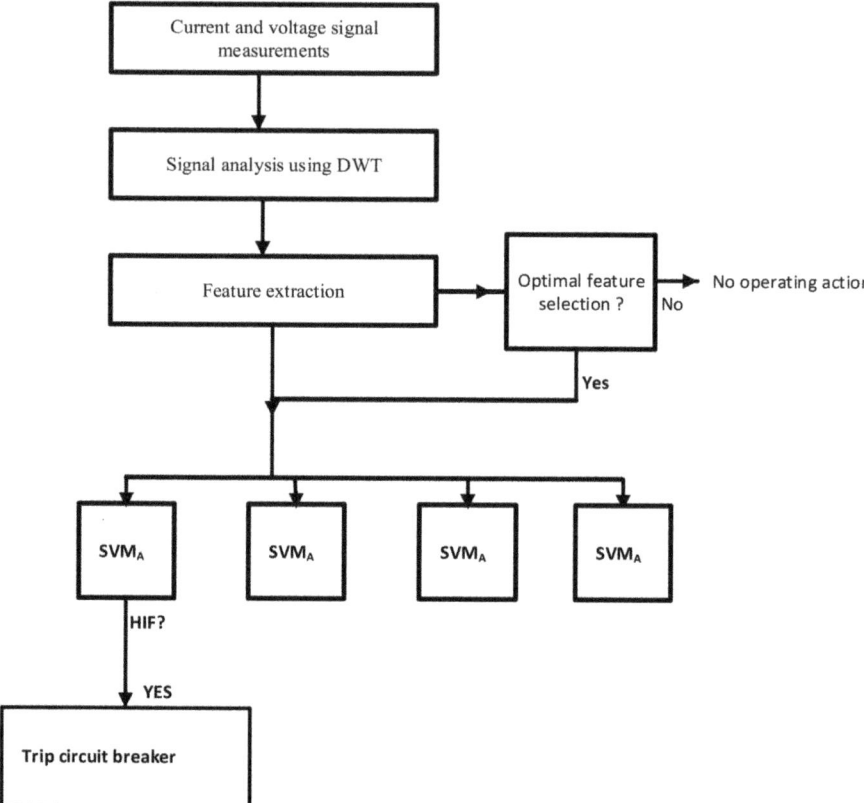

Figure 2. Wavelet transform and support vector machine fault classification scheme.

4.1. Selection of Mother Wavelet

The selection of a mother wavelet is an important aspect of utilizing DWT for signal processing. Essentially, DWT has been used effectively to decompose and extract features from non-stationary signals. In the present work, six (6) mother wavelets were tested using the statistical measures to validate the choice of selection. The comparison was acquired by computing the standard deviation, mean deviation and median absolute deviation. In Table 2, the mother wavelet selection is depicted. From the obtained analysis, the Daubechies (db4) yielded the best results compared to the other mother wavelets and thus was selected for the purposes of the current study.

Table 2. Selection of mother wavelet.

Mother Wavelet	Standard Deviation	Mean Deviation	Median Absolute Deviation
Db4	2.112	2.505	1.955
Db7	3.450	2.913	2.085
Db14	3.551	2.588	2.910
Sym4	2.941	3.528	3.201
Sym7	2.887	3.415	2.580
Sym14	3.815	3.117	3.155

4.2. SVM Implementation

SVMs have been widely used to solve both the classification and the regression problems in power systems. To minimize the computational and design complexity, four (4) SVMs are used and each SVM is trained to classify and detect HIF and other power system operating conditions. The other power system operating conditions include the capacitor switching (CS), load switching (LS) and normal operation (NO). The four (4) SVMs, are arranged chronologically as: SVM_A for HIF, SVM_B for CS, SVM_C for LS, and SVM_D for NO, respectively. The output of each SVM is either +1 or −1, where +1 indicates that an operation has occurred in the corresponding SVM, and −1 means there is no operation in any of the SVMs. In Table 3, the SVM training matrix is depicted. The matrix would then be used to send a trip pulse to a circuit breaker in a case of an HIF. For instance, the output $[+1, -1, -1, -1]$ would correspond to the presence of HIF leading to a decision to operate the circuit breaker.

Table 3. SVM classification matrix.

Type of Incident	SVM_A	SVM_B	SVM_C	SVM_D
HIF	+1	−1	−1	−1
CS	−1	+1	+1	−1
LS	−1	−1	+1	−1
NO	−1	−1	−1	+1

5. Power Distribution System: Case Study

To demonstrate the validity of the proposed method, an Eskom power distribution system is studied. The model is carried out using the DIgSILENT PowerFactory engineering software tool. Eskom is South Africa's dominant electric utility responsible for over 95% of power generation and exporting to some neighbouring countries in Southern Africa. The reduced network segment consists of a substation at 132/22 kV with three outgoing 22 kV feeders named (Siyabuswa 22kV, Verena 22 kV, and Amanda 22kV). The substation parameters and distribution line parameters are shown Tables 4 and 5, respectively. Additionally, in this paper an HIF case is modelled based on an improved model proposed in [30]. The model consists of a sending node model based on an AC source supply, two impedances based on the capacitance, resistance, and inductance of the power system. The other distinctive property regarding HIF cases is the non-linearity of the current magnitude at each cycle of the signal. One major challenge with HIF detection is the similarity in nature with other power system operation signals such as capacitor switching, non-linear load switching, and inductive load switching. It is thus imperative, to develop a scheme that can distinguish HIF from other power system operations to minimize the incorrect tripping of the breaker. Consequently, in this paper, such operations are considered for the validation of the scheme. Moreover, low impedance fault such as single line, double line, and three phase faults are included to improve the validity of the scheme.

Table 4. Substation source parameters.

Substation Source	Short Circuit Power (MVA)	Short Circuit Current (kA)	X/R	X_1/X_0	R_1/R_0
Parameters	1140	15.3	132.1	0.55	0.61

Table 5. Distribution line parameters.

Feeder (22 kV)	Length (km)	Pos = Neg. Seq. $R_1 = R_2$ (Ω/km)	Pos = Neg. Seq. $X_1 = X_2$ (Ω/km)	Zero. Seq. R_0 (Ω/km)	Zero Seq. X_0 (Ω/km)
Siyabuswa	15.3	0.119	0.168	0.145	1.850
Amandla	12.8	0.119	0.168	0.145	1.850
Verena	19.6	0.119	0.167	0.145	1.850

HIF Modelling

The accurate modelling of HIF has been a challenge to many engineers over the years, although several models have been proposed. In the current study we adopted a high resistance and dynamic model to formulate HIF. We assumed that the HIFs exit because of the tree contacting the energized medium voltage line. The dynamic model is given by:

$$\frac{dg}{dt} = \frac{1}{\tau}(G - g) \tag{19}$$

$$G = \frac{|i|}{V_{arc}} \tag{20}$$

$$\tau = Ae^{Bg} \tag{21}$$

where, g, G, represents the time-varying conductance and stationary arc conductance, respectively, the arc current absolute is given by $|i|$, the time constant is given by τ, V_{arc} is the arc voltage parameters, A and B are the arc constants.

6. Results and Discussion

This section discusses the simulation results of HIF and other power system cases. The signal processing of a fault current improves the classification scheme. The DWT has been successfully used to analyze and track points of interest withing a range of a signal. However, the selection of a mother wavelet is an essential part of using DWT efficiently. As depicted in Table 2, a db4 mother wavelet was selected for purposes of signal decomposition. Some of the fourth level decomposition coefficients are depicted in Figure 3. The simulations were performed using the sampling frequency of 12.5 kHz.

The HIF current signal obtained from the simulation platform is shown in Figure 4. The results emphasize the random behavior of HIF signals. As indicated by the signal, the positive and negative cycles exhibit different current magnitudes for a similar fault. In addition, the fault magnitude depletes with time resulting in difficulties of detection. Generally, HIFs are associated with arching. This phenomenon of arching has been widely used to develop the possible detection schemes for HIFs. The HIF arc voltage is shown in Figure 5. The signal indicates a significant increase on the voltage at the incipient of the fault. However, the signal is not uniformly distributed and decreases in magnitude over time.

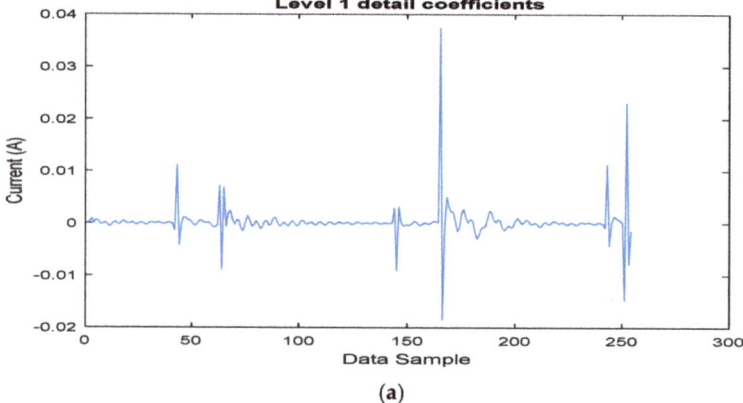

Figure 3. *Cont.*

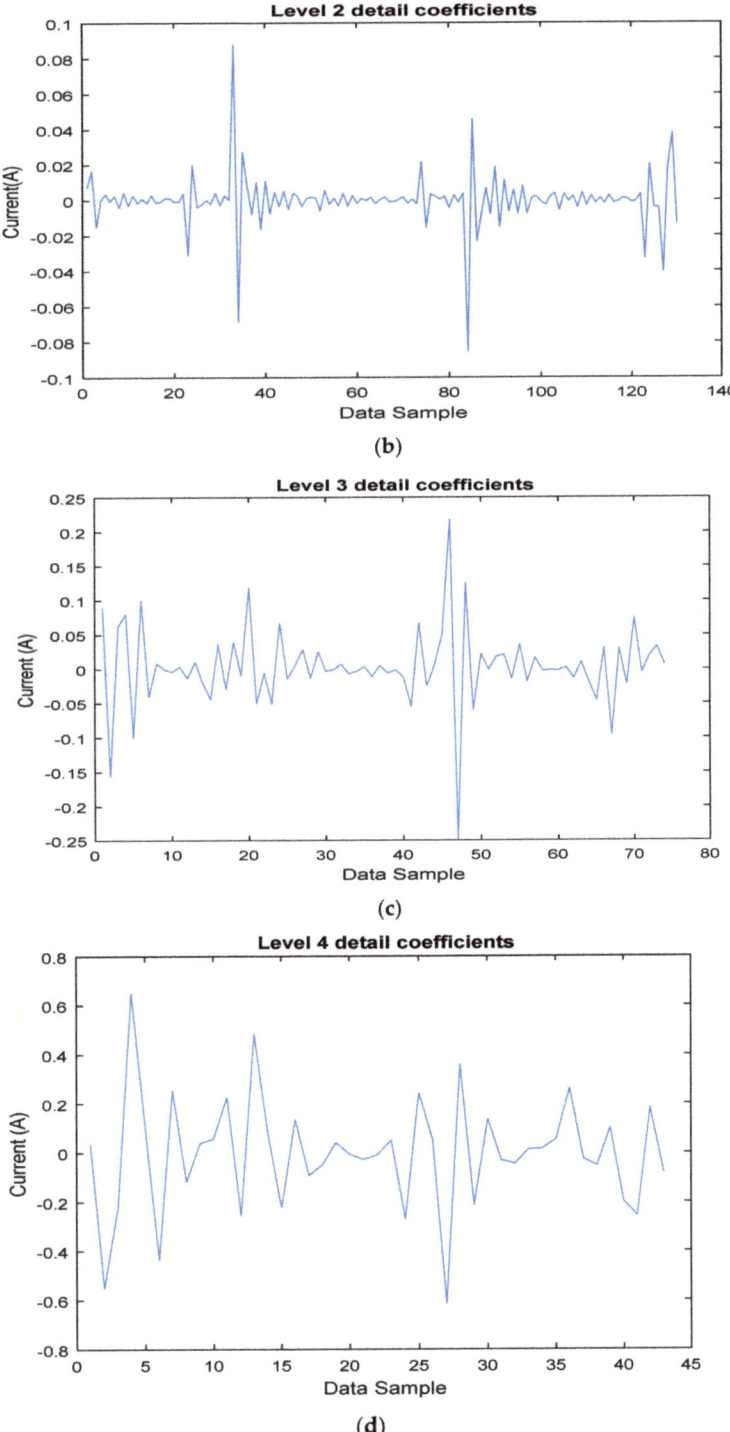

Figure 3. (**a–d**) level decomposition.

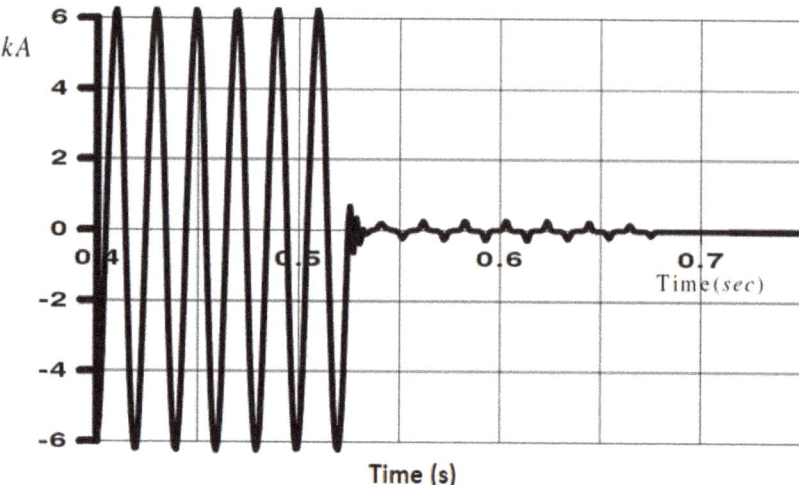

Figure 4. HIF current signal.

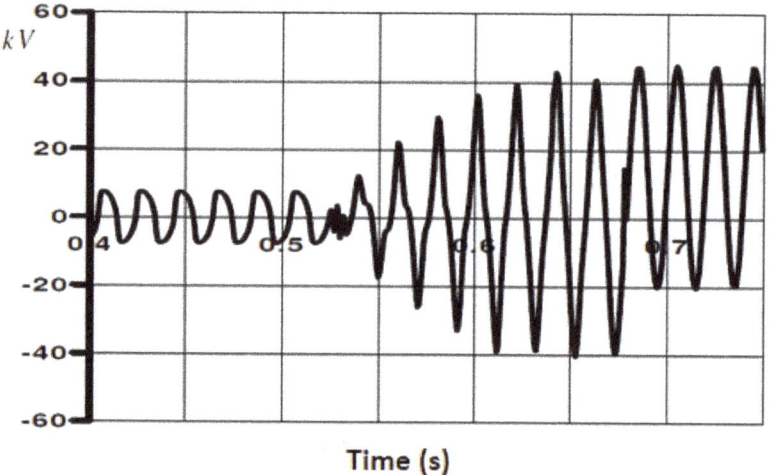

Figure 5. HIF arc voltage.

6.1. Classification of HIF

The art of protection engineering encompasses the segment of classification. In technical terms, classification can be defined as a systematic method of organizing features according to their category. In the present work, the SVM, DT, and NN mathematical techniques are used primarily to classify HIF and other technical operations on the power system. The description of the different signal events considered for the classification schemes is shown in Table 6.

The classification accuracy of signal events (A1–A5) and (A6–A10) using the SVM and DT techniques are shown in Tables 7 and 8, respectively. The results show 97.3% accuracy using SVM for signal events (A1–A5) and 95.5% accuracy using the DT technique, as well as an accuracy level of 98.5% and 97.8% using the SVM and DT techniques for (A1–A5) and (A6–A10), respectively. To demonstrate the effectiveness of using the GA technique for optimal feature selection, a comparison between the optimal features and non-optimal features used for classification is shown in Table 9.

Table 6. Description of different signal events.

Signal Events	Class Identification
High impedance fault	A1
Normal current	A2
Capacitor switching	A3
Inductor switching	A4
Load switching	A5
Line to ground fault	A6
Line to line fault	A7
Line to line to ground fault	A8
Three phase fault	A9
Three phase fault to ground	A10

Table 7. Classification of A1–A5 signal events.

	SVM					DT				
	A1	A2	A3	A4	A5	A1	A2	A3	A4	A5
A1	96	1	0	0	1	95	2	0	0	1
A2	2	94	2	1	0	1	95	0	2	1
A3	2	0	95	1	2	1	0	96	2	2
A4	1	1	0	98	1	1	1	0	97	0
A5	0	2	1	2	96	0	1	2	1	98
Accuracy (%) = 97.3						Accuracy (%) = 95.5				

Table 8. Classification of A6–A10 signal events.

	SVM					DT				
	A6	A7	A8	A9	A10	A6	A7	A8	A9	A10
A6	98	1	0	0	1	97	1	2	0	1
A7	1	93	1	1	0	1	98	0	1	0
A8	1	0	97	1	1	0	1	98	0	0
A9	1	1	0	92	1	0	0	2	95	1
A10	0	1	1	0	98	1	2	1	0	95
Accuracy (%) = 98.5						Accuracy (%) = 97.8				

The performance of the SVM, DT and NN classifiers is presented Tables 9–11, respectively. The results show that SVM has a high accuracy level compared to the DT and NN classifiers. The accuracy precision of the SVM, DT, and NN is given by 97.6%, 96.5%, and 95.4%, respectively.

Table 9. SVM classification performance.

Class	TP	FP	Precision	Recall	F-Measure	ROC
A1	0.952	0.000	0.975	0.966	0.983	0.988
A2	0.919	0.000	1.000	0.965	0.960	0.967
A3	0.966	0.000	0.952	0.958	0.983	0.988
A4	0.935	0.002	0.955	0.938	0.966	0.985
A5	0.953	0.001	0.985	0.933	0.980	0.976
A6	0.950	0.000	1.000	0.960	0.966	0.958
A7	0.961	0.004	0.998	0.955	0.952	0.960
A8	0.911	0.001	0.952	0.961	0.955	0.976
A9	0.915	0.004	0.965	0.979	0.982	0.971
A10	0.973	0.043	0.980	0.982	0.904	0.988
Avg	0.944	0.006	0.976	0.961	0.961	0.976

Table 10. DT classification performance.

Class	TP	FP	Precision	Recall	F-Measure	ROC
A1	0.991	0.000	0.960	1.000	0.991	0.970
A2	0.920	0.000	0.991	0.990	0.990	0.966
A3	0.933	0.000	0.990	0.980	0.995	0.985
A4	0.985	0.000	0.961	0.985	0.977	0.990
A5	0.968	0.003	0.950	0.930	0.988	0.900
A6	0.987	0.000	0.910	0.975	0.970	0.988
A7	0.980	0.014	0.960	0.957	0.966	0.955
A8	0.981	0.010	0.988	0.911	0.960	0.975
A9	0.971	0.035	0.991	0.980	0.991	0.961
A10	0.981	0.043	0.950	0.990	0.955	0.992
Avg	0.969	0.011	0.965	0.970	0.978	0.968

Table 11. NN classification performance.

Class	TP	FP	Precision	Recall	F-Measure	ROC
A1	0.902	0.000	0.915	0.990	0.905	0.955
A2	0.915	0.000	0.943	0.955	0.922	0.950
A3	0.945	0.000	0.955	0.965	0.959	0.980
A4	0.955	0.000	0.940	0.952	0.960	0.970
A5	0.915	0.003	0.933	0.911	0.977	0.911
A6	0.920	0.010	0.990	0.933	0.950	0.916
A7	0.991	0.000	0.982	0.960	0.960	0.958
A8	0.900	0.010	0.922	0.915	0.955	0.988
A9	0.980	0.050	0.985	0.930	0.965	0.966
A10	0.975	0.045	0.975	0.945	0.977	0.911
Avg	0.941	0.012	0.954	0.946	0.953	0.951

The screenshots demonstrating the impact of the kernel function to maximize the hyperplane and thus, improving classification accuracy between HIF and LS, and HIF and CS are depicted in Figures 6 and 7, respectively. As shown in both figures, the hyperplane maximally separates the two different classes of data events and thus improving the accuracy of the protection scheme.

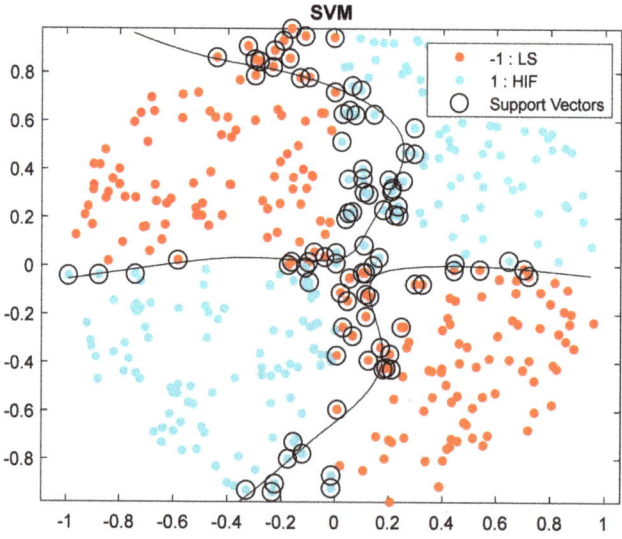

Figure 6. SVM classification between HIF and LS.

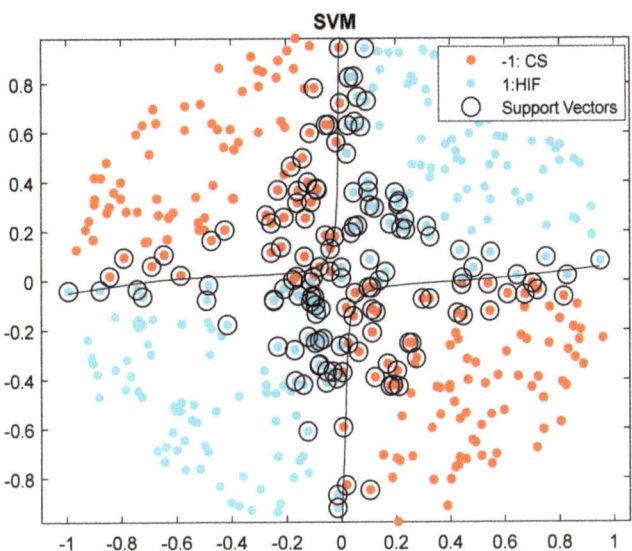

Figure 7. SVM classification between HIF and CS.

6.2. Experimental Analysis

To validate the proposed HIF detection scheme, an experimental setup was conducted. The experiment was conducted at a high voltage engineering lab in Mpumalanga province in South Africa. The experimental data are presented in Table 12.

The electrical circuit and experimental setup for HIF detection is shown in Figures 8 and 9, respectively. The instrument used to perform the experiment is the ICM8 power analyzer. The HIF voltage and current magnitude measured from the experimental setup are shown in Figures 10 and 11, respectively. To accurately measure the experimental parameters, the experimental error of margin must be minimized. These errors occur as a result of the measuring instruments such the voltage transformer, the current transformer, and the ICM8 power analyzer. Another element which may increase the error is the saturation of both the voltage and current transformers. However, in the case of HIF the current magnitude is minimum and such fault cannot lead to saturation. Thus, during the HIF, the current transformer operates linearly.

Table 12. Experimental setup data.

Description	Parameters
Source	5 A, 50 Hz, 2.5% short circuit impedance using transformer
Transformer	10 kVA, 110/132 kV, 4.5%
Capacitor	High Voltage 100 pF, 100 kV, Low voltage 100 nF
Atmospheric conditions	T = 31 °C

The experiment was conducted in two stages. In the first stage, the tree resistance was measured when there was physical contact between the tree and the energized conductor. The voltage applied on the tree was then increased steadily in steps to measure the current and the voltage values and thus, estimated the tree impedance as shown in Figure 12. The dotted line was obtained using the experimental data and the solid line gives the initial resistance value. The resistance value also depends on the atmospheric conditions and the location on the tree where the measurement is obtained. In the second stage of the experiment, the voltage was applied to the conductor while moving the tree very close to the energized conductor resulting in an arc. The movement of the tree from the energized conductor varied between 2–5 cm; in these instances, the arc current was established.

Figure 8. HIF experimental setup.

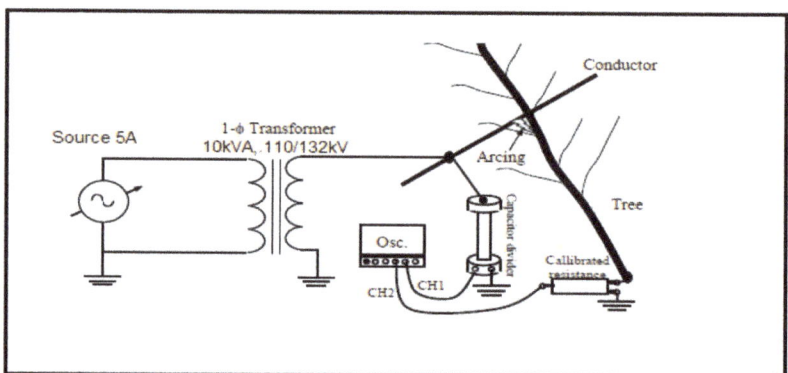

Figure 9. HIF electrical circuit.

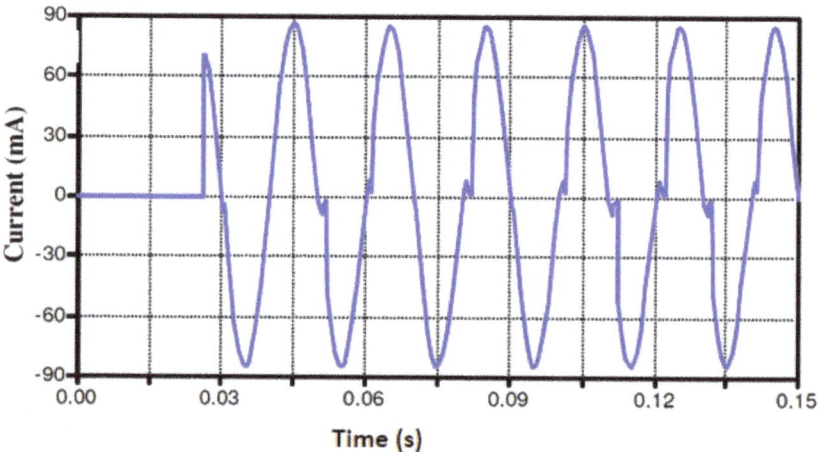

Figure 10. HIF current and voltage measurements.

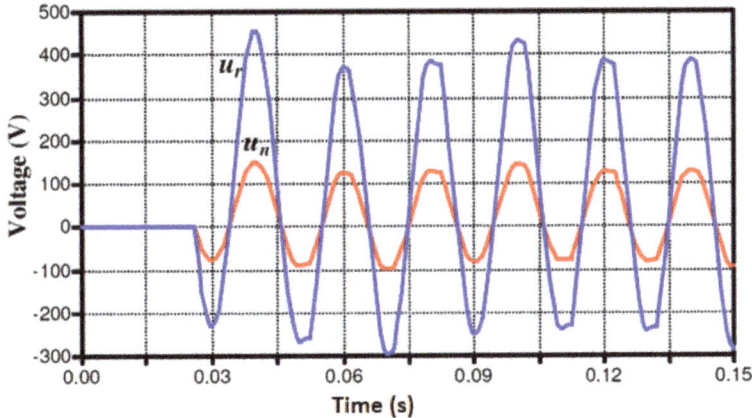

Figure 11. Measure neutral (U_n), and residual voltage (U_r).

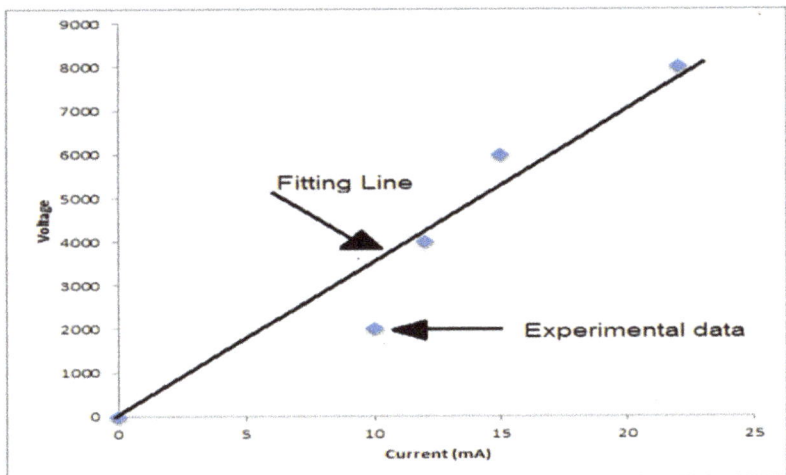

Figure 12. Tree resistance measurement.

The results of the classification validation of the proposed scheme for the simulation and experimental based analysis are presented in Tables 13 and 14, respectively. From the results presented in Table 13, the classification accuracy of SVM is reported to be 97.6% compared to the DT and NN classification results of 96.5% and 95.4%, respectively. In addition, the time required to classify the fault using SVM is reported to be 0.90ms compared to the time required to classify the fault by the DT and NN classifiers. These results emphasize the importance of computational efficiency and processing time reduction. Thus, our proposed model uses less time to detect the fault with high accuracy. The classification accuracy is determined by calculating the ratio between the correctly classified instances and the total instance tested.

Table 13. Classification accuracy of different classifiers using simulation data with GA.

Classifiers and Wavelet	Time Required to Build Model (min)	Time Required to Classify the Fault (ms)	Total Number of Instances	Total Number of Correctly Classified Instances	Total Number of Incorrectly Classified Instances	Accuracy (%)
SVM & DWT	2.51	0.90	4000	3904	96	97.6
DT & DWT	10.3	0.92	4000	3860	140	96.5
NN & DWT	15.3	0.95	4000	3816	184	95.4

Table 14. Classification accuracy of different classifiers using experimental data with GA.

Classifiers	Time Required to Build Model (min)	Time Required to Classify the Fault (ms)	Total Number of Instances	Total Number of Correctly Classified Instances	Total Number of Incorrectly Classified Instances	Accuracy (%)
SVM	1.25	0.75	100	87	13	87.0
DT	3.22	0.88	100	85	15	85.0
NN	4.50	0.91	100	83	17	83.0

7. Conclusions

In this study, we proposed an HIF detection technique. The technique uses mathematical models for signal analysis, feature extraction, optimization, and classification to detect the HIF. The study of HIF has been conducted over many years to find an optimal solution. Generally, HIF can cause severe consequences such as infrastructure damage and possible human fatalities. It therefore is imperative to design a protection scheme that will effectively detect HIF. In the present work, a hybrid mathematical protection scheme is proposed; the scheme uses DWT to decompose and analyze different fault signals using the db4 mother wavelet. Thereafter, the statistical features from the decomposed signals are extracted to build a feature matrix. Consequently, the feature matrix is used to test, train, and validate the SVM classifier. The GA is used to improve the performance of the classifier. The results presented depict that HIF can be detected with an accuracy level of 97.6% using SVM compared to the DT and NN classifier with accuracy levels of 96.5% and 95.4%, respectively, for simulation-based data instances. Finally, we conducted an experiment to test the validity of the proposed method. The experimental results show that the HIF can be classified correctly using SVM with an accuracy of 84% compared to the classification accuracy 85% and 83% when using the DT and NN classifiers, respectively. The development of a hybrid protection for HIF detection is made possible by using mathematical models. In future studies, the HIF detection with high penetration levels of renewable distributed generation will be considered.

Author Contributions: Conceptualization, K.M. and I.D.; methodology, K.M.; software, K.M.; validation, K.M and I.D.; formal analysis, K.M.; investigation, K.M.; resources, I.D., data curation, K.M.; writing—original draft preparation, K.M.; writing—review and editing, I.D.; visualization, I.D.; supervision, I.D.; project administration, K.M.; funding acquisition, I.D. All authors have read and agreed to the published version of the manuscript.

Funding: This research received no external funding.

Institutional Review Board Statement: Not applicable.

Informed Consent Statement: Not applicable.

Data Availability Statement: Not applicable.

Conflicts of Interest: The authors declare no conflict of interest.

Appendix A

In this section, the GA, and SVM parameters are given in Tables A1 and A2, respectively.

Table A1. GA parameters.

Parameter	Value
Probability of mutation (pm)	0.005
Probability of cross over (pc)	0.010
Population size (N)	1000

Table A2. SVM parameters.

SVM	RBF Kernel Parameters
SVM_A	$\gamma = 50\ \sigma^2$
SVM_B	$\gamma = 50\ \sigma^2$
SVM_C	$\gamma = 50\ \sigma^2$
SVM_D	$\gamma = 50\ \sigma^2$

References

1. Moloi, K.; Yusuff, A.A. A Support Vector Machine Based Fault Diagnostic Technique in Power Distribution Networks. In Proceedings of the 2019 Southern African Universities Power Engineering Conference/Robotics and Mechatronics/Pattern Recognition Association of South Africa (SAUPEC/RobMech/PRASA), Cape Town, South Africa, 28–30 January 2019.
2. Magagula, X.G.; Hamam, Y.; Jordaan, J.A.; Yusuff, A.A. A Fault Classification and Localization Method in A Power Distribution Network. In Proceeding of the IEEE Africon, Cape Town, South Africa, 18–20 September 2017.
3. Aucoin, B.M.; Jones, R.H. High Impedance Fault Implementation. *IEEE Trans. Power Deliv.* **1996**, *11*, 139–148. [CrossRef]
4. Silva, S.; Costa, P.; Gouvea, M.; Lacerda, A.; Alves, F.; Leite, D. High impedance fault detection in power distribution systems using wavelet transform and evolving neural network. *Electr. Power Syst. Res.* **2018**, *154*, 474–483. [CrossRef]
5. Nikander, A.; Jarventausta, P. Methods for earth fault identification and distance estimation in a compensated medium voltage distribution network. In Proceedings of the Energy Management and Power Delivery, Singapore, 3–5 March 1998.
6. AL-Dabbagh; Technisearch, M. High Impedance Fault Detector. Australian Patent PL3451, July 1992.
7. Jeerings, I.; Linders, I.R. A Practical Protective Relay for Down Conductor Faults. *IEEE Trans. Power Deliv.* **1991**, *6*, 565–574. [CrossRef]
8. Reason, J. Relay Detects Down Wires by Fault Current Harmonics. *Electr. World* **1994**, *208*, 58–59.
9. Li, K.K.; Chan, W.L. Novel Methods for High-Impedance Ground Fault Protection in Low-Voltage Supply Systems. *Electr. Power Compon. Syst.* **2003**, *33*, 1133–1150. [CrossRef]
10. Cui, Q.; El-Arroudi, K.; Weng, Y. A Feature Selection Method for High Impedance Fault Detection. *IEEE Trans. Power Deliv.* **2019**, *34*, 1203–1215. [CrossRef]
11. Langeroudi, A.T.; Abdelaziz, M.M. Preventative high impedance fault detection using distribution system state estimation. *Electr. Power Syst. Res.* **2020**, *186*, 1–11. [CrossRef]
12. Chen, J.C.; Phung, B.T.; Wu, H.W.; Zhang, D.M.; Blackburn, T. Detection of High Impedance Faults using wavelet transform. In Proceedings of the Australasian Universities Power Engineering Conference (AUPEC), Perth, Australia, 28 September–1 October 2014.
13. Lai, T.; Snider, L.; Lo, E.; Sutanto, D. High-Impedance Fault Detection Using Discrete Wavelet Transform and Frequency Range and RMS Conversion. *IEEE Trans. Power Deliv.* **2005**, *20*, 397–407. [CrossRef]
14. Sedighi, A.-R.; Haghifam, M.-R.; Malik, O.; Ghassemian, M.-H. High Impedance Fault Detection Based on Wavelet Transform and Statistical Pattern Recognition. *IEEE Trans. Power Deliv.* **2005**, *20*, 2414–2421. [CrossRef]
15. Baqui, I.; Zamora, I.; Mazón, J.; Buigues, G. High impedance fault detection methodology using wavelet transform and artificial neural networks. *Electr. Power Syst. Res.* **2011**, *81*, 1325–1333. [CrossRef]
16. Huang, S.-J.; Hsieh, C.-T. High-impedance fault detection utilizing a Morlet wavelet transform approach. *IEEE Trans. Power Deliv.* **1999**, *14*, 1401–1410. [CrossRef]
17. Elkalashy, N.I.; Lehtonen, M.; Darwish, H.A.; Taalab, A.-M.I.; Izzularab, M.A. DWT-based extraction of residual currents throughout unearthed MV networks for detecting high-impedance faults due to leaning trees. *Eur. Trans. Electr. Power* **2007**, *17*, 597–614. [CrossRef]
18. Tyska, W.; Russell, B.D.; Aucoin, B.M. A Microprocessor-Based Digital Feeder Monitor with High Impedance Fault Detection. In Proceedings of the 47th Annual Texas A&M Relay Conference, College Station, TX, USA, 21–23 March 1994.
19. Butler, K.; Momoh, J. Neural network-based classification of arcing faults in a power distribution system. In Proceedings of the Protection of the North American Power symposium, Washington, DC, USA, 11–12 October 1993.
20. Snider, L.; Yuen, Y. The artificial neural-networks-based relay algorithm for the detection of stochastic high impedance faults. *Neurocomputing* **1998**, *23*, 243–254. [CrossRef]
21. Lai, T.M.; Snider, L.A.; Lo, E.; Cheung, C.H.; Chan, K.W. High impedance faults detection using artificial neural network. In Proceedings of the Sixth International Conference on Advances in Power System Control, Operation and Management, Hong Kong, China, 11–14 November 2003.
22. Moloi, K.; Jordaan, J.A.; Hamam, Y. A hybrid method for high impedance fault classification and detection. In Proceedings of the Southern African Universities Power Engineering Conference/Robotics Mechatronics/Pattern Recognition Association of South Africa, Bloemfontein, South Africa, 28–30 January 2019.
23. Sheng, Y.; Rovnyak, S. Decision Tree-Based Methodology for High Impedance Fault Detection. *IEEE Trans. Power Deliv.* **2004**, *19*, 533–536. [CrossRef]

24. Shahrtash, S.M.; Sarlak, M. High Impedance Fault Detection Using Harmonics Energy Decision Tree Algorithm. In Proceedings of the International Conference on Power System Technology, Chongqing, China, 28 September–1 October 2006.
25. Rai, K.; Hojatpanah, F.; Ajaei, F.B.; Grolinger, K. Deep Learning for High-Impedance Fault Detection: Convolutional Autoencoders. *Energies* **2021**, *14*, 3623. [CrossRef]
26. Sulaiman, M.B.; Tawafan, A.H.; Ibrahim, Z.B. Detection of High Impedance Fault Using Probabilistic Neural Network Classifier. *J. Theor. Appl. Inf. Technol.* **2013**, *53*, 180–191.
27. Wang, S.; Dehghanian, P. On the Use of Artificial Intelligence for High Impedance Fault Detection and Electrical Safety. *IEEE Trans. Ind. Appl.* **2020**, *56*, 7208–7216. [CrossRef]
28. Lala, H.; Karmakar, S. Detection and Experimental Validation of High Impedance Arc Fault in Distribution System Using Empirical Mode Decomposition. *IEEE Syst. J.* **2020**, *14*, 3494–3505. [CrossRef]
29. Yang, M.T.; Gu, J.C.; Guan, J.L.; Cheng, C.Y. Detection of high impedance faults in distribution system. In Proceedings of the 2005 IEEE/PES Transmission & Distribution Conference & Exposition: Asia and Pacific, Dalian, China, 18 August 2005.
30. Wang, S.; Dehghanian, P.; Li, L.; Wang, B. A machine learning approach to detection of geomagnetically induced currents in power grids. *IEEE Trans. Ind. Appl.* **2019**, *56*, 1098–1106. [CrossRef]
31. Wang, S.; Dehghanian, P.; Li, L. Power grid online surveillance through PMU-embedded convolutional neural networks. *IEEE Trans. Ind. Appl.* **2019**, *56*, 1146–1155. [CrossRef]
32. Yusuff, A.A.; Fei, C.; Jimoh, A.; Munda, J.L. Fault location in a series compensated transmission line based on wavelet packet decomposition and support vector regression. *Electr. Power Syst. Res.* **2011**, *81*, 1258–1265. [CrossRef]
33. Jadhav, A.; Thakur, K. Fault Detection and Classification in Transmission Lines Based on Wavelet Transform. *Int. J. Sci. Eng. Res.* **2015**, *3*, 14–19.
34. Keerthana, G.; Umayal, S.P. Analysis of faults in transmission line with the help of discrete wavelet transform. In Proceedings of the International Conference on Current Research in Engineering Science and Technology, Hyderabad, India, 5–27 October 2016.
35. Ray, P.K.; Mohanty, S.R.; Kishor, N.; Catalão, J.P.S. Optimal Feature and Decision Tree-Based Classification of Power Quality Disturbances in Distributed Generation System. *IEEE Trans. Sustain. Energy* **2014**, *5*, 200–208. [CrossRef]
36. Moloi, K.; Jordaan, J.A.; Hamam, Y. High impedance fault detection technique based on Discrete Wavelet Transform and support vector machine in power distribution networks. In Proceedings of the IEEE Africon, Cape Town, South Africa, 18–20 September 2017.
37. HLivani, H.; Evrenosoglu, C.Y. A Fault Classification and Localization Method for Three-Terminal Circuits Using Machine Learning. *IEEE Trans. Power Deliv.* **2013**, *28*, 2282–2290. [CrossRef]
38. Nakho, A.; Moloi, K.; Hamam, Y. Hig Impedance Fault Detection Based on HS-Transform and Decision Tree Techniques. In Proceedings of the IEEE Southern African Universities Power Engineering Conference/Robotics and Mechatronics/Pattern Recognition Association of South Africa, Potchefstroom, South Africa, 27–29 January 2021.
39. Sedighizadeh, M.; Rezazadeh, A.; Elkalashy, N.I. Approaches in High Impedance Fault Detection—A Chronological Review. *Adv. Electr. Comput. Eng.* **2010**, *10*, 114–128. [CrossRef]

Article

A 2D Multi-Layer Model to Study the External Magnetic Field Generated by a Polymer Exchange Membrane Fuel Cell

Antony Plait and Frédéric Dubas *

Département ENERGIE, FEMTO-ST, CNRS, Univ. Bourgogne Franche-Comté, F-90000 Belfort, France
* Correspondence: frederic.dubas@univ-fcomte.fr; Tel.: +33-38-458-3648

Abstract: An original innovative two-dimensional (2D) multi-layer model based on the Maxwell–Fourier method for the diagnosis of a polymer exchange membrane (PEM) fuel cell (FC) stack is presented. It is possible to determine the magnetic field distribution generated around the PEMFC stack from the (non-)homogenous current density distribution inside the PEMFC stack. Analysis of the magnetic field distribution can indicate whether the FC is healthy or faulty. In this way, an explicit, accurate and fast analytical model can allow the health state of an FC to be studied. To evaluate the capacity and the efficiency of the 2D analytical model, the distribution of local quantities (i.e., magnetic vector potential and magnetic field) in a PEMFC stack has been validated with those obtained by the 2D finite-element analysis (FEA). The comparisons demonstrate excellent results both in terms of amplitude and waveform. The validation of this 2D analytical model is essential for the subsequent generation of an inverse model useful for the diagnosis of a PEMFC.

Keywords: diagnosis; fuel cell; magneto-tomography; Maxwell–Fourier method; multi-layer model; finite-element analysis

MSC: 35A24; 34M25

1. Introduction

1.1. Context of the Paper

Currently, the fuel cell (FC) system presents the advantage of combining high-energy efficiency (e.g., FC efficiency about 50% and heat engine efficiency about 32%) and environmental friendliness compared to other more conventional technologies [1]. This presents an interesting combination for conventional power generation systems, especially in stationary and portable FC applications [2–4]. Furthermore, a polymer exchange membrane fuel cell (PEMFC), which is the most commonly used FC in transport applications, is an interesting alternative to the internal combustion engine [5].

In order to transform fuel cells in to a mass-market technology, many challenges must be overcome. For that, it is necessary to estimate and comprehend the possible performance of PEMFC in automotive applications as analyzed in [6]. Hydrogen is a promising energy resource. In order to obtain an overview, various studies are based on economics and storage, while recommendations [7] or a critical view can be given on technologies, applications, trends and challenges [8].

The main application of FCs is in the automotive industry. A comprehensive review and research on FC electrical vehicles are presented in [9]. Many topics such as topologies, power electronic converters, energy management methods, technical challenges, marketing and future aspects are discussed.

Some research studies are conducted on the degradation mechanism and impacts on FC degradation with a focus in hybrid transport applications [10]. Similarly, FC diagnosis methods for embedded automotive applications are detailed in [11].

Many factors such as temperature, reactant humidity, partial pressures of feed gases, obstruction of gas diffusion layers, degradation of the membrane electrode assembly

(MEA), flow field structure in combination with MEA, etc., could affect the current density distribution and consequently the PEMFC performance [12]. In order to diagnose the PEMFC, an abundance of methods have been developed from physical models [13–15] or experimental data information [16]. These can be divided into two categories [17], viz., intrusive and non-intrusive methods. Often, in order to study the (non-)homogeneous current density distribution inside a PEMFC reflecting an FC failure or a healthy FC, intrusive and inflexible methods are applied. Furthermore, the feedback effect of these methods on the measured current density distribution is uncertain. Therefore, a non-destructive control method for the diagnosis, at reasonable costs, will help in PEMFC deployment [18]. Magneto-tomography is one of the most efficient and easiest non-intrusive methods that comprises the mapping of an external magnetic field produced by any electrical device. This measurement technique allows researchers to deduce the current density distribution inside the device [19] by solving an inverse problem [20,21]. In [12], magneto-tomography is for the first time applied to the analysis and diagnosis of a PEMFC stack. The current density distribution can be evaluated by the Biot–Savart law [12,21–31], the varying-network electric circuit [30,31] or the heuristic search method [32,33] where the external magnetic field is determined by

- numerical method [e.g., finite-element analysis (FEA), volume finite method, etc.], viz., (i) two-dimensional (2D) [27,28], or (ii) three-dimensional (3D) [25–28,32,33];
- and, experimental data from magnetic sensors, viz., (i) uniaxial [22,23,25,26,30,31,33], or (ii) triaxial [12,24,34].

Experimental validations have been realized on

- FC emulator/simulator [25,30,32,35,36];
- single cell [12,22–24,27];
- and, real FC stack [24–26,30,31,33,34].

Moreover, studies were performed on the external magnetic field determination by imposing the (non-)homogeneous current density distribution inside a PEMFC with 3D numerical [35–37] and experimental [35] validations. It is interesting to note that a ferromagnetic circuit placed around a PEMFC emulator permits concentration of the external magnetic field and, consequently, the detection amplification of an FC failure [35]. The actual methods to determine the current density distribution can be globally described as redundant, time-consuming and have variable accuracy.

1.2. Objective of the Paper

In the literature, to the best of the authors' knowledge, there is no 2D multi-layer model based on the Maxwell–Fourier method for the diagnosis of a PEMFC stack. This method is based on the formal resolution of Maxwell's magnetostatic equations in Cartesian coordinates (x, y) by using the separation of variables method, the Fourier's series and the superposition principle. This purely analytical multi-layer model could give a perfect theoretical study of the magnetic field distribution with a rapid computation time. Moreover, this model could present more precision than the usual methods. Here, the magneto-tomography study is realized on a real bipolar plate of a PEMFC stack (without ferromagnetic circuit placed around the PEMFC stack) in order to diagnose the health state of this one, viz., (i) healthy FC, and (ii) FC failure. As reported in [35–37], the developed analytical model will allow the magnetic field distribution generated around the PEMFC to be determined by imposing the (non-)homogeneous current density distribution inside the FC. The reverse study, which is the second step of this scientific work, could then be easily performed on the 2D multi-layer model. However, this is beyond the scope of this paper. To evaluate the efficiency of the proposed analytical model, the distribution of local quantities (i.e., magnetic vector potential and magnetic field) for a healthy FC and an FC failure has been validated with those obtained by the 2D FEA [38]. The comparisons demonstrate excellent results both in terms of amplitude and waveform. The validation of this 2D analytical model is essential for the subsequent generation of an inverse model. This kind of model, which has never been previously developed for FC diagnosis application, must first be validated (e.g., by numerical validation) in order to finally obtain a powerful diagnostic tool of the FC operating mode.

2. Mathematical Development of 2D Multi-Layer Model

2.1. Problem Description

2.1.1. Bipolar Plate of a PEMFC Stack

Usually, a PEMFC stack (of dimension $w_{fc} \times h_{fc} \times L_{fc}$), with a related constant current of I_{fc}, consists of elements in parallel, i.e., compression of two end-plates, two current collectors and several unit cells. Each unit cell is composed of two bipolar plates with an inactive zone (seal) and an electrochemical active cell (of dimension $w_a \times h_a$), two gas diffusion layers and MEA. In electrochemical active cells, at the MEA level, the current density $J = \{0; 0; J_z\}$, which is perpendicular to the cells with a constant value along the z-axis, is directly related to the flow of hydrogen protons travelling through the electrolyte membrane. Outside of the electrochemical active cell, J is neglected. An FC failure is characterized by a non-homogeneous distribution of J [12]. The considered bipolar plate for the diagnosis of a PEMFC stack, with the geometrical and physical parameters, is illustrated in Figure 1a. This system is surrounded by a vacuum via an infinite box (of dimension $w \times h$).

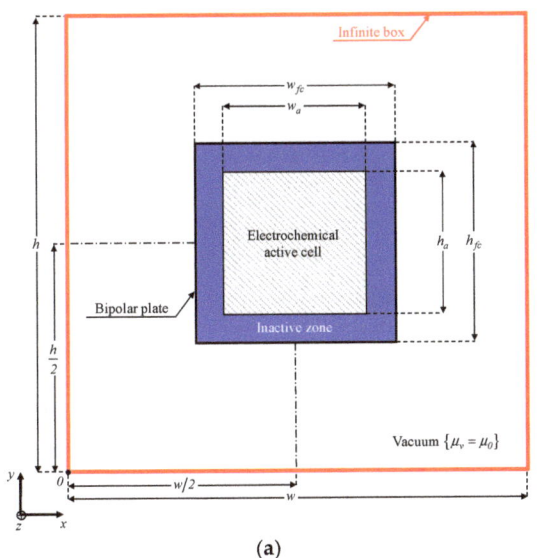
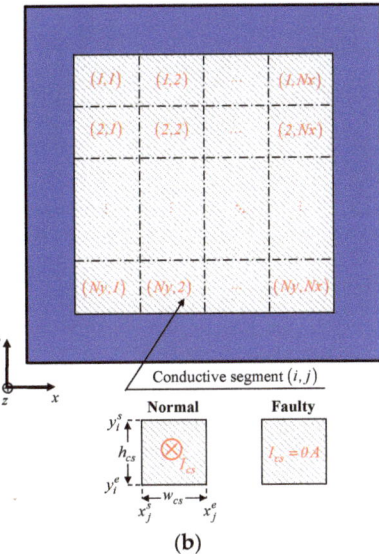

(a) (b)

Figure 1. (a) Bipolar plate and (b) Electrochemical active cell modeled by $Ny \times Nx$ conductive segments for the diagnosis of a PEMFC stack (see Table 1 for the various parameters).

Table 1. Geometrical and physical parameters of a PEMFC stack.

Symbol	Quantity	Values
$S_{fc} = w_{fc} \times h_{fc}$	FC surface	$14 \times 14 = 196$ cm^2
$S = w \times h$	Infinite box surface	$42 \times 42 = 1764$ cm^2
$S_a = w_a \times h_a$	Electrochemical active cell surface	$10 \times 10 = 100$ cm^2
$Ny \times Nx$	Number of conductive segments	$25 \times 20 = 500$
$S_{cs} = w_{cs} \times h_{cs}$	Conductive segment surface	$5 \times 4 = 20$ mm^2
L_{fc}	FC depth	10 cm

2.1.2. Normal and Faulty Conductive Segments

To simulate (non-)homogeneous distribution of J, this electrochemical active cell of bipolar plates can be decomposed into $Ny \times Nx$ regular conductive segments (of dimension $S_{cs} = w_{cs} \times h_{cs}$ with $w_{cs} = w_a/Nx$ and $h_{cs} = h_a/Nx$), as shown in Figure 1b, where $Ny \in \mathbb{N}^*$ and $Nx \in \mathbb{N}^*$ are the segmentation number in the y- and x-axis, respectively. In Cartesian coordinates (x, y), the starting and ending points of conductive segments in both axes can be defined by

$$x_j^s = \frac{(w - w_a)}{2} + (j - 1) \cdot w_{cs} \text{ and } x_j^e = x_j^s + w_{cs}, \tag{1a}$$

$$y_i^s = \frac{(h + h_a)}{2} - (i - 1) \cdot h_{cs} \text{ and } y_i^e = y_i^s - h_{cs}, \tag{1b}$$

with $j \in \{\mathbb{N}^*, Nx\}$ and $i \in \{\mathbb{N}^*, Ny\}$.

Any conductive segments (i, j) may be faulty (not supplied) or normal (supplied). For a non-homogeneous distribution of J reflecting an FC failure, some conductive segments are not supplied (i.e., the current are assumed to be null in the conductive segment). For a homogenous distribution of J reflecting a healthy FC, all conductive segments are supplied by direct current I_{cs} symbolized by $\otimes$ equal to

$$I_{cs} = \frac{I_{fc}}{N_{cs}^n}, \tag{2}$$

with N_{cs}^n the number of conductive segments in a normal condition defined by

$$N_{cs}^n = \sum_{i,j} M_{i,j}^{cs}, \tag{3}$$

where $M_{i,j}^{cs}$ represents the element (i, j) of the power supply matrix of the conductive segments M^{cs} (of dimension $Ny \times Nx$), viz.,

$$M_{i,j}^{cs} = \begin{cases} 0 & \text{if \textbf{faulty} condition,} \\ 1 & \text{if \textbf{normal} condition.} \end{cases} \tag{4}$$

2.1.3. Modeling of Spatial Current Density Distribution in any Conductive Segments (i, j)

Figure 2 illustrates the waveform of the spatial distribution of J_{zij} in any conductive segments (i, j), which can be expressed as a Fourier's series with $x \in [0, w]$:

$$J_{zij} = \sum_k J_{i,j,k}^{cs} \cdot \sin(\beta_k \cdot x), \tag{5a}$$

$$J_{i,j,k}^{cs} = \frac{2 \cdot J_{max}^{cs}}{\beta_k \cdot w} \cdot F_{i,j,k}^{cs}, \tag{5b}$$

with

$$F_{i,j,k}^{cs} = M_{i,j}^{cs} \cdot \left[\cos\left(\beta_k \cdot x_j^s\right) - \cos\left(\beta_k \cdot x_j^e\right)\right] \tag{5c}$$

where $J_{max}^{cs} = I_{cs}/S_{cs}$ is the maximum current density in each conductive segment, $\beta_k = k\pi/w$ is the spatial frequency (or periodicity) of J_{zij} with $k \in \{\mathbb{N}^*, K_{max}\}$ the spatial harmonic orders in which K_{max} is the finite number of spatial harmonics terms.

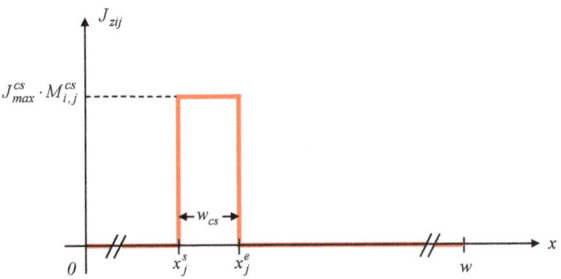

Figure 2. Waveform of the spatial distribution of J_{zij} in any conductive segments (i, j).

2.2. List of Assumptions

The 2D magnetic field distribution generated around the PEMFC stack has been studied from multi-layer model by solving Maxwell's magnetostatic equations in Cartesian coordinates (x, y). In this analysis, the simplifying assumptions of the analytical model are:

- the end-effects are neglected (i.e., the magnetic variables are independent of z);
- the magnetic vector potential and current density have only one component along the z-axis, i.e., $A = \{0; 0; A_z\}$ and $J = \{0; 0; J_z\}$;
- the skin depth effect is not considered;
- many industrial PEMFC stacks use graphite bipolar plates and non-magnetic stainless steels for rods and other mechanical parts [35]. Thus, the absolute magnetic permeability μ of all components used in bipolar plates of PEMFC stack are very close to the vacuum magnetic permeability $\mu_v = \mu_0 = 4\pi \times 10^{-7}\ H/m$.

2.3. Problem Discretization in Regions

As shown in Figure 3, the bipolar plate for the diagnosis of a PEMFC stack is divided into three regions $\forall x \in [0, w]$, viz.,

- Vacuum above the electrochemical active cell: Region 1 (R1) for $y \in [y_1^s, h]$ with $\mu_1 = \mu_v = \mu_0$;
- The electrochemical active cell: Region 2 (R2) for $y \in \left[y_{Ny}^e, y_1^s\right]$, which is divided in the y-axis into sub-regions i (R2$_i$) for $y \in [y_i^e, y_i^s]$, with $\mu_2 = \mu_v = \mu_0$;
- Vacuum below the electrochemical active cell: Region 3 (R3) for $y \in \left[0, y_{Ny}^e\right]$ with $\mu_3 = \mu_v = \mu_0$.

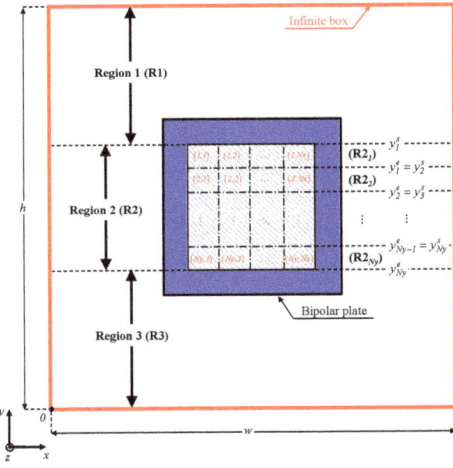

Figure 3. Definition of regions in the bipolar plate for the diagnosis of a PEMFC stack.

2.4. Governing Partial Differential Equations (PDEs) in Cartesian Coordinates

In quasi-stationary, the magnetostatic Maxwell's equations are represented by Maxwell–Ampère [39]

$$\nabla \times \boldsymbol{H} = \boldsymbol{J} \ (with \ \boldsymbol{J} = 0 \ for \ the \ no-load \ \text{operation}), \quad (6)$$

Maxwell–Thomson

$$\nabla \cdot \boldsymbol{B} = 0 \ (Magnetic \ flux \ conservation), \quad (7a)$$

$$\boldsymbol{B} = \nabla \times \boldsymbol{A} \ with \ \nabla \cdot \boldsymbol{A} = 0 \ (Coulomb's \ gauge), \quad (7b)$$

and the magnetic material equation

$$\boldsymbol{B} = \mu \cdot \boldsymbol{H} + \mu_0 \cdot \boldsymbol{Mr} \quad (8)$$

where $\boldsymbol{B}$, $\boldsymbol{H}$ and $\boldsymbol{Mr}$ are the magnetic flux density, the magnetic field and the remanent (or residual) magnetization vector (with $\boldsymbol{Mr} = 0$ for other materials or $\boldsymbol{Mr} \neq 0$ for the magnets), respectively.

Using (6)~(8) with the general assumptions of the study, in Cartesian coordinates (x, y), the general PDEs of magnetostatic in terms of $\boldsymbol{A} = \{0; 0; A_z\}$ inside an isotropic and uniform material (i.e., $\mu = C^{ste}$) can be expressed by the Laplace's equation in (R1)

$$\Delta A_{z1} = \frac{\partial^2 A_{z1}}{\partial x^2} + \frac{\partial^2 A_{z1}}{\partial y^2} = 0, \quad (9a)$$

and in (R3)

$$\Delta A_{z3} = \frac{\partial^2 A_{z3}}{\partial x^2} + \frac{\partial^2 A_{z3}}{\partial y^2} = 0, \quad (9b)$$

and the Poisson's equation in (R2$_i$)

$$\Delta A_{z2_i} = \frac{\partial^2 A_{z2_i}}{\partial x^2} + \frac{\partial^2 A_{z2_i}}{\partial y^2} = -\mu_0 \cdot J_{z2_i} \quad (9c)$$

where J_{z2_i} is the spatial distribution of $\boldsymbol{J}$ along the x-axis for each i, i.e., in (R2$_i$), which is defined by

$$J_{z2_i} = \sum_j J_{zij} = \sum_k J^{cs}_{i,k} \cdot \sin(\beta_k \cdot x), \quad (10a)$$

$$J^{cs}_{i,k} = \sum_j J^{cs}_{i,j,k} = \frac{2 \cdot J^{cs}_{max}}{\beta_k \cdot w} \cdot \sum_j F^{cs}_{i,j,k}. \quad (10b)$$

For example, with $Ny \times Nx = 2 \times 10$ conductive segments, Figure 4 illustrates the waveforms of J_{z2_i} for each i in (R2$_i$) obtained by (10) and for

$$M^{cs} = \begin{bmatrix} 1 & 1 & 1 & 1 & 1 & 1 & 1 & 1 & 1 \\ 1 & 0 & 1 & 1 & 1 & 0 & 0 & 1 & 0 & 1 \end{bmatrix}. \quad (11)$$

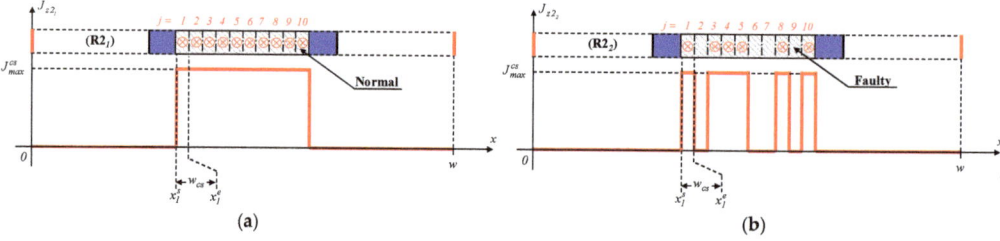

Figure 4. Spatial distribution of: (**a**) J_{z2_1} in (R2$_1$), and (**b**) J_{z2_2} in (R2$_2$).

Using (7b), the components of $\boldsymbol{B} = \{B_x; B_y; 0\}$ whatever the region can be deduced from A_z by

$$B_x = \frac{\partial A_z}{\partial y} \text{ and } B_y = -\frac{\partial A_z}{\partial x}. \tag{12}$$

From (8), and according to the materials assumptions, the components of $\boldsymbol{H} = \{H_x; H_y; 0\}$ are defined by $\boldsymbol{H} = \boldsymbol{B}/\mu_0$.

2.5. Boundary Conditions (BCs)

In electromagnetic field problems, the general solutions of regions depend on boundary conditions (BCs) at the interface between two surfaces, which are defined by the continuity of parallel magnetic field $H_\parallel$ and A [40]. On the infinite box [see Figure 1], A_z satisfies the Dirichlet's BC, viz., $A_z = 0$. Figure 5 represents the respective BCs at the interface between the various regions.

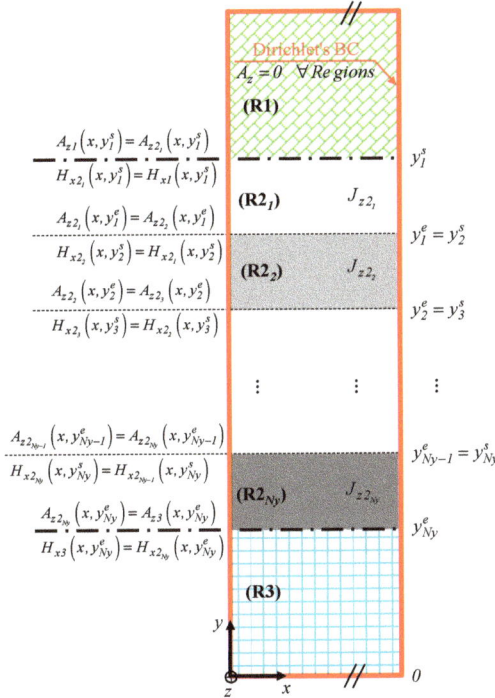

Figure 5. Boundary conditions at interface between (R1), (R2$_i$) and (R3).

2.6. General Solution of Magnetic Field in Each Region

Using the separation of variables method, the 2D general solution of A in each region can be defined by Fourier's series. The unknown coefficients (or integration constants) of series are determined analytically from a linear matrix system satisfying the BCs of Figure 5. To simplify the formal resolution of the Cramer's system, the superposition principle on each sub-region i (R2$_i$) has been applied.

In (R1) for $y \in [y_1^s, h]$ and $\forall x \in [0, w]$, the magnetic vector potential A_{z1}, which is a solution of (9a) satisfying the various BCs, is defined by

$$A_{z1} = \sum_{i,k} \mu_0 \cdot \frac{J_{i,k}^{cs}}{\beta_k^2} \cdot D1_{i,k} \cdot f1_{i,k}(y) \cdot \sin(\beta_k \cdot x) \tag{13}$$

and the components of H_1 (i.e., H_{x1} and H_{y1}) by

$$H_{x1} = -\sum_{i,k} \frac{J_{i,k}^{cs}}{\beta_k} \cdot D1_{i,k} \cdot g1_{i,k}(y) \cdot \sin(\beta_k \cdot x), \tag{14a}$$

$$H_{y1} = -\sum_{i,k} \frac{J_{i,k}^{cs}}{\beta_k} \cdot D1_{i,k} \cdot f1_{i,k}(y) \cdot \cos(\beta_k \cdot x), \tag{14b}$$

where

$$f1_{i,k}(y) = sh[\beta_k \cdot (h-y)]/sh[\beta_k \cdot (h-y_i^s)], \tag{15}$$

$$g1_{i,k}(y) = ch[\beta_k \cdot (h-y)]/sh[\beta_k \cdot (h-y_i^s)], \tag{16}$$

$$D1_{i,k} = NumD1_{i,k}/Den_{i,k}, \tag{17}$$

with

$$Den_{i,k} = a1_k \cdot (a3_{i,k} - a4_{i,k}) + a3_{i,k} \cdot a4_{i,k} \cdot \left(a1_k^2 + a2_k^2\right) - 1, \tag{18}$$

$$NumD1_{i,k} = -(a2_k + a1_k \cdot a4_{i,k} + 1), \tag{19}$$

in which

$$a1_k = -sh(\beta_k \cdot h_{cs})/ch(\beta_k \cdot h_{cs}), \tag{20}$$

$$a2_k = -1/ch(\beta_k \cdot h_{cs}), \tag{21}$$

$$a3_{i,k} = ch[\beta_k \cdot (h-y_i^s)]/sh[\beta_k \cdot (h-y_i^s)], \tag{22}$$

$$a4_{i,k} = -sh(\beta_k \cdot y_i^e)/ch(\beta_k \cdot y_i^e). \tag{23}$$

In (R3) for $y \in \left[0, y_{Ny}^e\right]$ and $\forall x \in [0, w]$, the magnetic vector potential A_{z3}, which is a solution of (9b) satisfying the various BCs, is defined by

$$A_{z3} = \sum_{i,k} \mu_0 \cdot \frac{J_{i,k}^{cs}}{\beta_k^2} \cdot D3_{i,k} \cdot f3_{i,k}(y) \cdot \sin(\beta_k \cdot x) \tag{24}$$

and the components of H_3 (i.e., H_{x3} and H_{y3}) by

$$H_{x3} = \sum_{i,k} \frac{J_{i,k}^{cs}}{\beta_k} \cdot D3_{i,k} \cdot g3_{i,k}(y) \cdot \sin(\beta_k \cdot x), \tag{25a}$$

$$H_{y3} = -\sum_{i,k} \frac{J_{i,k}^{cs}}{\beta_k} \cdot D3_{i,k} \cdot f3_{i,k}(y) \cdot \cos(\beta_k \cdot x), \tag{25b}$$

where

$$f3_{i,k}(y) = sh(\beta_k \cdot y)/ch(\beta_k \cdot y_i^e), \tag{26}$$

$$g3_{i,k}(y) = ch(\beta_k \cdot y)/ch(\beta_k \cdot y_i^e), \tag{27}$$

$$D3_{i,k} = NumD3_{i,k}/Den_{i,k}, \tag{28}$$

with

$$NumD3_{i,k} = -\left[a3_{i,k} \cdot \left(a1_k^2 + a2_k^2\right) + a3_{i,k} \cdot a2_k - a1_k\right]. \tag{29}$$

In (R2$_i$) for $y \in [y_i^e, y_i^s]$ and $\forall x \in [0, w]$, the magnetic vector potential A_{z2_i}, which is a solution of (9c) satisfying the various BCs, is defined by

$$A_{z2_i} = \sum_k \left\{ \sum_{n=1}^{i-1} \mu_0 \cdot \frac{J_{n,k}^{cs}}{\beta_k^2} \cdot D3_{n,k} \cdot f3_{n,k}(y) + \mu_0 \cdot \frac{J_{i,k}^{cs}}{\beta_k^2} \cdot \begin{bmatrix} C2_{i,k} \cdot f2c_{i,k}(y) \\ \cdots + D2_{i,k} \cdot f2d_{i,k}(y) \\ \cdots + 1 \end{bmatrix} + \sum_{v=i+1}^{Ny} \mu_0 \cdot \frac{J_{v,k}^{cs}}{\beta_k^2} \cdot D1_{v,k} \cdot f1_{v,k}(y) \right\} \cdot \sin(\beta_k \cdot x) \tag{30}$$

and the components of H_{2_i} (i.e., H_{x2_i} and H_{y2_i}) by

$$H_{x2_i} = \sum_k \left\{ \sum_{n=1}^{i-1} \frac{J_{n,k}^{cs}}{\beta_k} \cdot D3_{n,k} \cdot g3_{n,k}(y) + \frac{J_{i,k}^{cs}}{\beta_k} \cdot \left[\begin{array}{c} C2_{i,k} \cdot g2c_{i,k}(y) \\ \cdots - D2_{i,k} \cdot g2d_{i,k}(y) \end{array} \right] - \sum_{v=i+1}^{Ny} \frac{J_{v,k}^{cs}}{\beta_k} \cdot D1_{v,k} \cdot g1_{v,k}(y) \right\} \cdot \sin(\beta_k \cdot x), \quad (31a)$$

$$H_{y2_i} = -\sum_k \left\{ \sum_{n=1}^{i-1} \frac{J_{n,k}^{cs}}{\beta_k} \cdot D3_{n,k} \cdot f3_{n,k}(y) - \frac{J_{i,k}^{cs}}{\beta_k} \cdot \left[\begin{array}{c} C2_{i,k} \cdot f2c_{i,k}(y) \\ \cdots + D2_{i,k} \cdot f2d_{i,k}(y) \\ \cdots + 1 \end{array} \right] + \sum_{v=i+1}^{Ny} \frac{J_{v,k}^{cs}}{\beta_k} \cdot D1_{v,k} \cdot f1_{v,k}(y) \right\} \cdot \cos(\beta_k \cdot x), \quad (31b)$$

where

$$f2c_{i,k}(y) = sh[\beta_k \cdot (y - y_i^e)]/ch(\beta_k \cdot h_{cs}), \quad (32)$$

$$f2d_{i,k}(y) = ch[\beta_k \cdot (y_i^s - y)]/ch(\beta_k \cdot h_{cs}), \quad (33)$$

$$g2c_{i,k}(y) = ch[\beta_k \cdot (y - y_i^e)]/ch(\beta_k \cdot h_{cs}), \quad (34)$$

$$g2d_{i,k}(y) = sh[\beta_k \cdot (y_i^s - y)]/ch(\beta_k \cdot h_{cs}), \quad (35)$$

$$C2_{i,k} = NumC2_{i,k}/Den_{i,k}, \quad (36)$$

$$D2_{i,k} = NumD2_{i,k}/Den_{i,k}, \quad (37)$$

with

$$NumC2_{i,k} = a3_{i,k} \cdot (a2_k + a1_k \cdot a4_{i,k} + 1), \quad (38)$$

$$NumD2_{i,k} = (a2_k \cdot a3_{i,k} \cdot a4_{i,k} - a1_k \cdot a3_{i,k} + 1). \quad (39)$$

3. Comparison of Analytical and Numerical Calculations

3.1. Problem Description

The real bipolar plate of a PEMFC stack is illustrated in Figure 6. The electrochemical active cell is presented with the oxygen/hydrogen channels and surrounded by an inactive zone (seal). The geometrical and physical parameters of the PEMFC stack are reported in Table 1.

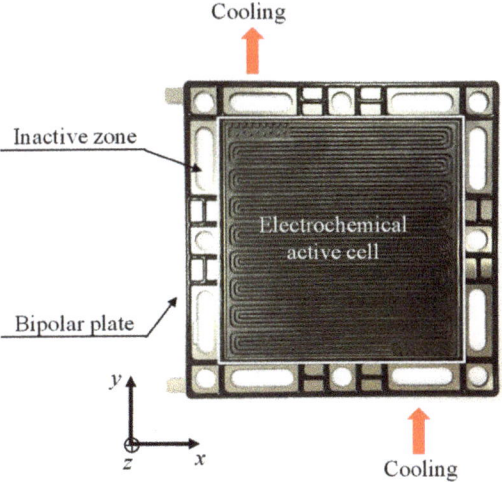

Figure 6. Real bipolar plate of a PEMFC stack.

To simulate the (non-)homogeneous distribution of J, this electrochemical active cell is decomposed into $Ny \times Nx$ regular conductive segments. For example, the discretization with $Ny \times Nx = 25 \times 20 = 500$ conductive segments is considered. This discretization of the electrochemical active cell to study health of the FC is illustrated in Figure 7.

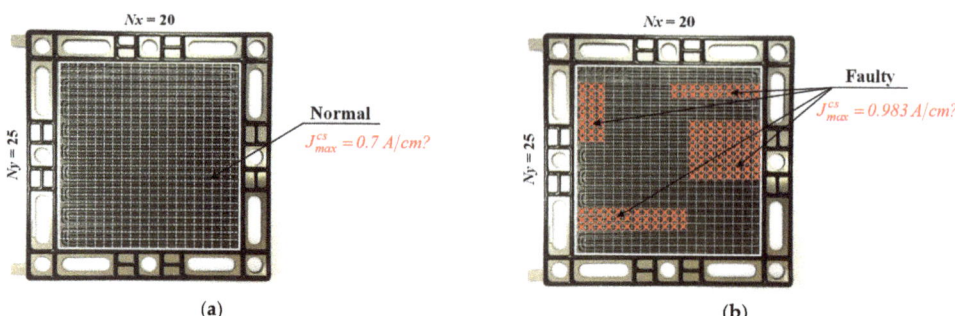

Figure 7. FC representation with a discretization of $Ny \times Nx = 25 \times 20 = 500$ conductive segments for: (**a**) a healthy FC, and (**b**) an FC failure.

Such an important discretization is necessary due to the channels' size (viz., ~0.8 mm) as shown in Figure 6. The total current flowing through the PEMFC stack (of surface $S_{fc} = 196$ cm^2) is considered constant and equal to $I_{fc} = 70$ A. With an electrochemical active cell surface of $S_a = 100$ cm^2, for a healthy FC [see Figure 7a], this corresponds to $J_{max}^{cs} = 0.7$ A/cm^2. An example of an FC failure, which can be a consequence of different operating problems, is exposed in Figure 7b. Considering 500 initial conductive segments, 144 segments are faulty (i.e., $J_{max}^{cs} = 0$ A/cm^2), so 356 conductive segments present a current density of $J_{max}^{cs} = 0.983$ A/cm^2.

To evaluate the capacity and the efficacy of the 2D analytical model, the two operating conditions (i.e., normal and faulty condition) are simulated and compared to 2D FEA. The geometrical and physical parameters described previously are simulated on FEMM software [38]. The FE calculations are performed with the same assumptions as the analytical model [see Section 2.2]. The infinite box surface has been imposed to $S = 3 \cdot S_{fc} = 1764$ cm^2 (viz., $w = 3 \cdot w_{fc}$ and $h = 3 \cdot h_{fc}$) so that the equipotential lines of A_z would not be perturbed by the Dirichlet's CB imposed on the edges. Figure 8 presents the mesh of the system (viz., 107,385 nodes) and different paths with their dimensions added around the PEMFC stack (in red, rated from A to D) in order to evaluate the magnetic field distribution. In this comparison, the analytical solution of A and H in each region have been computed with $K_{max} = 100$.

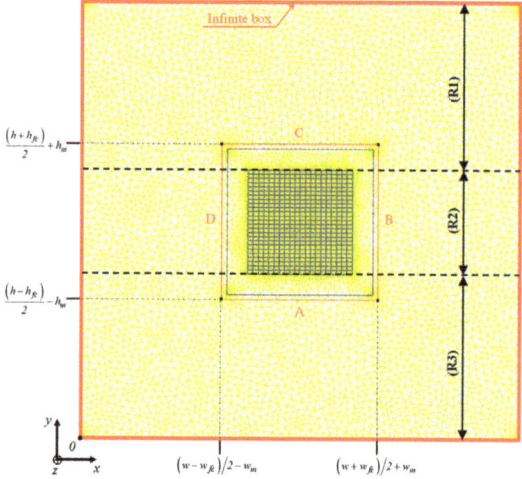

Figure 8. The 2D FEA mesh and paths of the magnetic field validation for comparison with $\{w_m; h_m\} = \{5; 5\}$ mm (see Table 1 for the various parameters).

3.2. Analysis of Equipotential Lines

3.2.1. Healthy FC

The 2D multi-layer model is implemented in order to obtain values of A_z in the bipolar plate for the diagnosis of a PEMFC stack. Figure 9 presents the equipotential lines ($\approx$30 lines) of A_z (or vector plot of **H**) for a healthy FC with the 2D analytical model and 2D FEA. As can be seen, a perfect accuracy between analytical and numerical results allows a first validation of the multi-layer model based on the Maxwell–Fourier method. Due to the homogenous distribution of **J**, and for a square FC, the field lines are symmetrical and circular along the centre of the bipolar plate.

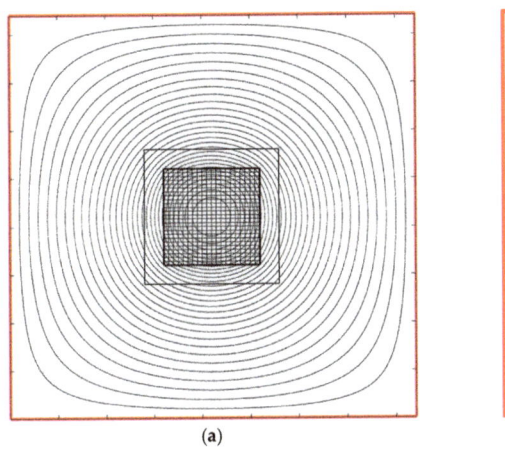
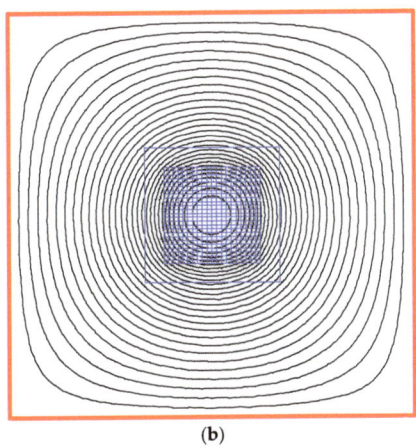

Figure 9. Equipotential lines of A_z for a healthy FC obtained by (**a**) 2D multi-layer model and (**b**) 2D FEA.

3.2.2. FC Failure

With a non-homogeneous distribution of **J** reflecting an FC failure, the generated magnetic field is impacted. In Figure 10, the deformation of the equipotential lines of A_z is clearly observed. It is also correctly predicted analytically. On the left, a distortion of the field lines (i.e., asymmetry of field lines) is widely visible due to the presence of a major defect on the right [see Figure 7b].

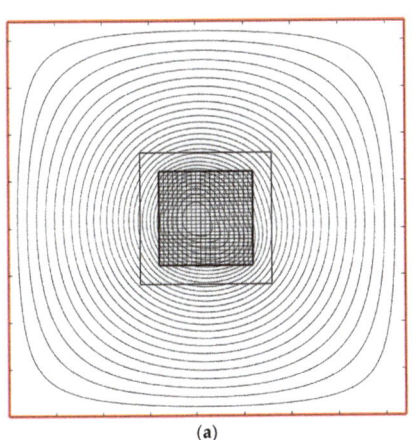
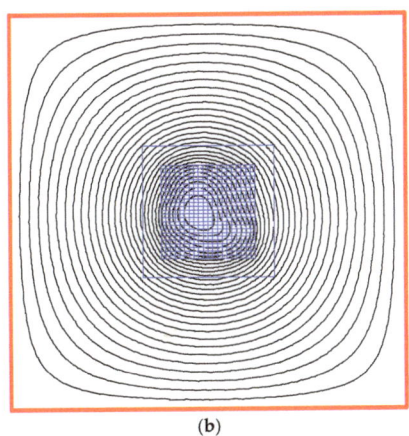

Figure 10. Equipotential lines of A_z for an FC failure obtained by (**a**) 2D multi-layer model and (**b**) 2D FEA.

3.3. Analysis of Magnetic Field Distributions

3.3.1. Healthy FC

Figure 11 represents the distribution of the components of $H = \{H_x; H_y; 0\}$ for a healthy FC computed by the 2D multi-layer model and verified by the 2D FEA on the various paths. Highly accurate results are achieved between the analytical and numerical model, regardless of the paths and the components of $H = \{H_x; H_y; 0\}$ in the different regions. It is interesting to note that the waveform and the amplitude of magnetic field components are exactly the same in the four paths (i.e., A to D in Figure 8) independently of the axis.

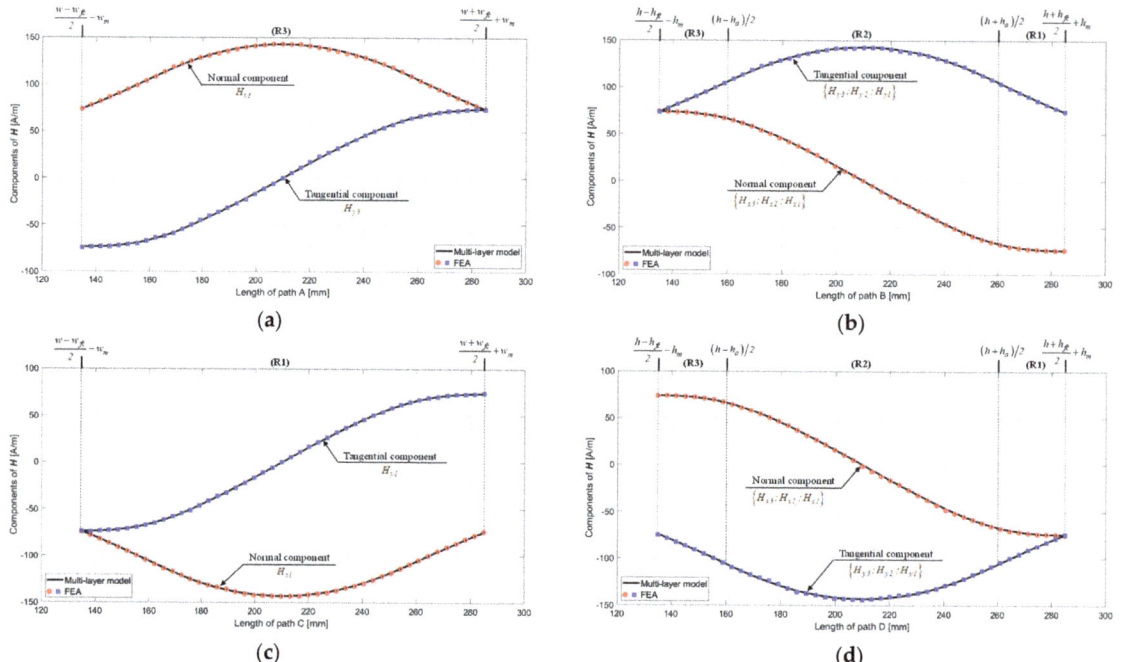

Figure 11. Components of H for a healthy FC calculated by the 2D multi-layer model and verified by 2D FEA on the path: (**a**) A, (**b**) B, (**c**) C and (**d**) D.

3.3.2. FC Failure

As for the case of a healthy FC, Figure 12 shows the distribution of the components of $H = \{H_x; H_y; 0\}$ for an FC failure obtained analytically and numerically on the different paths. Irrespective of the paths and regions, similar and accurate magnetic field results also permit validation of the 2D analytical model. A magnetic field variation reflects a failure in the FC. Around the most damaged part, the greatest decrease in magnetic field is observed (viz., on the path B and C), as shown in Figure 12b,c. As the overall current value I_{fc} is maintained, the current density J_{max}^{cs} increases in the rest of the FC [see Figure 7]. Thus, the magnetic field variation is less observable for small surface defaults.

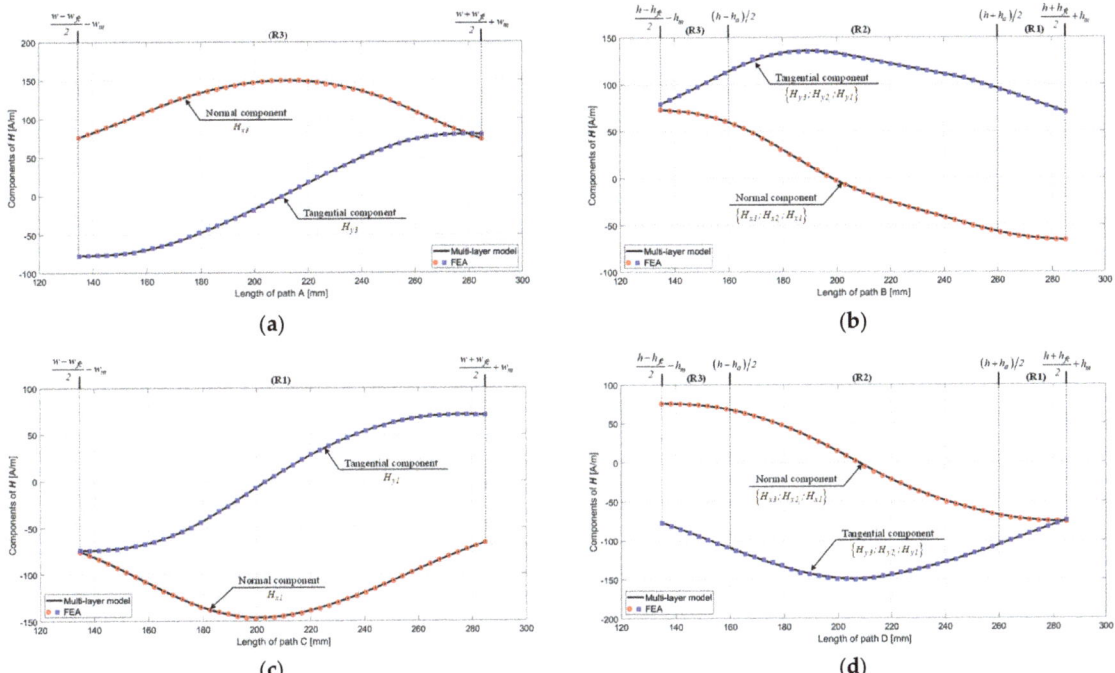

Figure 12. Components of H for an FC failure calculated by the 2D multi-layer model and verified by 2D FEA on the path: (**a**) A, (**b**) B, (**c**) C and (**d**) D.

4. Conclusions

In this paper, an original innovative 2D multi-layer model, based on the Maxwell–Fourier method, has been developed analytically and validated numerically to estimate the health state of FC. Using magneto-tomography, knowledge of the magnetic field distribution generated around the PEMFC stack allowed for the expression of the (non-)homogeneous current density distribution inside the PEMFC stack and its operation mode. Through FEA validation, the accuracy and efficiency of the analytical model are undeniable.

This 2D analytical model represents the first step towards a comprehensive FC diagnostic tool. For this, the inverse problem of the presented analytical model will have to be solved. From the magnetic field measured around the FC, the inverse model will allow us to know the current density distribution inside the FC. It will then be quick and easy to check the health state of an FC with the advantage of being able to accurately locate a potential fault.

Author Contributions: Conceptualization, A.P.; methodology, F.D. and A.P.; validation, A.P. and F.D.; writing—original draft preparation, A.P and F.D.; writing—review and editing, F.D. and A.P. All authors have read and agreed to the published version of the manuscript.

Funding: This research received no external funding.

Acknowledgments: This work was supported by the EIPHI Graduate School (contract ANR-17-EURE-0002), and the Region Bourgogne-Franche-Comté.

Conflicts of Interest: The authors declare no conflict of interest.

References

1. Dubas, F.; Espanet, C.; Miraoui, A. Design of a high-speed permanent magnet motor for the drive of a fuel cell air compressor. In Proceedings of the IEEE Vehicular Power and Propulsion Conference (VPPC), Chicago, IL, USA, 7–9 September 2005. [CrossRef]
2. Hart, D.; Lehner, F.; Jones, S.; Lewis, J. The Fuel Cell Industry Review 2019. E4tech (an ERM Group Company). Available online: http://www.fuelcellindustryreview.com (accessed on 9 January 2020).
3. Wilberforce, T.; Alaswad, A.; Palumbo, A.; Dassisti, M.; Olabi, A.G. Advances in stationary and portable fuel cell applications. *Int. J. Hydrogen Energy* **2016**, *41*, 16509–16522. [CrossRef]
4. Gurz, M.; Baltacioglu, E.; Hames, Y.; Kaya, K. The meeting of hydrogen and automotive: A review. *Int. J. Hydrogen Energy* **2017**, *42*, 23334–23346. [CrossRef]
5. Yue, M.; Jemei, S.; Gouriveau, R.; Zerhouni, N. Review on health-conscious energy management strategies for fuel cell hybrid electric vehicles: Degradation models and strategies. *Int. J. Hydrogen Energy* **2019**, *44*, 6844–6861. [CrossRef]
6. Pahon, E.; Bouquain, D.; Hissel, D.; Rouet, A.; Vacquier, C. Performance analysis of proton exchange membrane fuel cell in automotive applications. *J. Power Sources* **2021**, *510*, 230385. [CrossRef]
7. Abe, J.O.; Popoola, A.P.I.; Ajenifuja, E.; Popoola, O.M. Hydrogen energy, economy and storage: Review and recommendation. *Int. J. Hydrogen Energy* **2019**, *44*, 15072–15086. [CrossRef]
8. Yue, M.; Lambert, H.; Pahon, E.; Roche, R.; Jemei, S.; Hissel, D. Hydrogen energy systems: A critical review of technologies, applications, trends and challenges. *Renew. Sustain. Energy Rev.* **2021**, *146*, 111180. [CrossRef]
9. Inci, M.; Büyük, M.; Demir, M.H.; Ilbey, G. A review and research on fuel cell electric vehicles: Topologies, power electronic converters, energy management methods, technical challenges, marketing and future aspects. *Renew. Sustain. Energy Rev.* **2020**, *137*, 110648. [CrossRef]
10. Lorenzo, C.; Bouquain, D.; Hibon, S.; Hissel, D. Synthesis of degradation mechanisms and of their impacts on degradation rates on proton-exchange membrane fuel cells and lithium-ion nickel–manganese–cobalt batteries in hybrid transport applications. *Reliab. Eng. Syst. Saf.* **2021**, *212*, 107369. [CrossRef]
11. Aubry, J.; Steiner, N.Y.; Morando, S.; Zerhouni, N.; Hissel, D. Fuel cell diagnosis methods for embedded automotive applications. *Energy Rep.* **2022**, *8*, 6687–6706. [CrossRef]
12. Hauer, K.-H.; Potthast, R.; Wüster, T.; Stolten, D. Magnetotomography—A new method for analysing fuel cell performance and quality. *J. Power Sources* **2005**, *143*, 67–74. [CrossRef]
13. Buchholz, M.; Brebs, V. Dynamic modelling of a polymer electrolyte membrane fuel cell stack by nonlinear system identification. *Fuel Cell.* **2007**, *7*, 392–401. [CrossRef]
14. Rubio, M.; Urquia, A.; Dormido, S. Diagnosis of performance degradation phenomena in PEM fuel cells. *Int. J. Hydrogen Energy* **2010**, *35*, 2586–2590. [CrossRef]
15. Chevalier, S.; Trichet, D.; Auvity, B.; Olivier, J.; Josset, C.; Machmoum, M. Multiphysics DC and AC models of a PEMFC for the detection of degraded cell parameters. *Int. J. Hydrogen Energy* **2013**, *38*, 11609–11618. [CrossRef]
16. Zheng, Z.; Petrone, R.; Péra, M.; Hissel, D.; Becherif, M.; Pianese, C.; Steiner, N.Y.; Sorrentino, M. A review on non-model based diagnosis methodologies for PEM fuel cell stacks and systems. *Int. J. Hydrogen Energy* **2013**, *38*, 8914–8926. [CrossRef]
17. Geske, M.; Heuer, M.; Heideck, G.; Styczynski, Z.A. Current Density Distribution Mapping in PEM Fuel Cells as An Instrument for Operational Measurements. *Energies* **2010**, *3*, 770–783. [CrossRef]
18. Pei, P.; Chen, H. Main factors affecting the lifetime of Proton Exchange Membrane fuel cells in vehicle applications: A review. *Appl. Energy* **2014**, *125*, 60–75. [CrossRef]
19. Ghezzi, L.; Piva, D.; Di Rienzo, L. Current Density Reconstruction in Vacuum Arcs by Inverting Magnetic Field Data. *IEEE Trans. Magn.* **2012**, *48*, 2324–2333. [CrossRef]
20. Begot, S.; Voisin, E.; Hiebel, P.; Artioukhine, E.; Kauffmann, J.M. D-optimal experimental design applied to a linear magnetostatic inverse problem. *IEEE Trans. Magn.* **2002**, *38*, 1065–1068. [CrossRef]
21. Potthast, R.; Wannert, M. Uniqueness of Current Reconstructions for Magnetic Tomography in Multilayer Devices. *SIAM J. Appl. Math.* **2009**, *70*, 563–578. [CrossRef]
22. Izumi, M.; Gotoh, Y.; Yamanaka, T. Verification of Measurement Method of Current Distribution in Polymer Electrolyte Fuel Cells. *ECS Trans.* **2009**, *17*, 401–409. [CrossRef]
23. Lustfeld, H.; Reißel, M.; Schmidt, U.; Steffen, B. Reconstruction of Electric Currents in a Fuel Cell by Magnetic Field Measurements. *J. Fuel Cell Sci. Technol.* **2009**, *6*, 021012. [CrossRef]
24. Nasu, T.; Matsushita, Y.; Okano, J.; Okajima, K. Study of Current Distribution in PEMFC Stack Using Magnetic Sensor Probe. *J. Int. Counc. Electr. Eng.* **2012**, *2*, 391–396. [CrossRef]
25. Le Ny, M.; Chadebec, O.; Cauffet, G.; Dedulle, J.-M.; Bultel, Y.; Rosini, S.; Fourneron, Y.; Kuo-Peng, P. Current Distribution Identification in Fuel Cell Stacks From External Magnetic Field Measurements. *IEEE Trans. Magn.* **2013**, *49*, 1925–1928. [CrossRef]
26. Le Ny, M.; Chadebec, O.; Cauffet, G.; Rosini, S.; Bultel, Y. PEMFC stack diagnosis based on external magnetic field measurements. *J. Appl. Electrochem.* **2015**, *45*, 667–677. [CrossRef]
27. Nara, T.; Koike, M.; Ando, S.; Gotoh, Y.; Izumi, M. Estimation of localized current anomalies in polymer electrolyte fuel cells from magnetic flux density measurements. *AIP Adv.* **2016**, *6*, 056603. [CrossRef]
28. Ifrek, L.; Cauffet, G.; Chadebec, O.; Bultel, Y.; Rouveyre, L.; Rosini, S. 2D and 3D fault basis for fuel cell diagnosis by external magnetic field measurements. *Eur. Phys. J. Appl. Phys.* **2017**, *79*, 20901. [CrossRef]

29. Claycomb, J.R. Algorithms for the Magnetic Assessment of Proton Exchange Membrane (PEM) Fuel Cells. *Res. Nondestruct. Eval.* **2017**, *29*, 167–182. [CrossRef]
30. Ifrek, L.; Chadebec, O.; Rosini, S.; Cauffet, G.; Bultel, Y.; Bannwarth, B. Fault Identification on a Fuel Cell by 3-D Current Density Reconstruction From External Magnetic Field Measurements. *IEEE Trans. Magn.* **2019**, *55*, 2512271. [CrossRef]
31. Ifrek, L.; Rosini, S.; Cauffet, G.; Chadebec, O.; Rouveyre, L.; Bultel, Y. Fault detection for polymer electrolyte membrane fuel cell stack by external magnetic field. *Electrochim. Acta* **2019**, *313*, 141–150. [CrossRef]
32. Katou, T.; Gotoh, Y.; Takahashi, N.; Izumi, M. Measurement Technique of Distribution of Power Generation Current Using Static Magnetic Field around Polymer Electrolyte Fuel Cell by 3D Inverse Problem FEM. *Mater. Trans.* **2012**, *53*, 279–284. [CrossRef]
33. Yamanashi, R.; Gotoh, Y.; Izumi, M.; Nara, T. Evaluation of Generation Current inside Membrane Electrode Assembly in Polymer Electrolyte Fuel Cell Using Static Magnetic Field around Fuel Cell. *ECS Trans.* **2015**, *65*, 219–226. [CrossRef]
34. Akimoto, Y.; Okajima, K.; Uchiyama, Y. Evaluation of Current Distribution in a PEMFC using a Magnetic Sensor Probe. *Energy Procedia* **2015**, *75*, 2015–2020. [CrossRef]
35. Plait, A.; Giurgea, S.; Hissel, D.; Espanet, C. New magnetic field analyzer device dedicated for polymer electrolyte fuel cells noninvasive diagnostic. *Int. J. Hydrogen Energy* **2020**, *45*, 14071–14082. [CrossRef]
36. Plait, A.; Dubas, F. Experimental validation of a 2-D multi-layer model for fuel cell diagnosis using magneto-tomography. In Proceedings of the 23rd World Hydrogen Energy Conference (WHEC), Istanbul, Turkey, 26–30 June 2022.
37. Lustfeld, H.; Reißel, M.; Steffen, B. Magnetotomography and Electric Currents in a Fuel Cell. *Fuel Cells* **2009**, *9*, 474–481. [CrossRef]
38. Meeker, D.C. Finite Element Method Magnetics Ver. 4.2. Available online: http://www.femm.info (accessed on 10 October 2018).
39. Stoll, R.L. *The Analysis of Eddy Currents*; Clarendon Press: Oxfort, UK, 1974.
40. Dubas, F.; Boughrara, K. New Scientific Contribution on the 2-D Subdomain Technique in Cartesian Coordinates: Taking into Account of Iron Parts. *Math. Comput. Appl.* **2017**, *22*, 17. [CrossRef]

 *mathematics*

Article

A Novel Universal Torque Control of Switched Reluctance Motors for Electric Vehicles

Mahmoud Hamouda [1], Fahad Al-Amyal [2], Ismoil Odinaev [3], Mohamed N. Ibrahim [4,5,6,*] and László Számel [2]

[1] Electrical Engineering Department, Mansoura University, 35516 Mansoura, Egypt
[2] Department of Electric Power Engineering, Budapest University of Technology and Economics, H-1521 Budapest, Hungary
[3] Department of Automated Electrical Systems, Ural Power Engineering Institute, Ural Federal University, 620002 Yekaterinburg, Russia
[4] Electrical Engineering Department, Kafrelshiekh University, 33511 Kafr El-Sheikh, Egypt
[5] Department of Electromechanical, Systems and Metal Engineering, Ghent University, 9000 Ghent, Belgium
[6] FlandersMake@UGent—Corelab EEDT-MP, 3001 Leuven, Belgium
* Correspondence: mohamed.ibrahim@ugent.be

Abstract: Due to their advantages, switched reluctance motors (SRMs) are interesting solutions for electric vehicle (EV) propulsion. However, they have the main drawback of high torque ripple. This paper develops a universal torque control (UTC) technique for SRM that can fulfill all vehicle requirements under a wide range of speeds. The developed UTC involves two different control techniques. It utilizes the direct instantaneous torque control (DITC) strategy in a low speed region, and the average torque control (ATC) strategy in high speeds. The selection of DITC and ATC is made based on their performance regarding torque ripple, torque/current ratio, and efficiency. Moreover, a novel transition control between the two control techniques is introduced. The results show the ability of the proposed UTC to achieve vehicle requirements while obtaining all the benefits of torque control over the possible range of speeds. The proposed UTC provides the best performance regarding minimum torque ripple, maximum torque/current ratio, and maximum efficiency over the whole speed range. The transition control achieves a smooth operation without any disturbances. The transition control helps to simplify the overall control algorithm, aiming to have a feasible and practical UTC without a complicated control structure.

Keywords: switched reluctance motor; universal torque control; direct instantaneous torque control; average torque control; firing angles; optimization

MSC: 70Q05; 37N35; 93-10; 13P25; 65K10

1. Introduction

Energy sources, pollution, and noise are currently the key issues facing many countries as a result of using enormous numbers of fuel-powered vehicles in transportation. Electric vehicles (EVs), on the other hand, can provide an alternative solution. They are considered the way for establishing a green transportation system [1,2]. For this purpose, many efforts have been made to solve the challenges facing EVs, such as saving energy and obtaining the highest performance at a reasonable cost. Various components of EVs can be optimized to enhance their performance. The most effective parts within the vehicle are the motors and their drive circuits. Generally, all types of electric motors can be used for EVs. Each motor type is associated with a set of benefits and drawbacks [3,4]. The switched reluctance motors (SRMs) can be considered a perfect choice because they have many features that are relevant to the requirements of EVs, such as, reasonable manufacturing cost, high fault resistance, high starting torque, normal operation in a high temperature environment, and they can provide high efficiency under high speed operation. Furthermore, SRMs do not contain permanent magnets nor conductors in the rotor structure, and thus, a robust and reliable operation exists. However, the SRMs encounter acoustic noise and considerable torque ripple [5–7]. These drawbacks make less utilization of SRMs in several industrial applications including the EVs. Therefore, they must be treated. Researchers have recently been focusing on solving these issues, and many strategies have been introduced. Some researchers proposed the optimization of SRM design by the modification of its geometrical dimensions [8–11]. These design modifications can minimize the acoustic noise and torque ripple only to a small extent [8,10].

On the other hand, other researchers succeeded in significantly reducing the SRM torque ripple and the associated acoustic noise by implementing innovative control techniques. Generally, the current chopping control (CCC) [12–15], instantaneous torque control (ITC) [16–20], and average torque control (ATC) [21–25] are the most popular control strategies. Despite the fact that these strategies can reduce torque ripple and improve system performance, they have disadvantages. For instant, the CCC can provide effective, uncomplicated, and low cost drive. It causes some torque ripple because each phase current can only be controlled in the form of square waveforms [26]. Thus, only the excitation angles can be optimized for a given reference current [15]. Hence, this paper focuses on the ITC and ATC as they are the most promising torque control strategies that can provide a better overall performance [27].

The ITC strategies are able to provide the remarkable suppression of torque ripple since they control the torque instantaneously at each sample of the rotor position. The ITC strategies can be categorized into direct ITC (DITC) and indirect ITC (IITC) using the torque sharing function (TSF). In [17,18], the IITC is achieved based on an improved TSF. The proposed TSF can impose optimal profiling for the phase current to reach a minimized torque ripple with reduced copper losses. In [19], a modified firefly algorithm is used to optimize the PID parameters of the speed regulator to achieve a precise DITC with an improved speed-change response. In [20], a new DITC is implemented by replacing the commonly used hysteresis torque controller with a PWM controller. In this way, the switching frequency can be controlled. Additionally, the commutation region is divided into many sectors, and by modifying the duty cycle coefficients in these sectors, further torque ripple enhancement is obtained. In [16], an improved DITC is implemented by utilizing a real-time commutation methodology for both switching angles. The turn on angle (θ_{on}) is controlled basically to maximize the torque tracking ability of the hysteresis torque controller. At the same time, the turn off angle (θ_{off}) is controlled in order to minimize the generated negative torque to an effective-less value. In [28], an improved DITC method is introduced. This method uses a torque sharing function (TSF) as well as it introduces an adaptive turn on angle control to suppress torque ripple and improve efficiency. However, it has a complex control structure.

In [21], the ATC strategy is investigated for EV applications. Then, an improved ATC is presented by optimizing both excitation angles, and the objective function is proposed to increase the performance of the SRM drive. In [22], the ATC is proposed involving both flux and current controllers. Unlike the conventional ATC, it controls the estimated average torque only over a complete electrical cycle. The new ATC can control the average torque through smaller intervals and thus can achieve less torque ripple. In [23], a sharing look-up table is used to equally divide the torque reference among motor phases, and thus, a reference value is set for each phase current, and by analytical accumulating each phase torque, the average torque can be estimated instantaneously. In [24], a multi-objective constrained optimization problem is created to implement an optimized ATC strategy. The objective function includes two terms: the efficiency and torque ripple. The control parameters are the switching angles, the constraints are set mainly to keep high torque/ampere ratio, and thus, further improvement is attained in order to meet the EVs' requirements. In [29], after investigations, the ATC is selected over ITC for EV applications. The turn on and turn off angles are optimized to enhance driving efficiency and reduce torque ripple. Despite the simple structure of ATC, it has a high torque ripple at low speeds.

Unfortunately, the majority of the literature includes only a comparison between one basic torque control strategy and its modification or compares the same type of modified strategies. Nonetheless, very few have conducted a comparison between different types of techniques. Therefore, in the authors' previous work [27], a comprehensive analysis and comparison between an improved DITC, IITC, and ATC are included regarding the requirements of EVs. The results showed the superiority of the proposed DITC since it produced lower torque ripples, highest efficiency, and the highest torque/ampere ratio at a low speed operation. Furthermore, the comparison also showed an advantage when using the ATC technique at high speeds. Above the rated speed of the SRM, a higher back-electromotive force is generated, and thus, the instantaneous torque can be only partially or poorly tracked. Therefore, the ATC method is proposed to be utilized at a high speed SRM operation. The study provides a pre-overview to implement a UTC strategy for the SRMs in EV application.

The literature shows the superior performance of ITC strategies (DITC and IITC) regarding the torque ripple reduction for low speed operation. It also gives the advantages of the ATC strategy in high speed regions as it gives a better torque/ampere ratio and hence better driving efficiency. For the best overall performance, the ITC and the ATC strategies have to be combined to implement a universal torque control (UTC) that provides the benefits of both strategies. Hence, it will be the best choice for several applications including EVs. In [25], the DITC is used below the rated speed of the SRM, while the ATC is utilized above the base speed. Furthermore, the excitation angles are optimized through a particle swarm algorithm to decrease the fuel consumption of the EVs (by attaining higher efficiency) and to guarantee a comfortable ride (by attaining lower torque ripple). However, the optimization of the excitation angles is only applied to the DITC at a low speed. No smooth transient is implemented between the DITC and the ATC. In [30], a unified controller is implemented by combining the DITC with the CCC method to attain torque ripple reduction in the low, medium, and high speed operation of the SRM. The unified controller also adopts a phase current demagnetization strategy to prevent negative torque production, and thus, a higher torque/ampere ratio can be attained.

The novelty of this article is the development of a novel UTC strategy that provides the best overall performance of SRM over the whole range of speeds while maintaining a feasible and practical control algorithm that fits several industrial applications including EVs. The proposed UTC uses DITC for low speed operation and ATC for high speeds to gather their benefits. The paper helps to overcome the problems and unfeasibility of developing such control. First, the paper uses modified DITC and ATC strategies to ensure a simple structure of the overall control algorithm. The modifications are made basically for performance improvement considering the ability of their integrations. Second, the proposed control guarantees maximum torque per ampere production via turn on angle optimization. It optimizes a turn off angle to improve efficiency and reduce torque ripples. Third, the novelty also includes a proposed smooth transition control (STC) from DITC to ATC and vice versa. The STC is guaranteed in motoring and generation modes of operation. Hence, this paper gives a novel UTC strategy that provides the best overall performance of SRM over the whole range of speeds, while maintaining a feasible and practical control algorithm.

The proposed method suits EV applications as it can fulfill the vehicle requirements that include minimum torque ripple for better drivability, maximum torque per ampere (MTPA) and/or maximum efficiency for longer millage per battery charge, wide speed operation, and high reliability to avoid breakdown on the road.

The rest of this article is outlined as follows: Section 3 describes the conventional structure of torque control strategies. The proposed UTC is given in Section 4. The STC is also included in this section. The results and discussions are presented in Section 5. Finally, Section 7 presents the conclusions.

2. Modeling of SRM

2.1. Machine Model

The SRM has highly nonlinear magnetic characteristics because of the salient structure of both stator and rotor poles. Hence, the flux-linkage $\lambda(i,\theta)$, inductance $L(i,\theta)$, and torque $T(i,\theta)$ are functions of both the current (i) and position (θ). The voltage equation of kth phase is given by Equation (1). It also illustrates how the flux is estimated from phase voltage. Equation (2) gives the electromagnetic torque of kth phase (T_k) and total electromagnetic torque (T_e) with q-phases. Equation (3) shows the mechanical motion.

$$v_k = Ri_k + \frac{\partial \lambda_k(i_k,\theta)}{\partial t}; \quad \therefore \lambda_k(i_k,\theta) = \int (v_k - Ri_k)dt \tag{1}$$

$$T_k = \frac{1}{2}\frac{\partial L_k}{\partial \theta}i_k^2; \quad T_e = \sum_{k=1}^{q} T_k \tag{2}$$

$$T_e - T_L = B\omega + J\frac{d\omega}{dt} \tag{3}$$

where R, J, B, ω, and T_L are the phase resistance, inertia, frictional coefficient, rotor speed, and load torque.

The simulation of one phase of SRM is shown in Figure 1. The inputs are the rotor phase position (θ_{ph}) and the phase voltage (V_{ph}). The outputs are the phase current (i_{ph}) and phase torque (T_{ph}). The finite element method (FEM) is employed to generate the magnetic characteristics of the studied SRM. The studied motor is 4 kW, 1500 r/min, 600 V, 8/6 poles, and 4 phases SRM. The calculated flux linkage and torque characteristics using FEM are shown in Figure 2a,b, respectively. This figure shows only part of the curves for simplification, while the complete flux and torque characteristics are calculated for current [0:0.5:30]A and position of [0:0.5:30]°. Due to the big size of FEM data, the interpolation and extrapolation within LUTs can ensure sufficient accuracy.

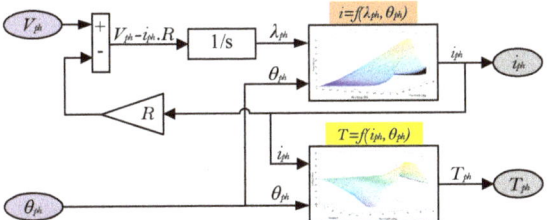

Figure 1. Simulation of one phase of SRM.

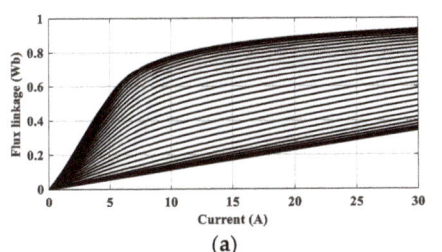
(a)

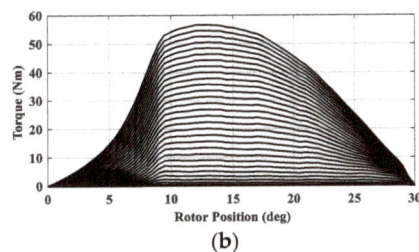
(b)

Figure 2. FEM-calculated characteristics: (**a**) flux linkage for [1:1:30]°; (**b**) torque for [1:1:30]A.

2.2. Performance Indices

The following indices are used for evaluations. The indices include average torque (T_{av}), mechanical output power (P_m), torque ripple (T_{rip}), efficiency (η), switching frequency (f_s) of power converter, average supply current (I_{av}), RMS supply current (I_{RMS}), and copper losses (P_{cu}). The average and RMS values are calculated over one electrical cycle (τ).

$$T_{av} = \frac{1}{\tau}\int_0^{\tau} T_e(t)dt, \quad P_m = \omega * T_{av} \tag{4}$$

$$T_{rip} = \frac{T_{max} - T_{min}}{T_{av}} \times 100 \tag{5}$$

$$\eta = \frac{\omega * T_{av}}{V_{dc} I_{av}}, \quad I_{av} = \frac{1}{\tau} \int_0^\tau i_s(t) dt \tag{6}$$

$$f_{sw} = \frac{1}{\tau} \int_0^\tau N_T dt \tag{7}$$

$$I_{RMS} = \sqrt{\tfrac{1}{\tau} \int_0^\tau i_s^2(t) dt}; \quad P_{cu} = q\, R\, I_{RMS}^2 \tag{8}$$

where T_{av} is the average torque; ω is the motor speed; T_{max} and T_{min} are the maximum and minimum instantaneous values of output motor torque (T_e) over one electrical cycle (τ); V_{dc} is the dc link voltage; I_{av} is the average supply current; i_s is instantaneous supply current; N_T is the total number of switching of IGBTs over one electric cycle.

3. The Conventional ATC and DITC Techniques for SRMs

As concluded from the literature, the best performance is achieved by DITC for low speeds and is achieved by ATC for high speeds. Hence, they are chosen and adopted for this research.

3.1. The Conventional ATC Technique

Figure 3 shows the block diagram of conventional ATC [22,29]. T_{ref} is the reference commanded torque; it is obtained from an outer loop speed controller. In Figure 3a, the average torque (T_{av-est}) is estimated online using a torque estimator. The torque error (ΔT) is processed through a PI torque controller that produces the reference current (i_{ref}). The mathematical representation of the PI controller is given generally in Equation (9). Its output (PI_{out}) depends on the input error and the constant gains K_p and K_I. In Figure 3a, the input error is ΔT and the output is i_{ref}. The closed loop direct current control is essential. The current controller is a hysteresis controller that outputs three voltage states from the power converter according to the current error ΔI and rotor position. If ΔI is greater than the current band, positive voltage is obtained. If ΔI is less than the current band, zero voltage is obtained. Finally, negative voltage is obtained if ΔI is less than the current band and the rotor position is after the turn off angle (θ_{off}). The torque estimator depends on the machine model as given in Figure 1.

In Figure 3b, the reference torque (T_{ref}) is converted directly to the reference current (i_{ref}) using the torque to current relationship as will be given later in Figure 7b. This helps to improve the dynamic performance [24]. In addition, the firing angles (θ_{on} and θ_{off}) must be optimized for better performance. The optimization is given below.

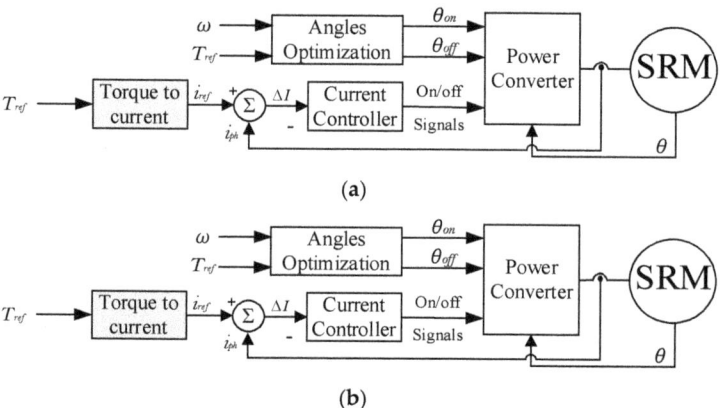

Figure 3. Block diagram of ATC with (**a**) torque controller, and (**b**) torque to current conversion.

$$PI_{out} = K_p * error + K_I * \int (error)\, dt \tag{9}$$

Optimization of Firing Angles (θ_{on} and θ_{off}) for ATC Technique

Due to the great effect of firing angles (θ_{on} and θ_{off}) on SRM performance as they affect torque production, amount of torque ripple, efficiency, and the range of speed control [15,31], the firing angles (θ_{on} and θ_{off}) are optimized based on a multi-objective problem to achieve the best overall performance. The multi-objective optimization includes the minimization of torque ripple (T_{rip}), reduction of copper losses (P_{cu}), increasing torque/current ratio, and improvement of motor efficiency (η).

A single-objective optimization problem can be obtained from the multi-objective problem by a linear combination of torque ripple (T_{rip}), copper losses (P_{cu}), and efficiency (η) as follows [27,29]:

$$F_{obj}\left(\theta_{on}, \theta_{off}\right) = minimum\left(w_r \frac{T_{rip}}{T_{rb}} + w_{cu} \frac{P_{cu}}{P_{cub}} + w_\eta \frac{\eta_b}{\eta}\right) \tag{10}$$

$$w_r + w_{cu} + w_\eta = 1 \tag{11}$$

$$\text{Subject to}: \theta_{on}^{min} \leq \theta_{on} \leq \theta_{on}^{max};\quad \theta_{off}^{min} \leq \theta_{off} \leq \theta_{off}^{max} \tag{12}$$

where F_{obj} is the objective function; T_{rb}, P_{cub}, and η_b are the base values for torque ripple, copper loss, and efficiency, respectively. w_r, w_{cu}, and w_η are the weight factors for torque ripple, copper loss, and efficiency, respectively. The weight factors can be chosen according to the desired level of optimization. θ_{on}^{min} and θ_{on}^{max} are the minimum and the maximum limits of the θ_{on}, respectively.

For every operating point, defined by motor speed and loading torque, the firing angles (θ_{on} and θ_{off}) are estimated to achieve Equation (10). The speed and torque are the inputs, and the firing angles (θ_{on} and θ_{off}) are the outputs that fulfill Equation (10). The inputs, speed, and torque are changed in small steps, and the corresponding optimal firing angles are estimated.

3.2. The Conventional DITC Technique

Figure 4 shows the block diagram of conventional DITC schemes [28,30,32].

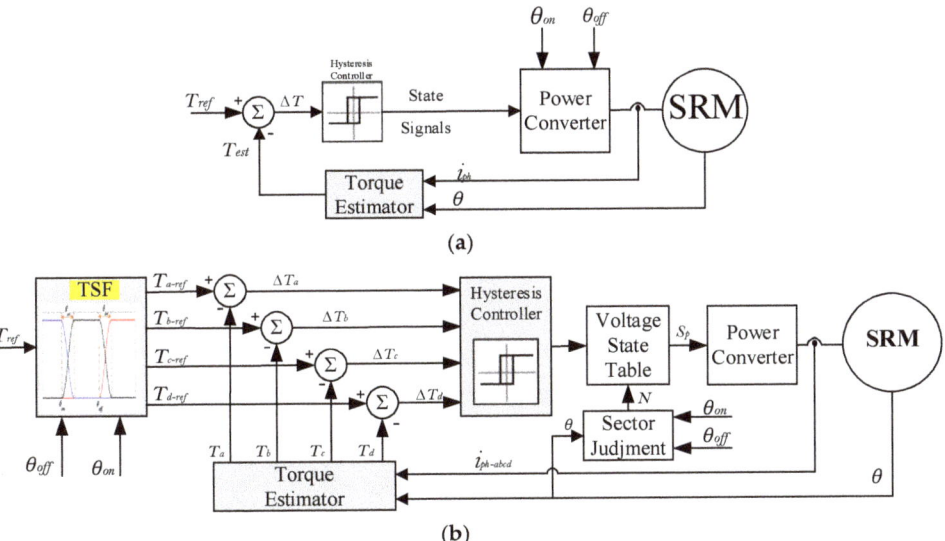

Figure 4. The block diagram of DITC, (**a**) conventional DITC, (**b**) TSF-based DITC.

T_{ref} is the reference commanded torque, it comes from an outer loop speed controller. In Figure 4a, the online torque estimator is essential to estimate the instantaneous motor torque (T_{est}). The torque error (ΔT) is processed through a hysteresis torque controller that outputs the state signals. The current control is included within the torque controller. Limiting the torque means limiting the maximum current value.

In Figure 4b, a torque sharing function (TSF) is introduced with the DITC algorithm [28]. The TSF is used to distribute T_{ref} between motor phases as described by Equation (13). The reference torque for each phase ($T_{a\text{-}ref}$, $T_{b\text{-}ref}$, $T_{c\text{-}ref}$, and $T_{d\text{-}ref}$) is compared to its actual measured value (T_a, T_b, T_c, and T_d). The torque errors (ΔT_a, ΔT_b, ΔT_c, and ΔT_d) are processed through a hysteresis torque controller that outputs the states. Sector judgment and voltage state table are employed to provide the proper control pulses. The sector judgment divides the control period into six sectors. The details are included in [28]. The voltage state table outputs the voltage state (S_p) (p could be phase a, b, c, or d) according to the torque error for each phase. For example, when ΔT_a > hysteresis band (ΔT), $S_a = 1$; when $\Delta T_a < -\Delta T$, $S_a = 1$; when phase is turned off, $S_a = -1$.

$$T_{ph-ref}(\theta) = \begin{cases} 0, & (0 \leq \theta \leq \theta_{on}) \\ T_{ref}\left(0.5 - 0.5\cos\left(\pi\frac{\theta - \theta_{on}}{\theta_{ov}}\right)\right), & (\theta_{on} \leq \theta \leq \theta_{on} + \theta_{ov}) \\ T_{ref}, & (\theta_{on} + \theta_{ov} \leq \theta \leq \theta_{off}) \end{cases} \quad (13)$$

where θ_{ov} is the overlap angle; θ_p depicts the rotor pitch angle.

Optimization of Firing Angles (θ_{on} and θ_{off}) for DITC Technique

For DITC, the firing angles (θ_{on} and θ_{off}) have to be optimized for better motor performance. The optimization is similar to that for ATC which is described by Equations (10)–(12). The difference is the structure of the control algorithm. Different weights can be chosen according to the required level of optimization.

4. The Proposed Universal Torque Control (UTC) of SRM for EVs

As each control technique provides a unique performance under a certain speed range, a combination of these techniques is the best solution in order to gain the benefits of each control technique. The best choice for the low speed range is the DITC. It provides a very low torque ripple, fast dynamic response, high efficiency, and the lowest RMS current. On the other hand, ATC is the best choice for high speeds. It provides high efficiency, a fast dynamic response, and the lowest RMS current. At high speeds, the vehicle inertial can filter the torque ripples [27].

A proposed universal control of SRM for EVs is given in Figure 5. It employs a DITC and ATC. The DITC is meant for low speeds and the ATC is meant for high speeds. The transition between the two techniques should be very smooth without transient or disturbances.

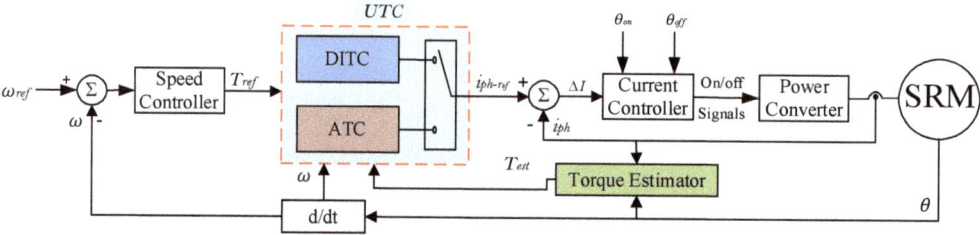

Figure 5. The universal torque control (UTC) of SRM.

For developing such universal control (Figure 5), several problems arise. First, the conventional structures of ATC and DITC, shown in Figures 3 and 4, are complicated to be combined in one control algorithm. The final UTC algorithm will be much more complicated. Second, the direct combination of conventional ATC and DITC seems to be infeasible. That is mainly because of the big differences between conventional ATC and conventional DITC regarding structure and operational principles. Third, these big differences could cause a non-smooth transition that could increase the noise and disturbance in a vehicle. The disturbance could be large enough to unstable the system operation. For these reasons, this paper improves the structure of conventional ATC and the structure of conventional DITC, in order to avoid the aforementioned issues, and aiming to have the best structure for the UTC strategy.

4.1. The Proposed DITC Technique

Figure 6a shows the block diagram of the modified DITC. The main difference is that it has a torque to current conversion as seen in Figure 6b. i_{ref} is obtained directly from T_{ref}. In addition, the torque error ($T_{ref} - T_{est}$) is compensated after being multiplied by K1.

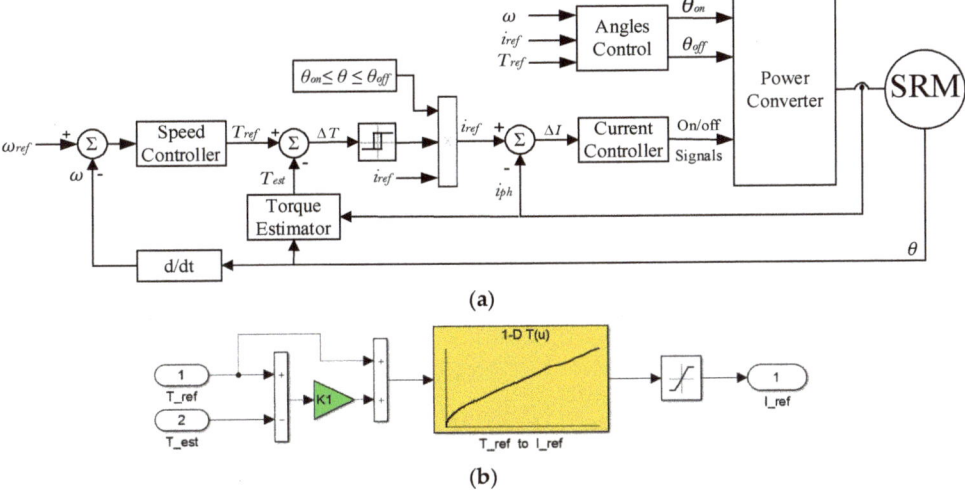

Figure 6. The adopted DITC (**a**) block diagram; (**b**) the torque to current conversion.

The torque to current conversion is achieved by the analyses of torque characteristics (Figure 7). The best torque production is achieved over the period [$\theta_m, \theta_m + \theta_c$] [33]. For each current magnitude, the average torque can be estimated. Then, a polynomial fitting can be simply applied, which is a very simple formulation as illustrated Figure 7b.

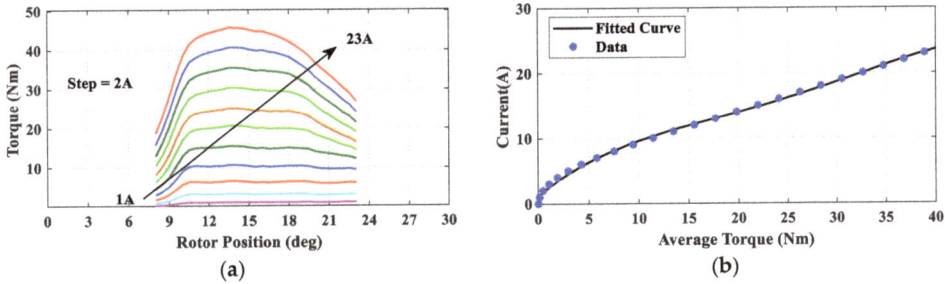

Figure 7. Torque to current conversion: (**a**) torque curves over 15°; (**b**) current versus average torque.

4.2. The Proposed ATC Technique

Figure 8 shows the block diagram of the simplified ATC (SATC). The reference current (i_{ref}) is obtained directly from the outer loop speed controller. In such a system, there is no need to estimate the average torque online as the optimization of firing angles at each operating point can be completed to ensure the motor can produce the required torque. Hence, the closed loop torque control is no longer needed. As a result, the structure of the overall system is reduced, which means a less complicated control algorithm [27].

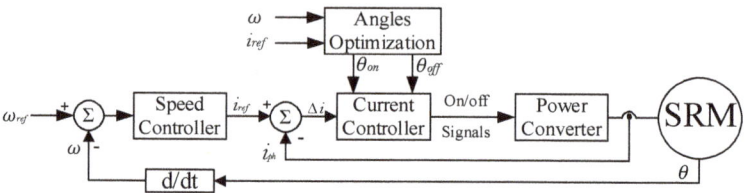

Figure 8. Block diagram of SATC.

4.3. The Final Intergrated UTC

Figure 9 shows the proposed UTC. It uses the modified versions of DITC and the SATC in order to have a feasible integration. The outer loop speed controller feeds the reference commanded torque to DITC and SATC. A switching unit determines which control algorithm is employed accordingly with motor speed. The switching value of speed is chosen to be 2000 r/min. That means below 2000 r/min, the DITC is utilized, while for higher speeds the ATC is used.

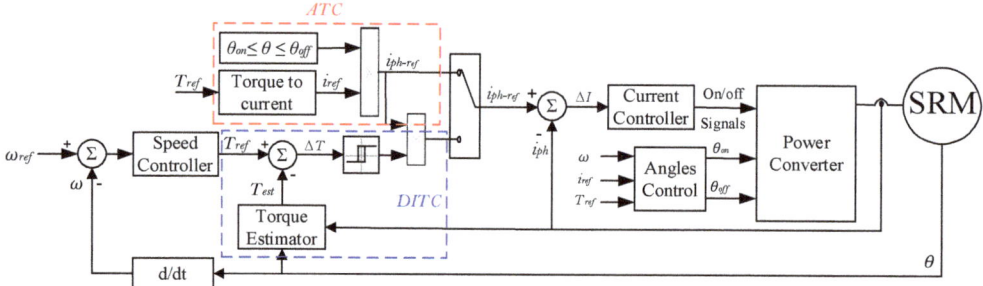

Figure 9. The proposed universal torque control (UTC) of SRM for EVs.

In order to have such an integration, a single torque to current conversion is needed. The firing angles are estimated online to improve the system performance. Furthermore, they are the keystone to have a smooth transition between the DITC and the ATC. The idea to guarantee a smooth transition is as follows.

Smooth Transition Control

The control will change from DITC to ATC after a certain speed (2000 r/min). At the changing instant, a smooth transition should occur. A smart solution without complications of the control algorithm could be the best. There is no need for cross-over control to transit from a state to another.

The UTC utilizes both the DITC and ATC. For each control strategy, the best performance is achieved with certain firing angles that achieve Equation (10). As the operation and structure of DITC is different from that of ATC, the solutions for the best firing angles that achieve Equation (10) for DITC and ATC will be large. Hence, the transition from one strategy (DITC) to the other (ATC), or vice versa, will lead to a large variation for the firing angles. This, in turn, gives rise to uncertainties and unpredicted performance for the system.

In this paper, the solution for a smooth transition is completed based on the solution of optimum excitation angles. A smooth transition from one technique to another can be guaranteed if small changes in the best firing angles can be achieved. Hence, this is the idea to achieve now.

Based on the observations, experience, and analysis of SRM behavior, the variations in the turn on angle (θ_{on}) with speed and loading torque are very large. The variations are also different from DITC to ATC. Hence, Equation (10) will not guarantee the best transition. For that reason, the authors propose the utilization of a single analytical formulation for the best turn on angle (θ_{on}) over the whole speed range for both the DITC and the ATC. The analytical solution for the turn on angle (θ_{on}) is given in detail in the previous authors' work [31]. This solution is illustrated in Equation (14). This equation provides a simple analytical solution for the optimum turn on angles over the whole range of speeds.

$$\theta_{on} = \theta_m + \frac{L(i,\theta)}{R+k_b\omega} \ln\left(1 - i_{ref}\frac{R+k_b\omega}{V_{DC}}\right) \tag{14}$$

where $L(i,\theta)$ is the phase inductance; V_{DC} is the supply voltage; and ω is the rotor speed. θ_m is the angle at which rotor poles start to overlap with stator poles.

The control variables are ω and i_{ref}. Hence, the optimum θ_{on} depends mainly on the inductance $L(i,\theta)$ and its slope $k_b = dL(i,\theta)/d\theta$. The inductance $L(i,\theta) = L(\theta)$ is the only position dependent over the minimum inductance zone $[-\theta_m, \theta_m]$. This fact simplifies the solution and implementation of Equation (4). The inductance can be easily fitted against the position as in Equation (15). The constants a, b, and c are easily estimated from the inductance data.

$$L = ae^{b|\theta|} + c, -\theta_m \leq \theta \leq \theta_m \tag{15}$$

For DITC, as it has an inherited capability for torque ripple reduction, its turn on angle can be optimized for MTPA using Equation (14). Its optimum turn off angle is estimated based on an optimization problem as in Equations (16)–(18). Greater importance is given to the efficiency due to the inherited capability of DITC for torque ripple reduction. Therefore, the weight factors are w_η = 0.6, w_r = 0.2, and w_{cu} = 0.2.

For SATC, the turn on angle is optimized for MTPA using Equation (14) too. As SATC is employed for high speeds, torque ripples are not of great interest as they could be filtered by inertia. The efficiency becomes of the greatest value. Therefore, the firing angles are optimized for higher efficiency focusing. A higher efficiency weighting factor of w_η = 0.6 is chosen, while the other weights are w_r = 0.2 and w_{cu} = 0.2.

The turn off angle is optimized based on the optimization problem that is described by Equations (16)–(19). Additional constraint is needed for SATC. It is given by Equation (19). This constraint guarantees the production of the commanded torque level within the obtained firing angles. Hence, the SATC has no need for a closed loop torque control.

$$F_{obj}\left(\theta_{off}\right) = minimum\left(w_r \frac{T_{rip}}{T_{rb}} + w_{cu}\frac{P_{cu}}{P_{cub}} + w_\eta \frac{\eta_b}{\eta}\right) \tag{16}$$

$$w_r + w_{cu} + w_\eta = 1 \tag{17}$$

$$\text{Subject to}: \theta_{on} = \theta_m + \frac{L(i,\theta)}{R+k_b\omega}\ln\left(1 - i_{ref}\frac{R+k_b\omega}{V_{DC}}\right); \quad \theta_{off}^{min} \leq \theta_{off} \leq \theta_{off}^{max} \tag{18}$$

$$T_e \leq T_{rated}\mid_{\omega,i_{ref}} \tag{19}$$

Figure 10 shows the flowchart of the developed searching algorithm. At each operating point, the turn off angle (θ_{off}) is changed in small steps. The simulation model is employed to calculate the torque ripple, copper loss, and efficiency at each step. At the end of the search, the minimum torque ripple, the minimum copper losses, and the maximum efficiency are defined as the base values (T_{rb}, P_{cub}, and η_b). The turn off angle (θ_{off}) is varied from $\theta_{off\text{-}min}$ = 15° to $\theta_{off\text{-}max}$ = 28° in steps of 0.2°, while the torque step is taken as 2 Nm.

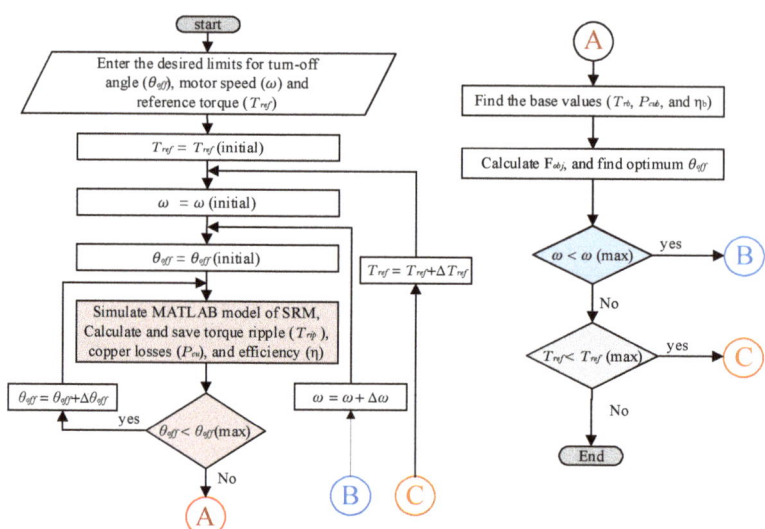

Figure 10. Flowchart of searching algorithm.

The optimized turn off angles for DITC and SATC are given in Figure 11. As noted, the turn off angles lie in a limited band. With firing angles' optimization, a smooth transition is guaranteed. Both DITC and SATC utilize Equation (14) for the optimum turn on angle, which means a perfect transition over the turn on. The changing band of the turn off angle is very limited as seen in Figure 11. Hence, the transition from DITC to SATC will be very smooth too. As a result, no change will happen over the turn on angle. Furthermore, no big change will happen over the turn off angle. That means a guaranteed STC.

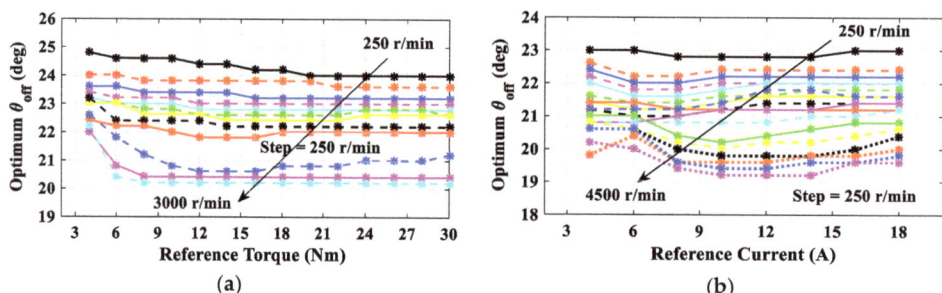

Figure 11. The optimized θoff angles: (**a**) DITC, (**b**) ATC.

5. Simulation Results and Discussion

A series of simulation results are conducted to provide a detailed analysis for the control performance. Hence, it is a powerful tool to validate the performance of a proposed control. The simulation results include a dynamic performance evaluation under a sudden change of reference speed and load torque. They also include an investigation of control behavior under EV loading in acceleration and deceleration regions.

5.1. Sudden Change in Reference Speed and Load Torque

Figure 12 gives the results of sudden changes in reference motor speed and its loading torque. The changing profiles for reference speed are shown in Figure 12a. The reference speed (ω_{ref}) is changed from 1500 r/min to 2500 r/min at the time of 1.5 s. The transition from DITC to SATC and vice versa is illustrated by signal (T_{sig}). The positive value of

T_{sig} means that UTC uses DITC, and the zero value of T_{sig} means that UTC uses SATC. The value of transition speed is 2000 r/min. Hence, the control starts with DITC until the speed of 2000 r/min, and then it changes to SATC for higher speeds than 2000 r/min. The loading torque profile is seen in Figure 12b, where a sudden change from 15 Nm to 12 Nm is observed at the time of 1.0 s. Figure 12c shows the reference current. The transition from DITC to SATC is obvious. The reference current is compensated under DITC (0–1.63 s) to reduce the torque ripple. That is why it is a thick line. On the other side, the reference current has a smooth profile under SATC (1.63–2.5 s).

Figure 12d,e illustrate the variation of firing angles, turn on, and turn off angles, respectively. They are changing accordingly with motor speed and the current level to provide the best performance. The estimation of firing angles is obtained based on the optimization problems that are described by Equations (16)–(19). A smooth behavior for the firing angles at the instant of transition is observed that ensures a smooth transition.

Figure 12f gives the output torque profile of the motor. As seen, the DITC (0–1.63 s) provides a smooth torque profile. The SATC (1.63–2.5 s) has higher torque ripples. The value of torque ripples is seen in Figure 12g. The DITC shows the torque ripple value of about 20%. On the contrary, the torque ripple under SATC reaches 45%. At the instant of transition at time 1.5 s, the curve of the torque ripple shows a noticeable notch that can be absolutely ignored. Regularly, the ripples are estimated under steady state conditions (constant speed and level currents), and the notch appears because of the transition.

Figure 12h illustrates the output mechanical power. In both regions of control, the motor can provide its rated power (4 kW) even with 50% higher. The switching frequency is seen by Figure 12i. It is changing with speed but has a limited band. The maximum frequency is about 10.6 kHz.

The efficiency curve is seen in Figure 12j. As noted, the efficiency increases with the motor speed. At the transition instant, the efficiency is very clear to be increased as the SATC provides higher efficiencies than DITC. The DITC provides a very good efficiency of almost 96.5%. On the other hand, the SATC provides a higher efficiency (almost 97.5%).

5.2. Acceleration with Electric Vehicle Loading

Figure 13 illustrates the results with EV as a loading under acceleration (motoring action).

Figure 13a shows the motor speed. ω_{ref} is changing from 1000 to 2000 at the time of 0.6 s, then it changes from 2000 to 3000 at the time of 1.3 s. The transition is illustrated by T_{sig}. The motor starts under DITC and converts to SATC at the time of 1.3 s. Figure 13b gives the load torque profile. As the load torque represents an EV, it increases with speed.

Figure 13c,d show the variation of turn on and turn off angles, respectively. A smooth and adaptive variation is observed even at the instant of transition that reflects the smooth transition control. Figure 13e illustrates the output torque profile of the motor. A smooth transition is obvious over the torque curve. Lower ripples are observed under DITC compared to SATC. Figure 13f shows the values of the torque ripple. The ripple is about 23% under DITC and reaches 60% under SATC.

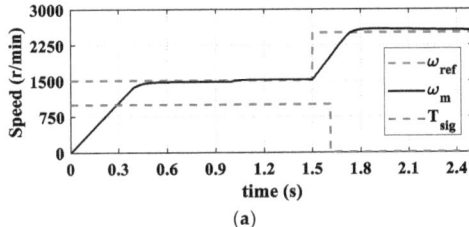
(a)

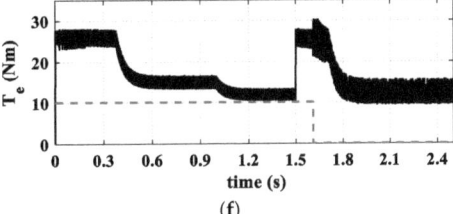
(f)

Figure 12. Cont.

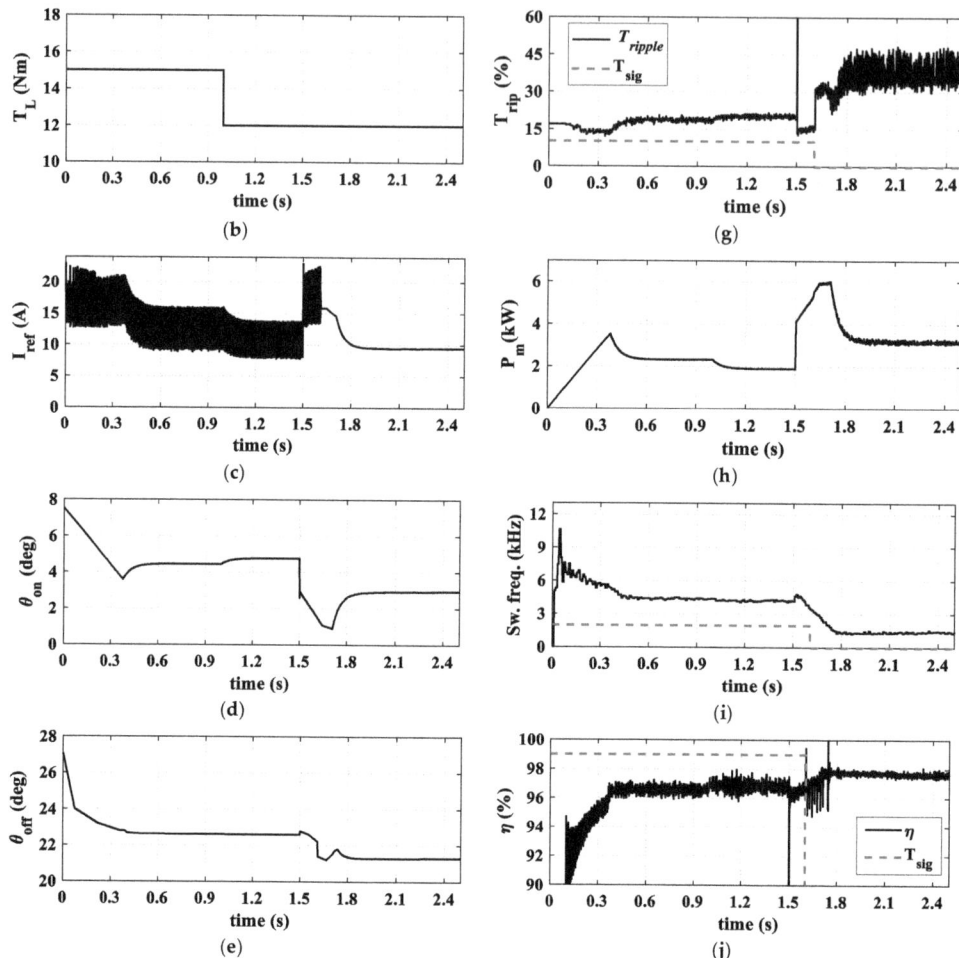

Figure 12. The simulation results under a sudden change in motor speed. (**a**) The motor speed; (**b**) the load torque; (**c**) the reference current; (**d**) the variation of the turn on angle; (**e**) the variation of the turn off angle; (**f**) the total electromagnetic torque; (**g**) the torque ripple; (**h**) the mechanical output power; (**i**) the switching frequency; (**j**) the efficiency.

The switching frequency is seen by Figure 13g. It has a maximum value of about 10.6 kHz. The switching frequency decreases with speed as the motor employs single pulse control under high speeds. Figure 13h shows the efficiency curve. The efficiency is observed to be around 97.5% under high speeds.

As a conclusion, the proposed UTC provides lower torque ripples at low speeds by using the DITC in order to reduce the oscillations of the vehicle body and to provide better drivability. Furthermore, the proposed UTC provides better efficiency at high speeds by using SATC. The torque ripples at high speeds can be filtered by vehicle inertia. Hence, the proposed UTC is a superior choice for EVs and several industrial applications.

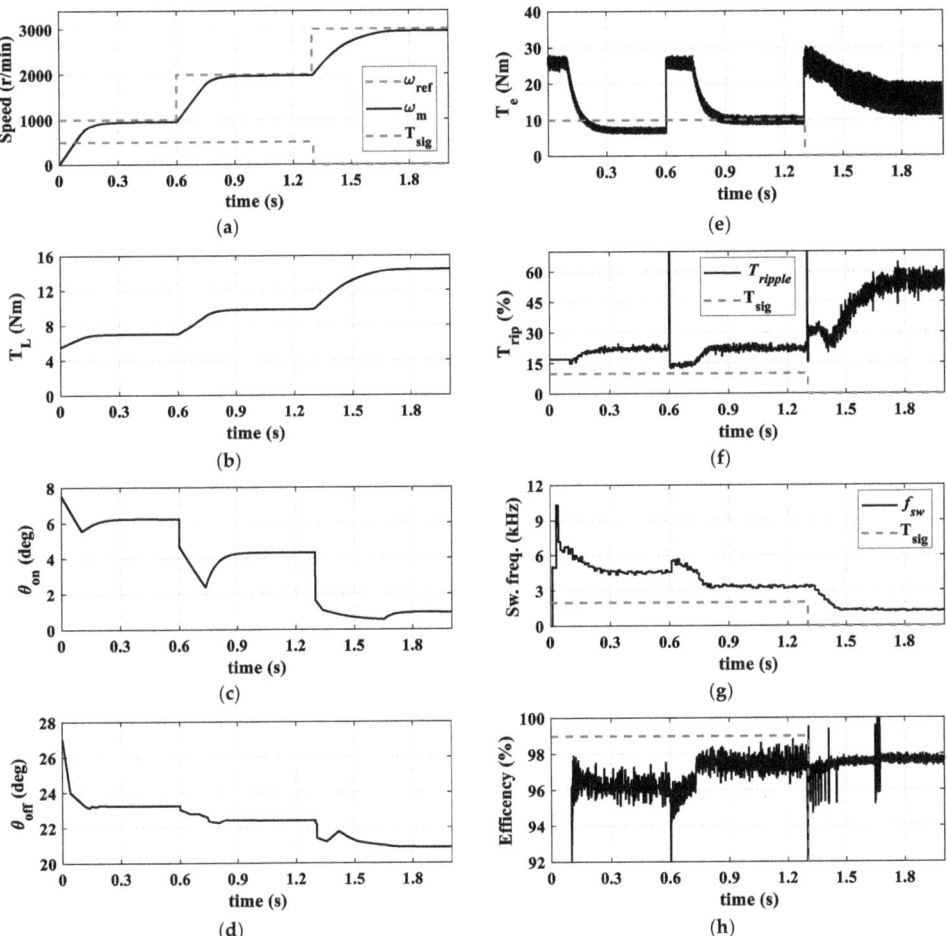

Figure 13. The simulation results under EV acceleration. (**a**) the motor speed; (**b**) the load torque on the motor side; (**c**) the variation of the turn on angle; (**d**) the variation of the turn off angle; (**e**) the total electromagnetic torque; (**f**) the torque ripple; (**g**) the switching frequency; (**h**) the efficiency.

5.3. Acceleration and Deceleration with Electric Vehicle Loading

Figure 14 is the results for the acceleration and deceleration with EV loading. The speed profile is shown in Figure 14a. From 1.0 s to 1,2 s, the vehicle deaccelerates or brakes. I_{ref} changes to negative at the beginning of the braking operation at 1.0 s as seen in Figure 14b. The negative I_{ref} means that the motor is in forward generating mode. Hence, the firing angles are changed from motoring (increasing inductance) to generating (decreasing inductance) as illustrated by Figure 14c,d. The firing angles are changing smoothly in motoring and generating modes.

In the generating (braking) region, the motor returns the current back to the supply/battery; hence, the average current becomes negative as shown in Figure 14e. The mechanical power and torque become negative in the braking region as seen in Figure 14f,g, respectively.

Zooming on the torque profile in the braking zone is shown in Figure 14h. First, the motor brakes under DITC, then under SATC, as seen, and a very smooth transition under generating is observed too. The torque ripple is seen in Figure 14i: it spikes because the machine torque crosses zero while changing from positive (motoring) to negative (generating) and vice versa. The switching frequency still has a limited value as seen in Figure 14j.

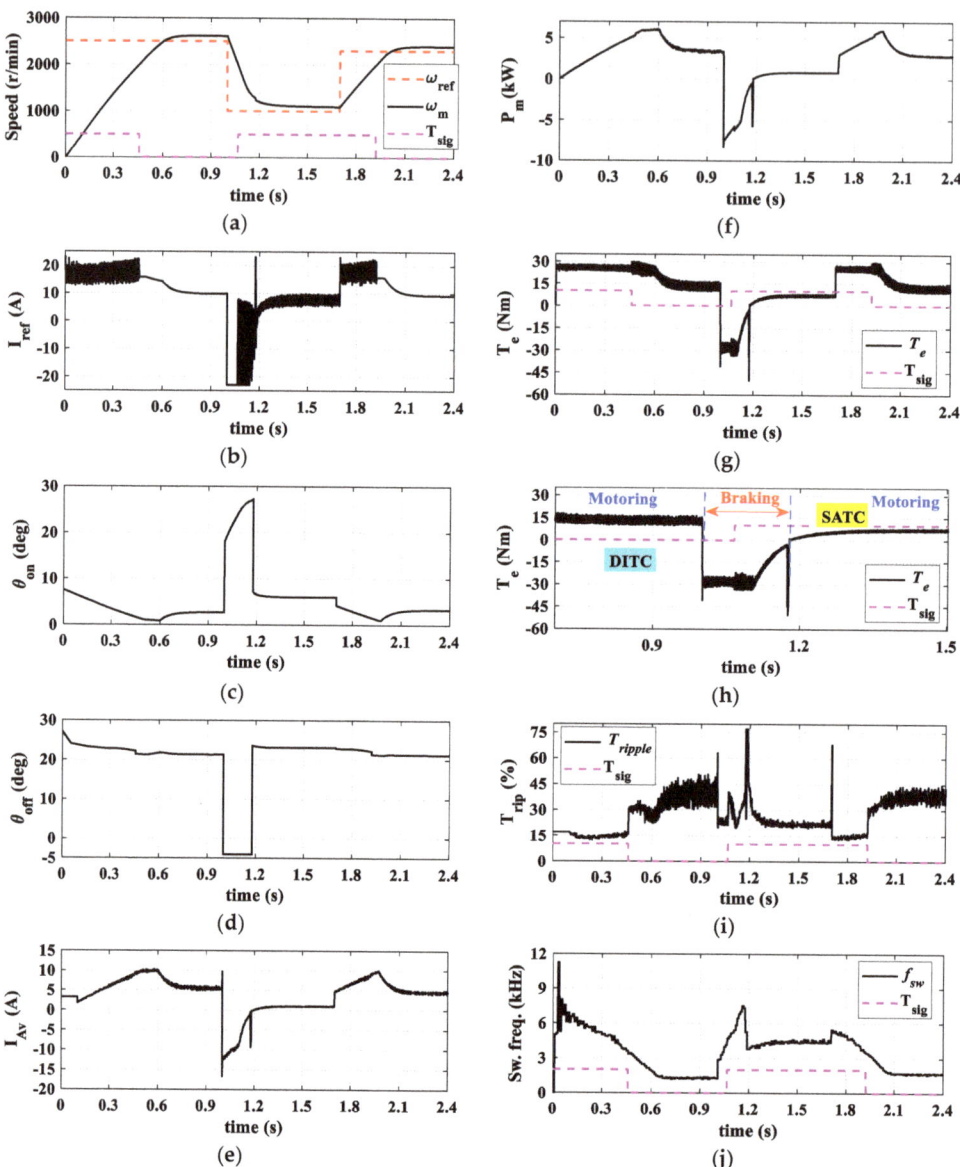

Figure 14. The simulation results under starting with EV acceleration and deceleration. (**a**) The motor speed; (**b**) the reference current; (**c**) the variation of the turn on angle; (**d**) the variation of the turn off angle; (**e**) the average current; (**f**) the mechanical output power; (**g**) the total electromagnetic torque; (**h**) zoom on electromagnetic torque; (**i**) the torque ripple; (**j**) the switching frequency.

Figure 15 shows one phase current and the phase inductance under the instant of braking. Before the time of 1.0 s, the machine is in motoring action. The current exists in the increasing inductance zone. Hence, it generates positive torque. On the other hand, after the instant of 1.0 s, the phase current exists in the decreasing inductance zone. Hence the generated torque is negative (braking torque). Note that the current is always positive as the SRM has a current that follows in one direction. The positive and negative torques are defined by the firing angles accordingly with the inductance profile. The generating current profile is seen to be a mirrored shape, with different amplitudes, to the motoring current profile.

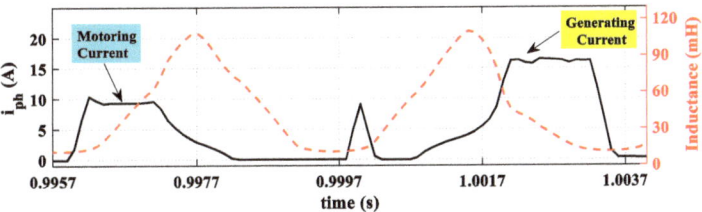

Figure 15. The phase current in motoring and generating modes.

6. Comparative Analysis

In order to show the effectiveness of the proposed UTC, a comparative analysis is conducted over a wide speed range. The proposed UTC is compared to two control techniques. The first is the ATC in [29] (Figure 3). The firing angles (θ_{on} and θ_{off}) of the ATC are optimized for the lowest torque ripple and the highest efficiency as in [29]. The optimization problem in Equations (16)–(19) is employed. The weights are set as $w_r = 0.4$, and $w_\eta = 0.6$. w_{cu} is not included in [29]; hence, it is set to zero ($w_{cu} = 0$).

The second technique for comparison is the DITC-based TSF in [28] (Figure 3b). In [28], the adaptive turn on angle control is introduced in [28] as seen in Figure 16. The conduction angle for TSF is constant: it is 15 for tested 8/6 SRM. Hence, the turn off angle is defined by $\theta_{off} = \theta_{on} + 15°$.

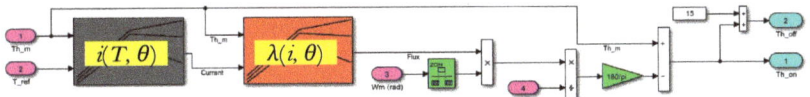

Figure 16. The simulation implementation of turn on and turn off angles of improved DITC of [28].

The comparative study is conducted under the dynamic loading condition, under the EV loading profile, and under the full load conditions as follows.

6.1. Under Dynamic Loading Conditions

Figure 17 shows the torque profiles, torque ripples, and efficiencies for the proposed UTC, ATC in [29], and DITC in [28]. The comparison is conducted for two speed levels: low speed of 1200 r/min, and high speed of 2400 r/min. This is mainly to illustrate the performance of the UTC in different control regions. The reference torque is changing between 10 Nm, 15 Nm, and 20 Nm.

Figure 17a shows the speed curve. The speed is changed from 1200 r/min to 2400 r/min at a time of 0.4 s. Figure 17b,c show the instantaneous motor torque for ATC, DITC, and UTC, respectively. As noticed, the three control techniques can track properly their refence torque (T_{ref}). The difference is the torque profile and amount of torque ripples.

Figure 17e shows the value of torque ripple. For low speed (0–0.4 s), the UTC shows the lowest torque ripple, followed by DITC, then SATC. For high speed (0.4–0.6 s), the UTC shows the lowest torque ripple followed by SATC. At high speeds, the DITC shows very high torque ripples.

Figure 17f gives the mean value of motor efficiency with a 60 Hz window. For low speed, the UTC shows the highest efficiency, followed by SATC, then DITC. For high speed, the UTC and SATC show the highest efficiencies. The DITC shows a lower efficiency.

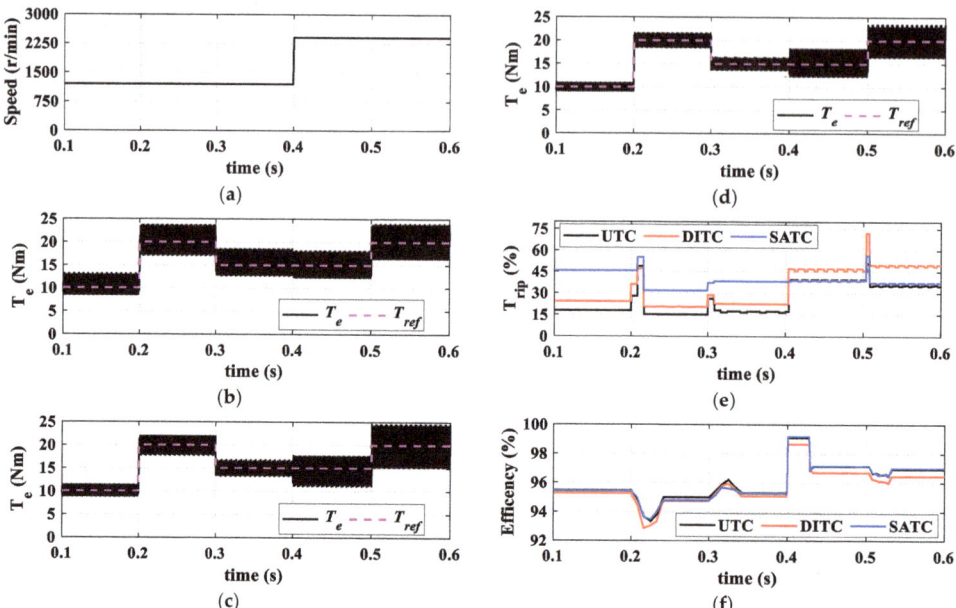

Figure 17. Comparative results under dynamic loading conditions. (**a**) Speed profile; (**b**) instantaneous torque of SATC in [29]; (**c**) instantaneous torque of DITC in [28]; (**d**) instantaneous torque of UTC; (**e**) torque ripple; (**f**) efficiency.

6.2. Under EV Loading Profile

The steady state results under the EV load profile, for three control techniques, are shown in Figure 18. The average torque is seen in Figure 18a, and the UTC and SATC can track their reference torque. The DITC fails to provide the commanded reference torque at high speeds (after 2000 r/min). The torque ripples are given in Figure 18b. The UTC provides the lowest torque ripple over the entire speed range. The SATC shows low ripples at a high speed, while obtaining high ripples at low speeds. The DITC shows low ripples under a low speed only.

Figure 18c shows the efficiency. The UTC and SATC show higher efficacies at high speeds compared to DITC. The average torque to RMS current ratio (T_{av}/I_{RMS}), in Figure 18d, is a very important index for electric machines, and the higher the better. As observed, the UTC shows a high T_{av}/I_{RMS} as the ATC. The DITC is the lowest. The switching frequency in Figure 18e lies in a limited band (<10 kHz) that fits well for experimental implementations without any constraints.

6.3. Under Full Load Conditions

The steady state results under the full load for three control techniques are shown in Figure 19. Figure 19a illustrates the full load average torque. The UTC and SATC can track their reference torque over the full speed range. The DITC fails tracking at high speeds (after 2000 r/min). Figure 19b gives the torque ripples. The UTC provides the lowest torque ripple over the entire speed range. The SATC shows high ripples at low speeds, and the DITC shows low ripples under low speed only. Figure 19c shows the efficiency. The UTC shows the highest efficiency over the full speed range.

Figure 19d shows the superior performance of the UTC to provide the highest T_{av}/I_{RMS} ratio over the full speed range. This in turn proves the MTPA operation and improved efficiency. The switching frequency, Figure 19e, is still lying in a limited band.

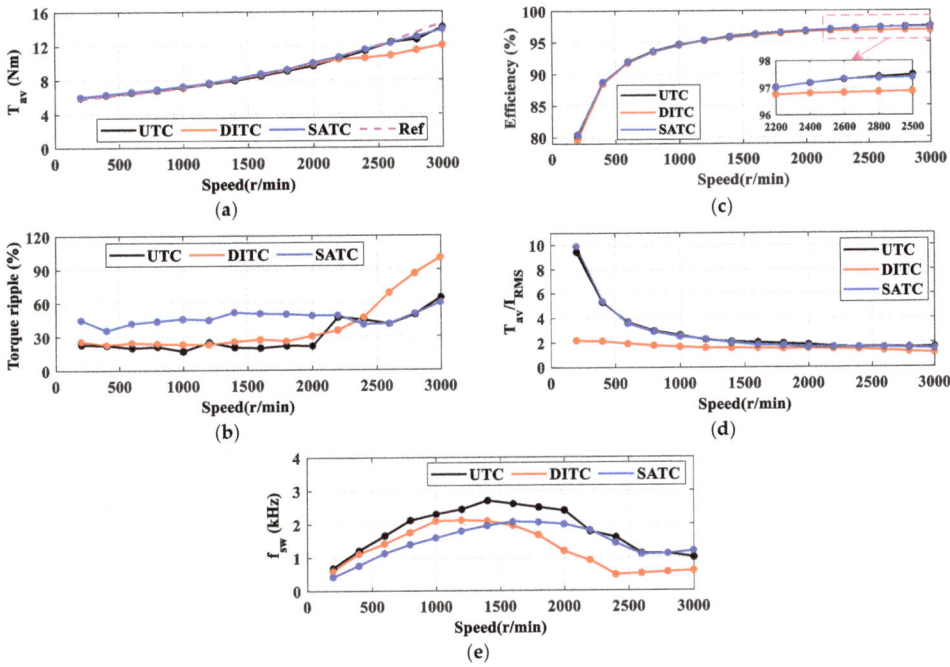

Figure 18. The steady state characteristics under EV loading profile. (**a**) Average torque; (**b**) torque ripple; (**c**) efficiency; (**d**) torque/current ratio; (**e**) switching frequency.

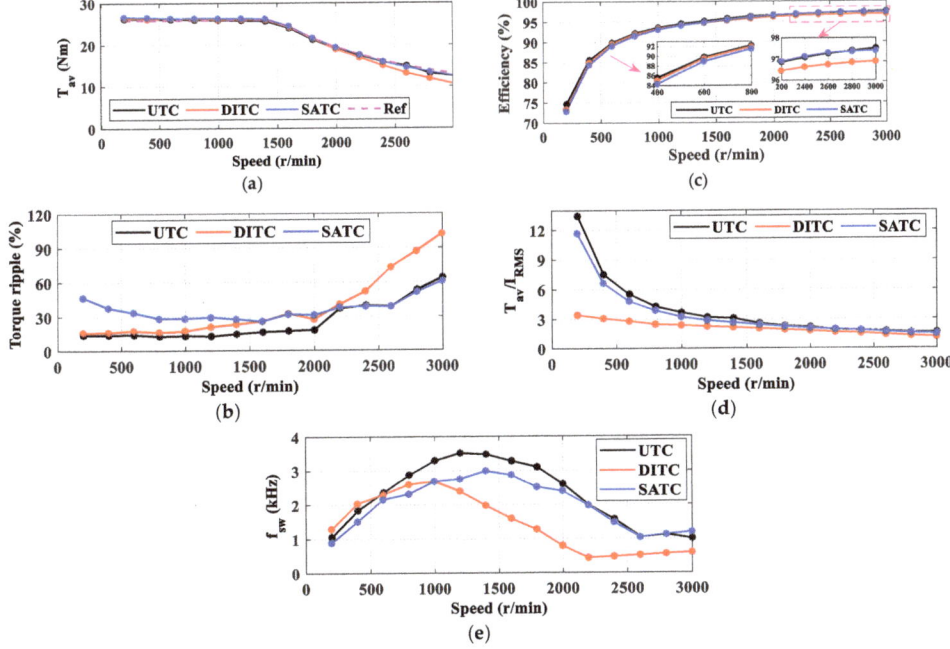

Figure 19. The steady state characteristics under full load conditions. (**a**) Average torque; (**b**) torque ripple; (**c**) efficiency; (**d**) torque/current ratio; (**e**) switching frequency.

6.4. Summary

Tables 1 and 2 show the detailed conclusion for comparative study. Table 1 shows a summary of the control performance over speed ranges. As a conclusion, the UTC provides the best performance regarding average torque production, torque ripples, T_{av}/I_{RMS}, and efficiency over the entire speed range. This is basically the contribution of this research; that is, to gather all the advantages in a single control method by combining two control strategies together. The proposed control method succeeded in gathering the advantages of the DITC for low speeds and the advantages of the ATC for high speeds. In addition, the proposed modification improved the performance of the proposed UTC to provide lower torque ripples and higher T_{av}/I_{RMS} compared to the DITC at low speeds.

Table 2 discusses the control structure and its feasibility for real time implementations. It shows the superior performance of the proposed UTC as it has a moderate complexity of the control algorithm. Hence, it is a feasible control method with no constrains in experimental implementations. Furthermore, the proposed UTC has fast dynamics and does not require a TSF.

Table 1. Summary of comparison results over speed ranges.

Indices	Low Speed			High Speed		
	ATC	DITC	UTC	ATC	DITC	UTC
Average torque	High	High	High	High	Low	High
Torque ripple	High	Low	Very Low	Low	High	Low
T_{av}/I_{RMS}	Medium	Low	High	High	Low	High
Switching frequency	Low	Low	Low	Low	Low	Low
Efficiency	Medium	High	High	High	Medium	High

Table 2. Summary of comparison results regarding control structure.

Indices	ATC [29]	DITC [28]	UTC
Algorithm complexity	Simple	Complex	Moderate
Dynamic torque response	Fast	Fast	Fast
Requirement of online torque estimation	Yes	Yes	Yes
Requirement of TSF algorithm	No	Yes	No
Required control period	Long	Short	Short

7. Conclusions

This paper presents the development of a UTC of SRM drives for EVs. The proposed UTC strategy is a combination of the DITC and SATC. The proposed technique utilizes a DITC for low speed operation and employs an SATC for high speeds. A very smart and smooth transition between the DITC and SATC is guaranteed within switching angles optimization. The results show the superior performance of the proposed UTC over the entire speed range. The proposed UTC can provide low torque ripples, even compared to conventional DITC, at low speeds. Furthermore, it shows the highest T_{av}/I_{RMS}, even compared to the ATC. For high speeds, the proposed UTC shows a similar performance to the ATC that is the best for SRMs regarding efficiency and torque ripples. The torque ripples at high speeds can be filtered by vehicle inertia. The very low torque ripples at low speeds reduce the oscillations of the vehicle body and provide better drivability. The high efficiency at high speeds increases the milage per charge. Hence, the proposed UTC is a superior choice for EVs and several industrial applications. Moreover, the proposed UTC provides a simple structure, high dynamics, extended constant power range, and reduced torque ripples. This, in turn, makes the proposed UTC feasible for easy implementations with high reliability. Future works could include the experimental implementations of the proposed control considering measurement errors and time delays.

Author Contributions: Conceptualization, M.H.; Formal analysis, M.H. and F.A.-A.; Investigation, M.H. and M.N.I.; Methodology, M.H. and F.A.-A.; Project administration, L.S.; Software, M.H. and F.A.-A.; Validation, M.H. and F.A.-A.; Visualization, I.O. and M.N.I.; Writing—original draft, M.H. and M.N.I.; Writing—review and editing, M.H., M.N.I., and L.S. All authors have read and agreed to the published version of the manuscript.

Funding: This research received no external funding.

Conflicts of Interest: The authors declare no conflict of interest.

References

1. Lan, Y.; Benomar, Y.; Deepak, K.; Aksoz, A.; El Baghdadi, M.; Bostanci, E.; Hegazy, O. Switched Reluctance Motors and Drive Systems for Electric Vehicle Powertrains: State of the Art Analysis and Future Trends. *Energies* **2021**, *14*, 2079. [CrossRef]
2. Yueying, Z.; Chuantian, Y.; Yuan, Y.; Weiyan, W.; Chengwen, Z. Design and optimisation of an In-wheel switched reluctance motor for electric vehicles. *IET Intell. Transp. Syst.* **2019**, *13*, 175–182. [CrossRef]
3. Nam, K.H. *AC Motor Control and Electrical Vehicle Applications*; CRC Press: Boca Raton, FL, USA, 2019.
4. Bostanci, E.; Moallem, M.; Parsapour, A.; Fahimi, B. Opportunities and Challenges of Switched Reluctance Motor Drives for Electric Propulsion: A Comparative Study. *IEEE Trans. Transp. Electrif.* **2017**, *3*, 58–75. [CrossRef]
5. Zhu, J.; Cheng, K.W.E.; Xue, X. Design and Analysis of a New Enhanced Torque Hybrid Switched Reluctance Motor. *IEEE Trans. Energy Convers.* **2018**, *33*, 1965–1977. [CrossRef]
6. Bramerdorfer, G.; Tapia, J.A.; Pyrhonen, J.J.; Cavagnino, A. Modern Electrical Machine Design Optimization: Techniques, Trends, and Best Practices. *IEEE Trans. Ind. Electron.* **2018**, *65*, 7672–7684. [CrossRef]
7. Chen, H.; Yan, W.; Gu, J.J.; Sun, M. Multiobjective Optimization Design of a Switched Reluctance Motor for Low-Speed Electric Vehicles with a Taguchi-CSO Algorithm. *IEEE/ASME Trans. Mechatron.* **2018**, *23*, 1762–1774. [CrossRef]
8. Mousavi-Aghdam, S.R.; Feyzi, M.R.; Bianchi, N.; Morandin, M. Design and Analysis of a Novel High-Torque Stator-Segmented SRM. *IEEE Trans. Ind. Electron.* **2016**, *63*, 1458–1466. [CrossRef]
9. Ding, W.; Yang, S.; Hu, Y.; Li, S.; Wang, T.; Yin, Z. Design Consideration and Evaluation of a 12/8 High-Torque Modular-Stator Hybrid Excitation Switched Reluctance Machine for EV Applications. *IEEE Trans. Ind. Electron.* **2017**, *64*, 9221–9232. [CrossRef]
10. Dmitrievskii, V.; Prakht, V.; Kazakbaev, V. Novel rotor design for high-speed flux reversal motor. *Energy Rep.* **2020**, *6*, 1544–1549. [CrossRef]
11. Fang, G.; Pinarello Scalcon, F.; Xiao, D.; Vieira, R.; Grundling, H.; Emadi, A. Advanced Control of Switched Reluctance Motors (SRMs): A Review on Current Regulation, Torque Control and Vibration Suppression. *IEEE Open J. Ind. Electron. Soc.* **2021**, *2*, 280–301. [CrossRef]
12. Al-Amyal, F.; Számel, L. Analytical Approach for the Turn-Off Angle in Switched Reluctance Motors. *Lect. Notes Networks Syst.* **2022**, *217*, 685–696. [CrossRef]
13. Al-Amyal, F.; Hamouda, M.; Számel, L. Torque Quality Improvement of Switched Reluctance Motor Using Ant Colony Algorithm. *Acta Polytech. Hung.* **2021**, *18*, 129–150. [CrossRef]
14. Shahabi, A.; Rashidi, A.; Afshoon, M.; Saghaıan Nejad, S.M. Commutation angles adjustment in SRM drives to reduce torque ripple below the motor base speed. *Turkısh J. Electr. Eng. Comput. Sci.* **2016**, *24*, 669–682. [CrossRef]
15. Hamouda, M.; Számel, L. Optimum Control Parameters of Switched Reluctance Motor for Torque Production Improvement over the Entire Speed Range. *Acta Polytech. Hung.* **2019**, *16*, 3. [CrossRef]
16. Al-Amyal, F.; Hamouda, M.; Számel, L. Performance improvement based on adaptive commutation strategy for switched reluctance motors using direct torque control. *Alex. Eng. J.* **2022**, *61*, 9219–9233. [CrossRef]
17. Li, H.; Bilgin, B.; Emadi, A. An Improved Torque Sharing Function for Torque Ripple Reduction in Switched Reluctance Machines. *IEEE Trans. Power Electron.* **2019**, *34*, 1635–1644. [CrossRef]
18. Li, C.; Zhang, C.; Liu, J.; Bian, D. A High-Performance Indirect Torque Control Strategy for Switched Reluctance Motor Drives. *Math. Probl. Eng.* **2021**, *2021*, 1–15. [CrossRef]
19. Liu, J.; Wang, L.; Yi, L.; Zhu, G.; Yin, X. Optimization of SRM Direct Instantaneous Torque Control Strategy based on Improved Firefly Algorithm. In Proceedings of the 2019 3rd IEEE Conference on Energy Internet and Energy System Integration: Ubiquitous Energy Network Connecting Everything, EI2 2019, Changsha, China, 8–10 November 2019; pp. 364–368. [CrossRef]
20. Wang, S.; Hu, Z.; Cui, X. Research on Novel Direct Instantaneous Torque Control Strategy for Switched Reluctance Motor. *IEEE Access* **2020**, *8*, 66910–66916. [CrossRef]
21. Pillai, A.; Anuradha, S.; Gangadharan, K.V.; Umesht, P.; Bhaktha, S. Modeling and Analysis of Average Torque Control Strategy on Switched Reluctance Motor for E-mobility. In Proceedings of the CONECCT 2021 7th IEEE International Conference on Electronics, Computing and Communication Technologies, Bengaluru, India, 9–11 July 2021. [CrossRef]
22. Fan, J.; Lee, Y. A Novel Average Torque Control of Switched Reluctance Motor Based on Flux-Current Locus Control. *Can. J. Electr. Comput. Eng.* **2020**, *43*, 273–281. [CrossRef]
23. Usman Jamil, M.; Kongprawechnon, W.; Chayopitak, N. Average Torque Control of a Switched Reluctance Motor Drive for Light Electric Vehicle Applications. *IFAC-PapersOnLine* **2017**, *50*, 11535–11540. [CrossRef]

24. Hamouda, M.; Abdel Menaem, A.; Rezk, H.; Ibrahim, M.N.; Számel, L. Numerical Estimation of Switched Reluctance Motor Excitation Parameters Based on a Simplified Structure Average Torque Control Strategy for Electric Vehicles. *Mathematics* **2020**, *8*, 1213. [CrossRef]
25. Fang, G.; Bauman, J. Optimized switching angle-based torque control of switched reluctance machines for electric vehicles. In Proceedings of the 2020 IEEE Transportation Electrification Conference and Expo, ITEC 2020, Chicago, IL, USA, 22–26 June 2020; pp. 186–191. [CrossRef]
26. Hamouda, M.; Szamel, L. Reduced Torque Ripple based on a Simplified Structure Average Torque Control of Switched Reluctance Motor for Electric Vehicles. In Proceedings of the 2018 International IEEE Conference and Workshop in Óbuda on Electrical and Power Engineering (CANDO-EPE), Budapest, Hungary, 20–21 November 2018; pp. 000109–000114. [CrossRef]
27. Hamouda, M.; Menaem, A.A.; Rezk, H.; Ibrahim, M.N.; Számel, L. Comparative Evaluation for an Improved Direct Instantaneous Torque Control Strategy of Switched Reluctance Motor Drives for Electric Vehicles. *Mathematics* **2021**, *9*, 302. [CrossRef]
28. Ren, P.; Zhu, J.; Jing, Z.; Guo, Z.; Xu, A. Improved DITC strategy of switched reluctance motor based on adaptive turn-on angle TSF. *Energy Rep.* **2022**, *8*, 1336–1343. [CrossRef]
29. Cheng, H.; Chen, H.; Yang, Z. Average torque control of switched reluctance machine drives for electric vehicles. *IET Electr. Power Appl.* **2015**, *9*, 459–468. [CrossRef]
30. Husain, T.; Elrayyah, A.; Sozer, Y.; Husain, I. Unified control for switched reluctance motors for wide speed operation. *IEEE Trans. Ind. Electron.* **2019**, *66*, 3401–3411. [CrossRef]
31. Hamouda, M.; Számel, L. A new technique for optimum excitation of switched reluctance motor drives over a wide speed range. *Turkısh J. Electr. Eng. Comput. Sci.* **2018**, *26*, 2753–2767. [CrossRef]
32. Liu, L.; Zhao, M.; Yuan, X.; Ruan, Y. Direct instantaneous torque control system for switched reluctance motor in electric vehicles. *J. Eng.* **2019**, *2019*, 1847–1852. [CrossRef]
33. Hamouda, M.; Menaem, A.A.; Rezk, H.; Ibrahim, M.N.; Számel, L. An improved indirect instantaneous torque control strategy of switched reluctance motor drives for light electric vehicles. *Energy Rep.* **2020**, *6*, 709–715. [CrossRef]

MDPI AG
Grosspeteranlage 5
4052 Basel
Switzerland
Tel.: +41 61 683 77 34

Mathematics Editorial Office
E-mail: mathematics@mdpi.com
www.mdpi.com/journal/mathematics

Disclaimer/Publisher's Note: The statements, opinions and data contained in all publications are solely those of the individual author(s) and contributor(s) and not of MDPI and/or the editor(s). MDPI and/or the editor(s) disclaim responsibility for any injury to people or property resulting from any ideas, methods, instructions or products referred to in the content.

www.ingramcontent.com/pod-product-compliance
Lightning Source LLC
LaVergne TN
LVHW070733100526
838202LV00013B/1227